D1484156

Electrochemical Aspects
of Ionic Liquids

Electrochemical Aspects of Ionic Liquids

Edited by

Hiroyuki Ohno

Department of Biotechnology,
Tokyo University of Agriculture and Technology,
Koganei, Tokyo,
Japan

WILEY-
INTERSCIENCE

A JOHN WILEY & SONS, INC., PUBLICATION

For general information on our other products and services please contact our Customer Care Department within the U.S. at 877-762-2974, outside the U.S. at 317-572-3993 or fax 317-572-4002.

Wiley also publishes its books in a variety of electronic formats. Some content that appears in print, however, may not be available in electronic format.

Library of Congress Cataloging-in-Publication Data:

Electrochemical aspects of ionic liquids / edited by Hiroyuki Ohno.
    p.   cm.
Includes bibliographical references.
ISBN 0-471-64851-5 (cloth)
1. Ionic solutions.   2. Electrochemistry.   3. Polymerization.
    I. Ohno, Hiroyuki, 1953-II. Title.

QD562.165E38   2005
541'.372–dc22   2004019936

Printed in the United States of America

10 9 8 7 6 5 4 3 2 1

Contents

Contributors

C. AUSTEN ANGELL, Department of Chemistry, Arizona State University caa@asu. edu

JEAN-PHILIPPE BELIERES, Department of Chemistry, Arizona State University jpbelieres@asu.edu

MARIA FORSYTH, School of Physics and Materials Engineering, Monash University maria.forsyth@spme.monash.edu.au

TOSHIO FUCHIGAMI, Department of Electronic Chemistry, Tokyo Institute of Technology fuchi@echem.titech.ac.jp

KYOKO FUJITA, School of Chemistry, Monash University kyoko.fujita@sci.-monash.edu.au

YUKINOBU FUKAYA, Department of Biotechnology, Tokyo University of Agriculture & Technology yuki-f@cc.tuat.ac.jp

RIKA HAGIWARA, Graduate School of Energy Science, Kyoto Universiry hagiwara@energy.kyoto-u.ac.jp

KENJI HANABUSA, Graduate School of Science and Technology, Shinshu University hanaken@giptc.shinshu-u.ac.jp

AKITOSHI HAYASHI, Department of Applied Materials Science, Graduate School of Engineering, Osaka Prefecture University hayashi@ams.osakafu-u.ac.jp

YASUSHI KATAYAMA, Department of Applied Chemistry, Faculty of Science and Technology, Keio University katayama@applc.keio.ac.jp

TAKASHI KATO, Department of Chemistry and Biotechnology, School of Engineering, The University of Tokyo kato@chiral.t.u-tokyo.ac.jp

NOBUO KIMIZUKA, Department of Chemistry and Biochemistry, Graduate School of Engineering, Kyushu University kimitcm@mbox.nc.kyushu-u.ac.jp

TOMOYA KITAZUME,   Department of Bioengineering, Tokyo Institute of Technology tkitazum@bio.titech.ac.jp

PIERRE LUCAS, Department of Chemistry, Arizona State University pierre@u.arizona.edu

DOUGLAS. R. MACFARLANE, School of Chemistry, Monash University Douglas.MacFARLANE@sci.monash.edu.au

HAJIME MATSUMOTO,   Research Institute for Ubiquitous Energy Devices, National Institute of Advanced Industrial Science and Technology h-matsumoto @aist.go.jp

KAZUHIKO MATSUMOTO, Graduate School of Energy Science, Kyoto University matsumoto@fchem.nucleng.kyoto-u.ac.jp

TOMONOBU MIZUMO,   Department of Biotechnology, Tokyo University of Agriculture & Technology mizmo@cc.tuat.ac.jp

TAKUYA NAKASHIMA,   Research and Education Center for Materials Science, Nara Institute of Science and Technology ntaka@ms.naist.jp

ASAKO NARITA,   Department of Biotechnology, Tokyo University of Agriculture & Technology nariasa@cc.tuat.ac.jp

NAOMI NISHIMURA,   Department of Biotechnology, Tokyo University of Agriculture & Technology ryutao@cc.tuat.ac.jp

AKIHIRO NODA,   Honda R&D Co., Ltd. akihiro_noda@n.t.rd.honda.co.jp

WATARU OGIHARA,   Department of Biotechnology, Tokyo University of Agriculture & Technology wogi@cc.tuat.ac.jp

HIROYUKI OHNO,   Department of Biotechnology, Tokyo University of Agriculture & Technology ohnoh@cc.tuat.ac.jp

JENNY PRINGLE,   School of Physics and Materials Engineering, Monash University Jenny.Pringle@sci.monash.edu.au

GUNZI SAITO,   Division of Chemistry, Graduate School of Science, Kyoto University saito@kuchem.kyoto-u.ac.jp

Md. ABU BIN HASAN SUSAN,   Department of Chemistry, Dhaka University abhsusan@yahoo.com

MAKOTO UE,   Mitsubishi Chemical Group, Science and Technology Research Center, Inc. Battery Materials Laboratory 3707052@cc.m-kagaku.co.jp

MARCELO VIDEA,   Department of Chemistry, Arizona State University mvidea@itesm.mx

GORDON. G. WALLACE,   Intelligent Polymer Research Institute, University of Wollongong gordon_wallace@uow.edu.au

MASAYOSHI WATANABE,   Department of Chemistry & Biotechnology, Yokohama National University mwatanab@ynu.ac.jp

HIKARI SAKAEBE,   Research Institute for Ubiquitous Energy Devices, National Institute of Advanced Industrial Science and Technology hikari.sakaebe@aist.go.jp

XU WU,   Department of Chemistry, Arizona State University wuxu@asu.edu

MASAFUMI YOSHIO,   Department of Chemistry and Biotechnology, School of Engineering, The University of Tokyo yoshio@chembio.t.u-tokyo.ac.jp

MASAHIRO YOSHIZAWA,   School of Chemistry, Monash University yoshi.fujita@sci.-monash.edu.au

Preface

This book introduces some basic and advanced studies on ionic liquids in the electrochemical field. Although ionic liquids are known by only a few scientists and engineers, their applications' potential in future technologies is unlimited. There are already many reports of basic and applied studies of ionic liquids as reaction solvents, but the reaction solvent is not the only brilliant future of the ionic liquids. Electrochemistry has become a big field covering several key ideas such as energy, environment, nanotechnology, and analysis. It is hoped that the contributions on ionic liquids in this book will open other areas of study as well as to inspire future aspects in the electrochemical field. The applications of ionic liquids in this book have been narrowed to the latest results of electrochemistry. For this reason only the results on room-temperature ionic liquids are presented, and not on high-temperature melts.

The reader of this book should have some basic knowledge of electrochemistry. Those who are engaged in work or study of electrochemistry will get to know the great advantages of using ionic liquids. Some readers may find the functionally designed ionic liquids to be helpful in developing novel materials not only in electrochemistry but also in other scientific fields. This book covers a wide range of subjects involving electrochemistry. Subjects such as the solubilization of biomolecules may not seem to be necessary for electrochemistry concerning ionic liquids, but some readers will recognize the significance of solubility control of functional molecules in ionic liquids even in an electrochemical field. Many more examples and topics on ionic liquids as solvents have been summarized and published elsewhere, and the interested reader will benefit from studying the references that are provided at the end of each chapter.

Acknowledgments

To prepare this book, many authors who are at the cutting edge of their areas of study have kindly agreed to write these chapters. This book should enable a practicing scientists to develop their own original research.

I would like to express my sincere thanks to all these authors. They were all very cooperative and summarized their progress despite their very busy schedules. This book therefore shares the fruits of their highly original, individual work.

I would also like to acknowledge Prof. Dr. K. R. Seddon of Queen's University of Belfast, England. He strongly urged me to publish this book when he saw the table of contents written in English. During his stay at our laboratory in January 2003, I explained to him the concept and the strong points of such a book. Without his urging, this book would not have been published.

I would like to extend my appreciation to all members of our laboratory in the Department of Biotechnology, Tokyo University of Agriculture and Technology. Especially Dr. M. Yoshizawa receives my most sincere thanks. This book would not have been finished without his sacrifices. Last I would like to thank Dr. Arza Seidel of John Wiley and Sons, Inc. for her kind support and encouragement.

HIROYUKI OHNO

Electrochemical Aspects
of Ionic Liquids

Chapter *1*

Importance and Possibility of Ionic Liquids

Hiroyuki Ohno

1.1 IONIC LIQUIDS

Ionic liquids are the salts having very low melting temperature. Ionic liquids have received great interests recently because of their unusual properties as liquids. These unique properties of ionic liquids have already been mentioned in some books, so we do not repeat them here more than simply summarize them in Table 1.1. Note these are entirely different properties from those of ordinary molecular liquids. The most important properties of electrolyte solutions are nonvolatility and high ion conductivity. These are essentially the properties of advanced (and safe) electrolyte solutions that are critical to energy devices put in outdoor use. Safety is more an issue than performance these days, and is taken into account in the trends in the materials developed for practical. Thus more developments in ionic liquids are expected to be seen in the future. The nonvolatile electrolyte solution will change the performance of electronic and ionic devices. These devices will become safer and have longer operational lives. But, more interesting, they will be composed of organic ions, and these organic compounds will have unlimited structural variations because of the easy preparation of many components. So there are unlimited possibilities open to the new field of ionic liquids. The most compelling idea is that ionic liquids are "designable" or "fine-tunable." So we can easily expect explosive developments in fields using these remarkable materials.

Electrochemical Aspects of Ionic Liquids Edited by Hiroyuki Ohno
ISBN 0-471-64851-5 Copyright © 2005 John Wiley & Sons, Inc.

TABLE 1.1   Basic Characteristics of Organic Ionic Liquids

Low melting point	• Treated as liquid at ambient temperature • Wide usable temperature range
Non-volatility	• Thermal stability • Nonflammability
Composed by Ions	• High ion density • High ion conductivity
Organic ions	• Various kinds of salts • Designable • Unlimited combination

1.2   IMPORTANCE OF IONIC LIQUIDS

Ionic liquids are salts that melt at ambient temperature. The principles of physical chemistry involved in the great difference between solution properties of molecular solvents and molten salts have already been introduced and summarized in a number of books. Thousands of papers have already been published on their outstanding characteristics and effectiveness for a variety of fields. Thus, as was mentioned above, in this book we take the most important point that these ionic salts are composed of organic ions and explore the unlimited possibility of creating extraordinary materials using molten salts.

Because ionic liquids are composed of only ions, they show very high ionic conductivity, nonvolatility, and nonflammability. The nonflammable liquids with high ionic conductivity are practical materials for use in electrochemistry. At the same time the nonflammability and nonvolatility inherent in ion conductive liquids open new possibilities in other fields as well. Because most energy devices can accidentally explode or ignite, for motor vehicles there is plenty of incentive to seek safe materials. Ionic liquids are being developed for energy devices. It is therefore important to have an understanding of the basic properties of these interesting materials. The ionic liquids are multi-purpose materials, so there should be considerable (and unexpected) applications. In this book we, however, will not venture into too many other areas. Our concern will be to assess the possible uses of ionic liquids in electrochemistry and allied research areas.

1.3   POTENTIAL OF IONIC LIQUIDS

At present most of the interest in ionic liquids is centered on the design of new solvents. While the development of "new solvents" has led the direction of possible applications for ionic liquids, there is more potential for development electrochemical applications.

Electrochemistry basically needs two materials: electro conductive materials and ion conductive materials. Ionic liquids open the possibility of improving ion conductive materials. The aqueous salt solution is one of the best electrolyte solutions for electrochemical studies. However, because water is volatile, it is impossible to use this at wide temperature range or on a very small scale. Many other organic polar solvents have been used instead of water to prepare electrolyte solutions. They, however, have more or less the same drawback, depending on the characteristics. The material known to be a nonvolatile ion conductor is the polymer electrolyte. Polymers do not vaporize but decompose at higher temperatures; the vapor pressure at ambient temperature is zero. Polymer electrolytes are considered a top class of electrolytes except for the one drawback: relatively low ionic conductivity.

Some of the literature has included statements that the ionic liquids are thermally stable and never decompose. This kind of statement has led to a misunderstanding that the ionic liquids are never vaporized and are stable even when on fire. Are the ionic liquids indestructable? The answer is no. However, while inorganic salts are entirely stable, the thermal stability of organic salts depends largely on their structure. Since most recently reported ionic liquids are organic compounds, their degradation begins at the weakest covalent bond. Nevertheless, ionic liquids are stable enough for ordinary use at temperatures of 200° to 300°C. So it is not difficult to design novel ionic liquids that can be decomposed at certain temperature or by certain trigger. It is also possible to design unique catalysts (or catalytic systems) that can decompose target ionic liquids. Some catalysts such as metal oxides or metal complexes have the potential to become excellent catalysts for the decomposition of certain ionic liquids under mild conditions. The post-treatment technologies of ionic liquids should therefore also be developed along with the work on the design of ionic liquids.

At the present time there has been little progress in this area. Although post-treatment technologies are beyond the scope of this book, we do attempt to give ideas on the various future developments in ionic liquid technologies as well as in electrochemistry. This book is dedicated to introducing, analyzing, and discussing ionic liquids as nonvolatile and highly ion conductive electrolyte solutions. The astute reader will find the future prospects for ionic liquids between the lines in all chapters of this book.

Chapter 2

Physical Chemistry of Ionic Liquids, Inorganic and Organic, Protic and Aprotic

**C. A. Angell, W. Xu, M. Yoshizawa, A. Hayashi,
J.-P. Belieres, P. Lucas, and M. Videa**

2.1 CLASSES OF IONIC LIQUIDS

Ionic liquids in their high temperature manifestations (liquid oxides, silicates, salts) have been studied for a long time, using quite sophisticated methods, and much of the physics is understood. By contrast, the low temperature ionic liquid field, the subject of the present volume, is very much in its formative stages. The many studies on the interesting transport properties of ionic liquids and thermodynamic properties have focused mainly on characterizing new systems for potential applications [1–5]. The job of placing ion conductive behavior within the wider phenomenology of liquid and amorphous solid electrolytes, and of the liquid state in general, has barely begun. In this chapter we review the collective physical properties of ionic liquids in an attempt to place them within this larger picture.

The first requirement of an ionic liquid is that, contrary to experience with most liquids consisting of ions, it must have a melting point that is not much above room temperature. The limit commonly suggested is 100°C [1b]. Given the cohesive energy of ionic liquids (about which more will be said below), ambient melting requires that the melting point occur at a temperature not too far above the glass

Electrochemical Aspects of Ionic Liquids   Edited by Hiroyuki Ohno
ISBN 0-471-64851-5   Copyright © 2005 John Wiley & Sons, Inc.

transition temperature, T_g, which provides the natural base for liquid-like behavior. Ionic liquids therefore nearly all melt within the range $T_g - 1.5T_g$ which we call the "low-temperature regime" of liquid behavior [6,7]. This means that they will, in most cases, supercool readily and will exhibit "super-Arrhenius" transport behavior near and below ambient temperature—as is nearly always reported.

Such liquids come in different classes. The most heavily researched class is the aprotic organic cation class [1–4, 8–15]. In this cation class the low melting point is a consequence of the problem of efficiently packing large irregular organic cations with small inorganic anions. More on this class is given in Section 2.3.

A second class [16] is one that may enjoy increased interest in the future because of the presence of one of its members in the first industrial IL process [1b], and also because of the new finding that its members can have aqueous solution-like conductivities [17] and can serve as novel electrolytes for fuel cells [18]. This class is closely related to the first but differs in that the cation has been formed by transfer of a proton from a Brønsted acid to a Brønsted base. The process is reversible, depending on how large the free energy of proton transfer is. When the gap across which

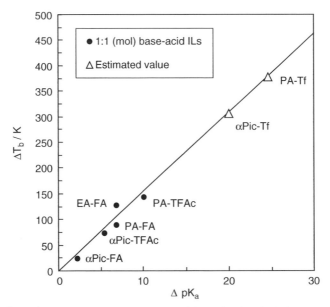

Figure 2.1 *Correlation of the excess boiling point (determined at the 1:1 composition) with the difference in aqueous solution pK$_a$ values for the component Brønsted acids and bases of the respective ionic liquids ΔpK$_a$. The ΔT$_b$ value is determined as the difference between the measured boiling point and the value, at 1:1, of the linear connection between pure acid and pure base boiling points. Note the very large excess boiling points extrapolated for the ionic liquids formed from the superacid HTf (open triangles). These values could not be determined experimentally because of prior decomposition. (Notation: EA = ethylammonium, PA = propy-propylammonium, αPic = α-picolinium = 2-methylpyridinium = 2MPy, FA = formate, TFAc = tri-trifluoroacetate, Tf = triflate = trifluoromethanesulfonate) (from Yoshizawa, Xu, and Angell [19]). Data for the three protic nitrates of ref. 17 (ethylammonium nitrate, dimethylammonium nitrate and methylammonium nitrate) fit precisely on this diagram.*

the proton must jump to reform the original molecular liquids is small, then the liquid will have a low conductivity and a high vapor pressure. These properties are not of great interest in an ionic liquid, though the liquid may be very fluid. If the gap is large, as in the case of ammonium nitrate, 87 kJ/mol (from ΔG_f^o data for $HNO_3 + NH_3 \rightarrow NH_4NO_3$), then the proton will remain largely on the cation and for many purposes the system is a molten salt. If the acid is a superacid, like triflic acid, HSO_3CF_3 or HTFSI [16], then the transfer of the proton will be very energetic, and the original acid will not be regenerated on heating before the organic cation decomposes. Such liquids will not be distinguishable in properties from the conventional aprotic salts in which some alkyl group, rather than a proton, has been transferred to the basic site. The stability of these systems has recently [19] been characterized in terms of the relation between the boiling point elevation (or excess boiling point) over the linear (or average value) of the components and the excess was shown to be a linear function of the difference in pK_a values determined in aqueous solutions.

This relation is shown in Figure 2.1. It has so far been found free of exceptions when base is an amine nitrogen. The protic ionic liquids as a class [16–18] are considerably more fluid than the aprotic ionic liquids [17], for reasons that are not clear at this time (see Section 2.6).

The third and distinct class of ionic liquid is the one comprised entirely of inorganic entities. These are mostly formed in consequence of the mismatch of large anions like tetrachloroaluminate or iodide with small cations like Li^+. The eutectic in the system $LiAlCl_4$–$LiAlI_4$ system, for instance, lies at 65°C, and the liquid is highly fluid, more fluid than the majority of aprotic ionic liquids. The phase diagram is shown in Figure 2.2 [20]. Also in this group is the more viscous system

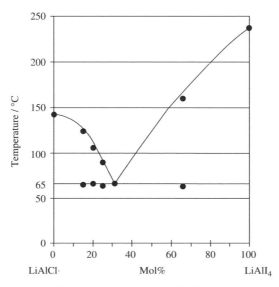

Figure 2.2 *Phase diagram of the system $LiAlCl_4 + LiAlI_4$ (from Lucas, Videa, and Angell [20]) showing ionic liquid domain ($T_l < 100°C$) in the mid-diagram.*

containing silver and alkali halides [21], which exhibits an ambient temperature electrical conductivity, $10^{-1.4}$ S/cm, that is higher than that of any aprotic ionic liquid.

A fourth class may be considered, though it contains non-ionic entities. This is the liquid state of various ionic solvates. In these systems molecules usually thought of as solvent molecules are tightly bound to high field cations and have no solvent function. Such "molten solvates" have very low vapor pressures at ambient, and only boil at temperatures near 200°C, for instance, $LiZnBr_4 \cdot 3H_2O$, has $T_b = 190°C$ while $T_g = -120°C$ [22].

To provide a better perspective on ionic liquids, we first make some observations on inorganic salts and the factors that make it possible to observe them as ionic liquids below 100°C.

2.2 LOW-TEMPERATURE LIQUID BEHAVIOR

Most inorganic salts, when they melt, are found to flow and conduct electricity according to a simple Arrhenius law at all temperatures down to their melting points. For instance, unless measurements of high precision are used, the alkali halides appear to remain obedient to the Arrhenius equation even down to the deep eutectic temperatures of their mixtures with other salts. LiCl and KCl form a eutectic mixture with a freezing point of 351°C, some 300 K below either pure salt freezing point, yet the viscosity of the melt barely departs from Arrhenius behavior before freezing.

In order to see viscosity behavior of the strongly super-Arrhenius type typical of almost any individual RTMS, it is necessary to avoid alkali halides altogether and examine salts that cannot form such symmetrical crystal lattices. For instance, alkali nitrates, like KNO_3, do not occupy much more volume in the liquid state than does KI, but they melt at much lower temperatures. However, even KNO_3 exhibits an Arrhenius-type viscosity temperature dependence according to any but the most precise measurements. It is only with deep eutectics like that for the ternary systems $LiNO_3$–$NaNO_3$–KNO_3 ($T_E = 143°C$), and $LiNO_3$–$NaNO_2$–KNO_3 ($T_E = 125°C$) that one starts to see clear deviations from the Arrhenius law [6a,7]. This stands in clear contrast with the behavior of the ILs, RTMS, or molten hydrates.

With all ionic liquids, deviations from the Arrhenius law, considerably in excess of those seen for the ternary LiNaK nitrate eutectic, are found well above their melting points. There is no need, with ionic liquids, to invoke eutectic mixtures to extend the stable liquid range to observe the "low-temperature domain" behavior. The low-temperature domain is typically seen in the temperature range at $T < 2T_0$ [6], where T_0 is the theoretical low temperature limit to the liquid state. At T_0, extrapolations of experimental data suggest that the excess entropy of the liquid (excess over that of the crystal) would vanish, and the time scale for fluid flow would diverge [7]. The practical low-temperature limit to the liquid state given by the glass temperature, T_g, is considerably higher than the theoretical T_0, by an amount that depends on the liquid fragility. Typically $T_g/T_0 = 1.2 - 1.3$, unless the liquid is very "strong". So the low-temperature domain is entered at

$T_g/T \approx 0.53 - 0.66$. For Arrhenius behavior, which represents the "strong" liquid limit of behavior and is rarely seen, $T_0 = 0$ K and $T_g/T_0 = \infty$. The upper end of this range, $T_g/T = 0.66$, is the number usually associated with the ratio of T_g/T_m for glassformers (the "2/3 rule"), though we have argued elsewhere [23] that this is not a rule but a tautology.

The behavior of ionic liquids is quite familiar to workers experienced with molten hydrates. With molten hydrates, the cation size is effectively increased by the shell of water molecules shielding the central cation from its anionic neighbors, such that the cation acquires a size not unlike those of cations in the typical IL. We show a selection of viscosity data for normal molten salts, molten hydrates, and ionic liquids in Figure 2.2, using a scaled Arrhenius plot to bring a wide range of data together on a single plot. A blurred distinction between "normal" and "low-temperature" domain behavior can be made by putting a vertical line at about $T_g/T = 0.625$.

The deviations from Arrhenius behavior seen in the low-temperature regime of Figure 2.3 are in most cases well accounted for by the three parameter

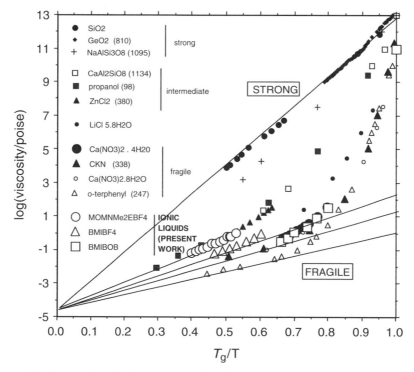

Figure 2.3 T_g-scaled Arrhenius plot showing data for molten salts ZnCl₂ and calcium potassium nitrate (CKN), with data for the calcium nitrate hydrate (CaNO₃·8H₂O) and the tetrafluoroborates of quaternary ammonium (MOMNM₂E, M = methyl, E = ethyl) and 1-n-butyl-3-methyl-imidazolium (BMI) cations, and the bis-oxalatoborate (BOB) of the latter cation, in relation to other liquids of varying fragility (from Xu, Cooper, and Angell [15]).

Vogel-Fulcher-Tammann (VFT) equation [24] in the modified form:

$$\eta = \eta_0 \exp\left(\frac{DT_0}{[T - T_0]}\right) \qquad (2.1)$$

where η_0, D, and T_0 are constants and T_0, the vanishing mobility temperature, lies below the glass transition temperature if a suitable range of data (>3 orders of magnitude) are included in the data fitting. The different curvatures in Figure 2.3 are then reproduced by variations in the parameter D of Eq. (2.1) [6b].

Figure 2.3 displays behavior that is almost universal to RTMS. Note that at their normal melting points, IL-RTMS are almost always in the "low-temperature region" of liquid behavior (defined by easily recognized "super-Arrhenius" transport behavior)—that is, they melt within the same range that is generally so difficult to access for uni-univalent inorganic molten salts systems and their mixtures. To understand how this can come about, it is helpful to give some consideration to the factors that decide where, in temperature, a given substance will melt (which we do in the next section).

Figure 2.4, on the other hand, shows behavior for some inorganic ILs [20] which, while superficially similar to that of Figure 2.3 (super-Arrhenius), is almost

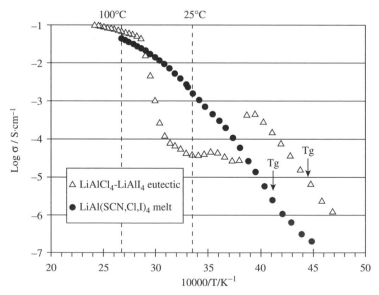

Figure 2.4 *Arrhenius plots for conductivity of lithium haloaluminate or pseudo-haloaluminate melts, showing super-Arrhenius conductivity that remains high at the glass temperature due to decoupling of Li^+ motion from its surroundings, in the "low-temperature regime." The break in the upper curve is due to crystallization of the supercooled liquid during re-heating. Note the change from curvilinear to Arrhenius behavior as T falls below T_g and the structure becomes fixed (from Lucas, Videa and Angell [20]).*

never seen in the organic cation ILs, either protic or aprotic, or in the molten hydrates. The distinction lies in the value of the conductivity near the glass temperature T_g. The conductivity of the low-melting eutectic of Figure 2.4 can easily be measured at temperatures near and below T_g if the melt is rapidly cooled to avoid crystallization. What is interesting is that the conductivity at T_g is about nine orders of magnitude higher than that of the typical IL measured at its glass temperature (which requires special equipment). The explanation for this dramatic difference lies in the ability of the small cation, Li^+, to slip through the holes in the vitrified structure and so establish a solid state ionic conductivity. These are known as decoupled systems, and are much sort after for solid state ionic devices. The only equivalent of this behavior in IL phenomenology is found in some little known protic salts in which the proton motion can apparently be slightly decoupled from the structural relaxation. A high degree of decoupling of protons, as in the Grotthus mechanism of aqueous solutions, is an urgent goal of research on protic ionic liquids. We discuss tests of this sort of behavior below.

2.3   MELTING POINTS AND THE LATTICE ENERGY

What is implied by the observations in the preceding section is that the crystal lattices of substances of the IL type, in common with salt hydrate (and salt solvate crystals in general), become thermodynamically unstable with respect to their liquid phases ($G_{liq} < G_{crys}$) at temperatures that are very low, relative to their cohesive energies.

The simplest explanation that can be given for this circumstance focuses on the difficulty of finding efficient packing modes for the more complex and size-mismatched ions characteristic of IL and inorganic salt hydrates. The idea is illustrated in Figure 2.5, which shows the Gibbs free energies of several crystalline forms of the same fictitious substance, along with that of the liquid phase, over a range of temperatures. The crystals are supposed to be nonconvertible except to the liquid, and to differ only in their lattice energies (lattice energy, $E_L = G$ at $T = 0$ K). There is only one liquid phase possible.

The diagram of Figure 2.5 shows that the lowest melting point must belong to the crystal phase with the lowest lattice energy. This can be verified using data for isomers of the same substance (e.g., xylene in which all three isomers o-, m-, and p- have essentially the same viscosity and boiling point) for which adequate thermodynamic data are available. We have shown elsewhere [25] how the difference in melting points of some 70 K between m- and p-isomers can be traced to a difference in lattice energy of $\sim$1 kcal/mol. Of course, the polymorph with the lowest melting point will be the one whose viscosity will be the highest at its melting point, and the one whose melting point will fall within, or closest to, the "low-temperature region" of the liquid state [26]. Most RTMSs therefore are characterized by low lattice energies that approach that of the glass formed by supercooling of the liquid to the vitreous state.

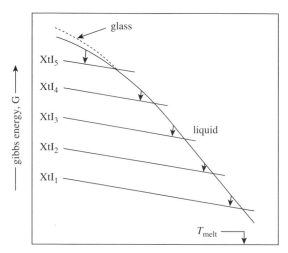

Figure 2.5 *Gibbs free energies for a system in which there are multiple noninterconvertable crystalline phases, each with a different lattice energy ($E_L = G$, at $T = 0\,K$) but all with the same entropy. All yield a liquid with the same free energy. Arrows show access to the different crystalline phases from the liquid. The crystal with the lowest melting point must be the one with the lowest lattice energy. It will be the one that melts to a liquid with the highest viscosity, and the one that therefore is the least likely to crystallize during cooling. When crystallization does not occur, the glassy state does.*

This conclusion is consistent with the usual strategy for making ILs. When the salt of a given organic cation does not melt at low enough temperature to qualify as an RTMS, the problem is usually rectified by addition of a side-group, or replacement of an existing side-group with a larger one or one that breaks a previous symmetry [1–5, 27, 28]. What has been done is, of course, to interfere with the possibility of achieving low energies by efficient packing in 3D order. This improves the competitive status of the disordered state; that is, it lowers the T/T_g value at the melting point.

2.4   RELATION BETWEEN ELECTRICAL CONDUCTIVITY AND LOW VAPOR PRESSURE

In order to vaporize an ionic liquid in which there is no possibility of forming molecular liquids by proton transfer (as in Figure 2.1), it is necessary to do work against electrical forces to remove an ion pair from the bulk of the liquid into the vacuum of space. If the ionic liquid were to consist mainly of ion pairs, this work would be the same as in a strongly dipolar liquid. Fortunately it is much larger. The reason that it is larger is that much of the energy of stabilization of an ionic liquid is gained by the formation of a quasi-lattice that is almost as efficient in minimizing the electrostatic energy of a system of ions as is the formation of a crystal lattice. After all, it is the enthalpy of vaporization that distinguishes ionic liquids from other ambient temperature liquids, not the enthalpy of fusion. The enthalpy of fusion is not exceptional at all.

The additional energy of stabilization of ionic *crystals* over dipolar crystals is gained by the uniform surrounding of ions of one charge by ions of the opposite charge. It is known as the Madelung energy [29]. It is rather difficult to estimate quantitatively, depending on details of crystal symmetry [29], and direct calculations for isotropic phases are not available, to the authors' knowledge. For simple molten salts these should be directly related to the three liquid radial distribution functions available from inelastic scattering studies. In any event, the low vapor pressure of ionic liquids is a direct reflection of the work that must be done against the Madelung potential to extract an ion pair from the bulk liquid. Quantitatively, the probability $p(h)$ of an enthalpy fluctuation h sufficient to permit an ion-pair to escape from the bulk liquid into the vapor is proportional to the Boltzmann factor:

$$p(h) \sim \exp{-\left(\frac{h}{k_B T}\right)} \tag{2.2}$$

where k_B is the Boltzmann constant and h is the enthalpy of extraction per molecule (ion-pair), which is dominated by the Madelung energy.

In an ideal quasi-lattice, no ion-pairs can be distinguished over statistically significant time intervals, since all sites are equivalent. This is also the condition that leads to the maximum ionic conductivity for a given fluidity (at least in the absence of decoupling, see below). This is because the presence of ion-pairs lowers the conductivity by permitting diffusion, hence the fluid flow without ionic current flow.

These relations are summed up in some of the classical equations of electrochemistry, which were derived by consideration of dilute aqueous solutions in which complete dissociation into *independently moving ions* could be assumed. Although these solutions present a rather different physical situation from that of solvent-free ionic liquids, the laws developed for their description remain very relevant to the description of the ionic liquid properties. The main difference is that the notion of "dissociation" is more obscure. In ionic liquids the "state of dissociation" must be decided by operational criteria, as we outline below.

The classical equations to which we refer are as follows:

1. The Nernst-Einstein equation connecting diffusion and partial equivalent ionic conductivity λ_i,

$$\lambda_i = \frac{RTz_i D_i}{F^2} \tag{2.3}$$

   where F is the Faraday (charge per equivalent).

2. The Stokes-Einstein equation connecting diffusivity D_i of ionic species i of charge z_i and radius r_i, with the viscosity η of the medium in which the diffusion is occurring,

$$D_i = \frac{k_B T}{6\pi \eta r_i} \tag{2.4}$$

3. The Walden rule connecting viscosity and equivalent conductivity $\Lambda = \Sigma\lambda_i$,

$$\Lambda\eta = \text{const.} \tag{2.5}$$

Equation (2.5) is particularly useful to us, and we will devote much attention to it in the following.

The Walden rule is interpreted in the same manner as the Stokes-Einstein relation. In each case it is supposed that the force impeding the motion of ions in the liquid is a viscous force due to the solvent through which the ions move. It is most appropriate for the case of large ions moving in a solvent of small molecules. However, we will see here that just as the Stokes-Einstein equation applies rather well to most pure nonviscous liquids [30], so does the Walden rule apply, rather well, to pure ionic liquids [15]. When the units for fluidity are chosen to be reciprocal poise and those for equivalent conductivity are $Smol^{-1}cm^2$, this plot has the particularly simple form shown in Figure 2.6.

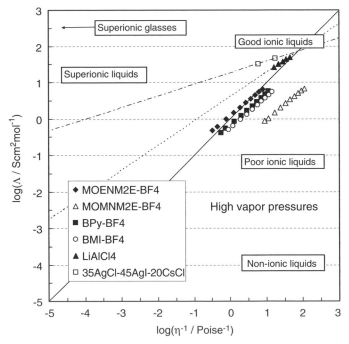

Figure 2.6 *Walden plot for tetrafluoroborate salts of various cations, showing subionic (or "supercoupled") behavior of the salt of the methoxymethyldimethylethyl ammonium cation $MOMNM_2E^+$. By contrast the salt of $MOENM_2E^+$ may become decoupled at low temperature. The dotted lines passing through the $LiAlCl_4$ points and the $AgCl–AgI–CsCl$ points are the fits of equation (2.6) to the data points. The slope, α, is inversely proportional to the log (decoupling index $R\tau$) of refs. [31, 35]. The plot is annotated to indicate how it may serve as a classification diagram for ionic liquids and other electrolytes. For notation, see the captions to Figures 2.1 and 2.3. (From Xu, Cooper, and Angell [15], by permission).*

In order to fix the position of the "ideal" Walden line in Figure 2.6, we use data for dilute aqueous KCl solutions in which the system is known to be fully dissociated and to have ions of equal mobility [31]. We have included some data for a glass-forming $Ca(NO_3)_2$ aqueous solution in which the fluidity has been measured over eight orders of magnitude [32]. For the units chosen, the ideal line runs from corner to corner of a square diagram. Figure 2.6 contains data for some of the salts of recent study [15] and some that were measured in 1983 to 1986 but remained unpublished until [15] (except for [15a] where the argument for unpolarizable anions like BF_4^- was given, and one example reported, in proof). It also includes data for some inorganic systems of interesting types, where the decoupling of conductivity from viscosity, which can be so large at T_g (Figure 2.6), can be seen in its earliest manifestation.

Because of the way Figure 2.6 reveals the various couplings and decouplings that can be encountered in ionic liquid media, we have used it as a basis for ionic liquid classification. We divide ionic liquids into ideal, "subionic" (or "poor" ionic) liquids, and "superionic" liquids. The subionic liquids may still be good conductors at ambient pressure because their fluidities are high, but the conductivity is much lower than if all the moving particles were cations or anions.

Among the subionic liquids are the cases of proton transfer salts in which the ΔpK_a value is small, and certain aprotic systems in which some special interionic locking mechanism appears to operate, causing a high proportion (some 90%) of the diffusive motions to be of the neutral pair type. In this regime also fall most non-aqueous lithium electrolyte solutions, where the failure to fully dissociate causes the low conductivities that plague these electrolytes. It is the progressive increase, with increasing temperature, in the population of nonconducting pairs that causes the less-than-unit slope of all the electrolytes that fall on the right-hand side of the Walden line.

Electrolytes that fall on the upper side of the ideal line are very desirable because of the high transport numbers. High transport numbers imply for the species that transports occur faster than expected by the Stokes-Einstein equation. The solutions whose conductivity are shown in Figure 2.4 provide a good example of this behavior, but unfortunately, fluidity data are not available. Instead, we show classic data for $LiAlCl_4$ [33] and also for the more striking case of the silver alkali halide ionic liquid for which fluidity and conductivity data were reported by McLin and Angell [34]. In this case it is the silver cation that can slip through the channels set up by the alkali halide sublattice [35]. These systems can retain their high conductivity in the glassy state, as illustrated in the case of the silver–alkali halide system in Figure 2.4b. For such systems a "decoupling index" has been defined from the ratio of the conductivity and fluidity relaxation times [32, 36], which can equally well be defined from the exponent α in the fractional Walden rule,

$$\Lambda\eta^\alpha = \text{const.} \tag{2.6}$$

which applies to such systems. α is the slope of the line in the Walden plot, and is unity for the ideal electrolyte.

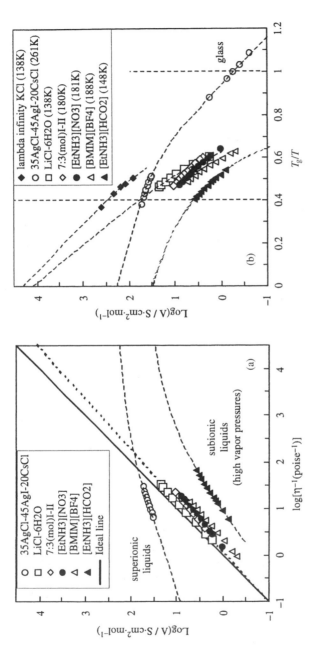

Figure 2.7 (a) Relation of equivalent conductivity to fluidity for various protic and aprotic ionic liquids. The heavy line is the ideal Walden line. Ideally the temperature dependence of conductivity is set by the value for fluidity, since the only force impeding the motion of an ion under fixed potential gradient is the viscous friction. The position of the ideal line is fixed by data for 1M aqueous KCl solution at ambient temperature. The data for LiCl· 6H$_2$O fall close to it. In most charge-concentrated systems, interionic friction causes loss of mobility, which is more important at high temperature. This gives rise to the below-ideal slope found for all the ILs, protic or aprotic, seen in the figure. A "fractional Walden rule" $\Lambda\eta^\alpha = const$, $(0 < \alpha < 1)$, applies. When there is a special mechanism for conductance, then the Walden plot falls above the ideal line, as for superionics, and the slope α provides a measure of the decoupling index [15].

(b) T_g-scaled Arrhenius plot to display the temperature dependence of the equivalent conductivity in relation to the temperature at which the fluidity reaches the glassy value of 10^{-11} poise (10^{-13} p for inorganic network glasses). The inorganic superionic systems have very high conductivities at their glass temperatures. Subionic (associated, ion-paired, etc.) systems have low conductivity at all temperatures. The ideal (ion interaction-free) behavior for conductivity is shown by the dashed line for infinite dilution conductivities for KCl in H$_2$O (data from Robinson and Stokes [31]). To include these data, we assign T_g as 138 K, since high-temperature viscosity fitting suggests it. The real value for water is controversial. Notation: I is [MeNH$_3$][NO$_3$], II is [Me$_2$NH$_2$][NO$_3$]. (From Xu and Angell [17], by permission)

We note in Figure 2.6 that while the majority of the IL systems studied lie close to the ideal Walden line, they all exhibit a slightly smaller slope, implying $\alpha < 1$ and also implying a smaller D parameter in equation (2.1) for conductivity than for fluidity. Since this behavior has been noted, historically, for glass-forming molten salts and liquid hydrate [24(d)], it appears to be fundamental to the physics of electrolytes, so it may prove useful in interpreting the manner in which the free space, introduced during configurational excitation, is distributed in ionic liquids. This effect, which leads all except infinitely dilute solutions to have limiting high temperature conductivities that are smaller than expected from their fluidities, is well illustrated in Figure 2.7.

Figure 2.7 is a composite representation of the transport properties of ionic liquids of different types intended to show the relation between Walden behavior and the temperature dependence of conductivity. In Figure 2.7a we show, in this Walden representation, an alternative set of data emphasizing proton transfer salts (protic ILs). The plot in this case terminates at the universal high T limit for fluidity implied by Figure 2.3, $10^{4.5}$ poise.

Figure 2.7b shows the corresponding behavior of the conductivity for these systems, to be discussed further below. In principle, the conductivity of the infinitely dilute aqueous solution, in which the ions move without any influence on each other, should have the same value at $1/T = 0 \text{ K}^{-1}$ as does the ideal Walden liquid at the high fluidity limit. Using data for aqueous KCl solutions at infinite dilution in the temperature range $0°$ to $100°C$ given by Robinson and Stokes [31], and applying the fact that all liquids obey an Arrhenius law for $T/T_g < 0.4$, we see that this correspondence is almost realized. Certainly the infinitely dilute solution (in which cation–anion friction is absent) has a much higher equivalent conductivity at the high-temperature limit than do any of the other systems.

Counted among these superionic electrolytes are the solutions of strong acids in water and other hydrogen-bonded solvents (e.g., glycerol, hydrogen peroxide) in which the proton has anomalously high mobility due to the Grotthus mechanism [38]. There is currently an urgent search for protic systems that can exhibit such behavior in the ambient temperature glassy state so that all-solid fuel cells can be realized. There is some hope that protonic superionic conductors might be found among the protic ionic liquids that have recently [17] been shown to rival aqueous solutions in their conductivity. In Figure 2.8 we show the conductivities of some of these systems, in comparison with the conductivities of their nearest aprotic relatives, and also with aqueous lithium chloride solutions of 1 and 7.7 molar concentration. The methyl ammonium nitrate–dimethyl ammonium nitrate mixture that is so highly conducting is known to obey the Walden rule, and indeed to fall close to the ideal Walden line, so its very high conductance is apparently not due to any "dry" proton mechanism. However, the extraordinary conductance found for ammonium bifluoride, which is seen to equal that of 7.7 M lithium chloride ($LiCl \cdot 6H_2O$) [39], suggests that some salts with protons in both cation and anions may have superprotonic properties. Nevertheless, even without superprotonic conduction, it seems that protic ionic liquids, which are neutral and noncorrosive, can serve as fuel cell electrolytes with unusual performance properties, as we now briefly describe.

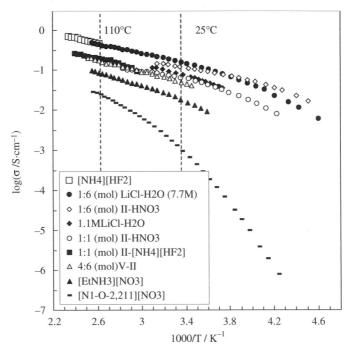

Figure 2.8 *Specific conductivities of protic vs. aprotic ionic liquids, showing matching of concentrated lithium chloride solution conductivity by solvent-free aprotic liquids. Note that at low temperature, the conductivity of protic nitrate in excess nitric acid is higher than that of the aqueous LiCl case with the same excess solvent. (From Xu and Angell [17] by permission)*

2.5 PROTON TRANSFER IONIC LIQUIDS AS NOVEL FUEL CELL ELECTROLYTES

Surprisingly, it is only a very recent recognition that protic ionic liquids can serve as proton transfer electrolytes in hydrogen–oxygen fuel cells. A current report from the laboratory of Watanabe [40] describes the performance of a hydrogen electrode utilizing, as the electrolyte, the salt formed by proton transfer from the acid form of *bis*-trifluoromethanesulfonyl imide (HTFSI) to the base imidazole.

Our own experience [18] has shown that by using certain favorable protic ILs, fuel cells can be made with short-term performance superior to that of the commercially feasible phosphoric acid high-temperature fuel cell. In refs. [18] we give comparisons of open circuit voltages produced, and short circuit currents flowing, in simple U-tube type hydrogen–oxygen fuel cells using protic IL electrolytes, spiral wound with platinum wire electrodes, with those produced when a phosphoric acid electrolyte is substituted in the same cell.

An example is shown in Figure 2.9. Because the bubbling of hydrogen gas was irregular, the current flowing was subject to occasional large fluctuations, which are

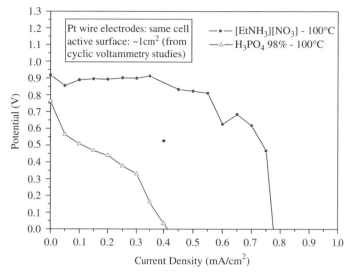

Figure 2.9 *Potential of the fuel cell under load using ethylammonium nitrate as electrolyte. Comparison is made with the behavior of the identical cell with phosphoric acid substituted for the ethylammonium nitrate electrolyte.*

not seen in more sophisticated gas diffusion cells that will be described separately [41]. Especially to be noted is the high potential of the cell relative to the value obtained when the IL is replaced by 98% H_3PO_4 with, (0.9 vs. 0.75 V), which is maintained under load when using ethylammonium nitrate as electrolyte. It is evident that with the protic ionic liquid, a much higher exchange current density at the oxygen electrode can be obtained than in other fuel cells.

Particularly when doped appropriately, the ionic liquid fuel cell (ILFC) gives clearly superior performance [18]. Using an ambient temperature version of the (technologically unacceptable) difluoride IL illustrated in Figure 2.8, equally good performance can be obtained at ambient temperature! It is expected that this application of the ionic liquid concept will receive considerable attention in the immediate future. Results using more efficient Teflon sandwich cells and colloidal Pt electrodes will be reported separately [41].

2.6 COHESION AND FLUIDITY: THE TRADE-OFF

The high cohesion of molten salts associated with electrostatic forces also makes them more viscous than other liquids. The smaller the ions, the higher is the cohesion. Since high cohesion means high viscosity, "good" ionic liquids need to achieve an optimum lowering of the cohesion if they are to serve as high-fluidity and high-conductivity media at low temperatures. It would seem that increasing the size of the ions, by increasingly lowering the coulombic attractions, would be the way to achieve high fluidities. Unfortunately, another effect comes into play as the particles are increased in size. This is the van der Waals interaction. It is important to keep this as low as possible by maintaining an unpolarizable outer surface on the anions. This is the reason for the prevalence of perfluorinated species among the ionic liquids used in practice. On the other hand, perfluorinated anions are very expensive to manufacture. And even perfluorinated species, begin to cause fluidity decreases when their size becomes large enough [12, 27].

One way of monitoring these effects is to use the glass temperature as a cohesion indicator. In Figure 2.10 we show the glass transition temperatures for a large number of different cation-anion pairs having in common that the species are chosen for low polarizabilities [15]. Thus aromatic cations are excluded, and anions are either fluorinated or, if oxyanions, the oxygen electrons are strongly polarized towards anion center species [Cl(VII), S(VI), or N(V)]. The plot is made against molar volume on the assumption that this is roughly the cube of the cation-anion separation. The plot shows a broad minimum around 250 cm^3 per mole, corresponding to a 4 molar ion concentration.

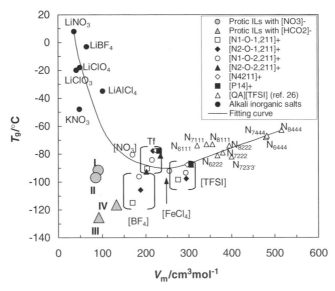

Figure 2.10 *Dependence of the cohesion of salts of weakly polarizable cations and anions, assessed by T_g value, on the ambient temperature molar volume V_m, hence on interionic spacing $[(r^+ + r^-) \sim V_m^{1/3}]$. A broad minimum in the ionic liquid cohesive energy is seen at a molar volume of 250 cm^{3}mol^{-1}, which corresponds to an interionic separation of about 0.6 nm, assuming face-centered cubic packing of anions about cations. The lowest T_g in the plot should probably be excluded from consideration because of the non-ideal Walden behavior for this IL (MOMNM$_2$E$^+$BF$_4^-$) [15]). The line through the points is a guide to the eye. The data for open triangles are from Sun, Forsyth, and MacFarlane [42].*

Since anions less polarizable than the perfluorinated species represented in Figure 2.6 cannot be made, it would seem that if fluorinated species are to be avoided for economic reasons, the number of anions that can be used in acceptable ionic liquids is rather limited. Nitrates, thiocyanates, nitrites, formates, dicyanamides, chlorosulfonates, and methanesulfonates would seem to be acceptable. Dicyanamides have been shown to have high fluidities with imidazolium and pyrrolidinium cations [28, 43].

The size of cations for use with these anions is already limited. The aprotic systems with the minimum glass temperatures in Figure 2.6 are quaternary ammonium cations with uniformly small alkyl groups attached. If we describe the alkyl groups using Wunderlich's "bead" notation [44] (a bead is taken as an atom or group that can participate in a change of configuration), then the methoxyethylethyldimethyl ammonium ($MOENM_2E$) cation has 8 beads. The lowest T_g's are obtained with the four-bead anion BF_4^- which is perfluorinated. If fluorine is avoided, then the lowest T_g's are obtained with three-bead and two-bead anions, NO_3^-, and formate HCO_2^-. Data on the latter are reported elsewhere [37].

Systems with cations smaller (and then more symmetrical) than this latter cation, seem to melt well above ambient, unless one uses protic cations [12]. Once protic cations are admitted, then a new range of high-fluidity (low T_g) systems opens up, and the maximum ambient temperature ionic conductivity jumps sharply as shown in Figure 2.7, largely because the T_g values go down at the same time as the charge concentration increases (to ~ 10 M).

2.7 CONCLUDING REMARKS

We have tried to cover important aspects of the physical chemistry of the ionic liquids currently under study, and to relate them to what is known about other types of low-melting ionic media. In concluding, we must emphasize that much of the success in their application, particularly in the Green Chemistry area where there is hope they will replace volatile solvents of environmentally hostile character, will depend on the important *chemical* properties of these media. These we have not addressed at all in this chapter. Properties such as donor and acceptor character, acidity and basicity, are in fact all within the capacity of physics to describe, though they are most commonly invoked in a more empirical manner based on experience, as described in [1–4]. An excellent treatment of acid base character of ionic liquids has recently been given by MacFarlane and Forsyth [45].

ACKNOWLEDGMENTS

This work was supported by the National Science Foundation under Solid State Chemistry grant DMR 0082535 and by the DOE-LANL high-temperature fuel cell program.

REFERENCES

1. (a) R. D. Rogers, K. R. Seddon, eds., *Ionic Liquids: Industrial Applications to Green Chemistry*, American Chemical Society, 2002. (b) K. R. Seddon, *Nat. Mater.*, **2**, 363 (2003).

2. R. D. Rogers, K. R. Seddon, S. Volkov, eds., *Green Industrial Applications of Ionic Liquids*, Kluwer Academic, 2003.

3. T. Welton, *Chem. Rev.*, **99**, 2071 (1999).

4. H. Ohno, ed., *Ionic Liquids: The Front and Future of Material Development*, High Tech. Info., 2003 (in Japanese).

5. M. Gaune-Escarde, ed., *Molten Salts: From Fundamentals to Applications*, Kluwer Academic, 2002.

6. (a) C. A. Angell, *J. Phys. Chem.*, **68**, 1917 (1964). (b) C. A. Angell, *J. Non-Cryst Sol.*, **13**, 131 (1991).

7. C. A. Angell, in *Molten Salts: From Fundamentals to Applications*, M. Gaune-Escarde, ed., Kluwer Academic, Delft, 2002, p. 305.

8. C. L. Hussey, *Adv. Molten Salt Chem.*, **5**, 185 (1983).

9. E. I. Cooper, E. J. M. Sullivan, *Proc. 8th. Int. Symp. Molten Salts*, Electrochem. Soc., 1992, p. 386.

10. J. S. Wilkes, M. J. Zawarotko, *J. Chem. Soc. Chem. Commun.*, 965 (1992).

11. R. T. Carlin, J. S. Wilkes, in *Chemistry of Nonaqueous Solutions*, G. Mamantov and A. I. Popov, ed., VCH Publishers, 1994.

12. Sun. J., Forsyth, M., MacFarlane, D. R., *Ionics*, **3**, 356 (1997).

13. P. Bonhote, A.-P. Dias, M. Armand, N. Papageorgiou, K. Kalyanasundaram, M. Graetzel, *Inorganic Chem.*, **35**, 1168 (1996).

14. (a) C. J. Bowles, D. W. Bruce, K. R. Seddon, *Chem. Commun.*, 1625 (1996). (b) J. D. Holbrey, K. R. Seddon, *J. Chem. Soc., Dalton Trans.*, 2133 (1999).

15. (a) W. Xu, E. I. Cooper, C. A. Angell, *J. Phys Chem. B*, **107**, 6170 (2003). (b) E. I. Cooper, C. A. Angell, *Solid State Ionics*, **9 & 10**, 617 (1983).

16. (a) M. Hirao, H. Sugimoto, H. Ohno, *J. Electrochem. Soc.*, **147**, 4168 (2000). (b) M. Yoshizawa, W. Ogihara, H. Ohno, *Electrochem. Solid State Lett.*, **4**, E25 (2001). (c) H. Ohno, M. Yoshizawa, *Solid State Ionics*, **154**, 303 (2002).

17. W. Xu, C. A. Angell, *Science*, **302**, (October 17) 422 (2003).

18. (a) J.-P. Belieres, W. Xu, C. A. Angell, Abstract 982, Fall ECS meeting, Orlando, Sept. 2003. (b) Ionic liquids by proton transfer, ΔpK_a, and the ionic liquid fuel cell. M. Yoshizawa, J.-P. Belieres, W. Xu, C. A. Angell, Abstracts of Papers of the American Chemical Society 226: U627-U627 083-IEC Part 1 Sep 2003. (c) C. A. Angell, W. Xu, J.-P. Belieres, M. Yoshizawa, Provisional patent application file No. 9138-0123.

19. M. Yoshizawa, W. Xu, C. A. Angell, *J. Am. Chem. Soc.*, **125**, 15411 (2003).

20. P. Lucas, M. Videa, C. A. Angell, to be published.

21. C.-L. Liu, H. G. K. Sundar, C. A. Angell, *Mat Res. Bull.*, **20**, 525 (1985).

22. A. Sivaraman, H. Senapati, C. A. Angell, *J. Phys. Chem., B* **103**, 4159 (1999).

23. C. Alba, J. Fan, C. A. Angell, *J. Chem. Phys.*, **110**, 5262 (1999).

24. (a) H. Vogel, *Phys. Z.*, **22**, 645 (1921). (b) G. S. Fulcher, *J. Am. Ceram. Soc.*, **8**, 339 (1925). (c) G. Tammann, W. Hesse, *Z. Anorg. Allg. Chem.*, **156**, 245 (1926). (d) This equation is frequently referred to in current literature as the VTF equation, which is idiosyncratic in view of the publication dates of the cited authors. The idiosyncrasy seems to have been introduced in the late 1960s to emphasize Tammann's contribution to the phenomenology of glass-forming liquids; see C. A. Angell (*J. Am. Ceram. Soc.*, **51**, 117 (1968) and C. A. Angell and C. T. Moynihan (in *Molten Salts: Analysis and Characterisation*, G. Mamantov, ed., Dekker, New York 1969, p. 315). See also G. W. Scherer, editor, *J. Am. Ceram. Soc.*, **75**(5), 1060 (1992).

25. J. Wong, C. A. Angell, *Glass: Structure by Spectroscopy*, Dekker, 1976, p. 12.

26. C. Alba, L. E. Busse, C. A. Angell. *J. Chem. Phys.*, **92**, 617 (1990).

27. J. Sun, M. Forsyth, D. R. MacFarlane, *J. Phys. Chem., B* **102**, 8858 (1998).

28. D. R. MacFarlane, J. Golding, S. Forsyth, M. Forsyth, *Chem. Comm.* (UK), **16**, 1430 (2001).

29. C. Kittel, *Solid State Physics*, 3rd ed., Wiley, 1967, ch. 3.

30. (a) F. Fujara, B. Geil, H. Sillescu, G. Fleischer, *Z. Phys. B. Condensed Matter*, **88**, 195 (1992). (b) I. Chang, H. Sillescu, *J. Phys. Chem., B* **101**, 8794 (1997). (c) M. T. Cicerone, M. D. Ediger, *J. Chem. Phys.*, **103**, 5684 (1995).

31. R. A. Robinson and R. H. Stokes, *Electrolyte Solutions*, 2nd ed., Butterworths, 1959, p. 465, app. 6.2.

32. (a) J. H. Ambrus, H. Dardy, C. T. Moynihan, *J. Phys. Chem.*, **76**, 3495 (1972). (b) J. H. Ambrus, C. T. Moynihan, P. B. Macedo, *J. Electrochem. Soc.*, **119**, 192 (1972).

33. C. R. Boston, L. F. Grantham, S. J. Yosim, *J. Electrochem. Soc.*, **117**, 28 (1970).

34. M. McLin, C. A. Angell, *J. Phys. Chem.*, **92**, 2083 (1988).

35. Changle Liu, H. G. K. Sundar, C. A. Angell, *Solid State Ionics*, **18** & **19**, 442 (1986).

36. C. A. Angell, *Chem. Rev.*, **90**, 523 (1990).

37. W. Xu, C. A. Angell (to be published).

38. C. J. T. von Grotthus, *Annales de Chemie*, **58**, 54 (1806).

39. C. T. Moynihan, N. Balitactac, L. Boone, T. A. Litovitz, *J. Chem. Phys.*, **55**, 3013 (1971).

40. A. B. H. Susan, A. Noda, S. Mıtsuchima, M. Watanabe, *Chem. Comm.*, 938 (2003).

41. J. P. Belieres, C. A. Angell, D. Gervasio (to be published).

42. J. Sun, M. Forsyth, D. R. MacFarlane, *J. Phys. Chem., B* **102**, 8858 (1998).

43. D. R. MacFarlane, J. Golding, S. Forsyth, G. B. Deacon, *Green Chem.*, **4**, 444 (2002).

44. W. B. Wunderlich, *J. Phys. Chem.*, **64**, 1052 (1960).

45. D. R. MacFarlane, S. A. Forsyth, ACS Symposium Series, in press.

Part I

Basic Electrochemistry

Chapter *3*

General Techniques

Yasushi Katayama

Electrochemical methods can be powerful tools. They can be used to reveal the chemical and physical properties of room-temperature ionic liquids. Most of existing electrochemical techniques [1] developed in aqueous solutions are applicable for the ionic liquids, as demonstrated in the chloroaluminate ionic liquids. However, there are several procedures that must be observed if one is to obtain reliable data in electrochemical measurements. This section describes the procedures that are important for the ionic liquids.

3.1 EQUIPMENT

Room-temperature ionic liquids are usually prepared from a variety of organic and inorganic salts. Since both ionic liquids and halide salts are hygroscopic in most cases, they must be dried under vacuum at elevated temperatures and handled in a dry atmosphere. It is convenient to construct a vacuum line for drying, as shown in Figure 3.1. The dried materials are usually handled in a glove box filled with a dry argon or nitrogen. These materials may be handled in dry air if the existence of oxygen is allowable. However, it should be noted that the solubility of oxygen in the ionic liquids is not negligible and the reduction of dissolved oxygen to superoxide will be observed in the absence of protic species [2–6]. The existence of oxygen is also unfavorable for the electrodeposition of base metals.

Electrochemical Aspects of Ionic Liquids   Edited by Hiroyuki Ohno
ISBN 0-471-64851-5   Copyright © 2005 John Wiley & Sons, Inc.

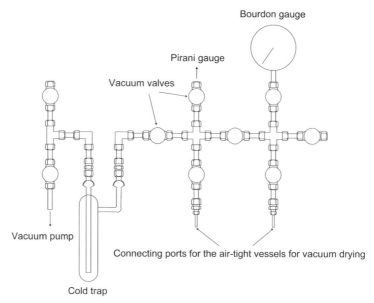

Figure 3.1 Schematic illustration of a metal vacuum line for drying materials.

3.2 PREPARATION OF ELECTROLYTES

Electrochemical methods are sensitive to the extent that it is possible to detect a trace of electroactive species in electrolyte solutions. Because of this distinctive feature, electrochemical methods have been developed and utilized for analytical purposes. The detection method used is known as polarography. For the electrochemical study purification of the electrolyte solutions is therefore important. As for most aqueous and organic electrolyte solutions, there are various well-established techniques for purifying both solvents and electrolytes. In the case of room-temperature ionic liquids, it is especially important to purify the starting materials used for preparing the ionic liquids.

Quaternary ammonium cations are frequently employed as the organic cations of the ionic liquids. They are usually prepared by the reaction of alkyl halides with such tertiary amines, as alkylamine, pyridine, alkylimidazole, and alkylpyrolidine. The purity of the resultant ionic liquids is often dependent on the purity of these starting materials. Therefore it is recommended that these starting materials are purified properly before the preparation of the organic cations. In addition to purification of the starting materials, it is also important to control the reaction when preparing the cations. When the reaction is exothermic, the reaction mixture becomes heated locally or wholly, resulting in the degradation of the starting materials or products. The starting materials can react even under mild conditions. So care must be taken in diluting the starting materials with aprotic solvents, stirring the reaction mixture thoroughly, keeping the temperature of the reaction mixture constant, adding the reactants very slowly, and so on. The organic salts are further purified by recrystallization, followed by evaporation of the solvents and vacuum drying at elevated temperatures.

The compounds that provide the anions of the ionic liquids should be checked carefully for purity, and further purified if necessary. In the case of chloroaluminate ionic liquids, it is essential to purify aluminum trichloride by sublimation under reduced pressure. The mixing of the organic chlorides with $AlCl_3$ can generate heat and cause the degradation of organic cations. Some nonchloroaluminate ionic liquids, BF_4^-, PF_6^-, and $CF_3SO_3^-$ ionic liquids, are prepared by the metathesis of the organic halides and silver, alkali metal, and ammonium salts of the desired anions in certain aprotic solvents [7–9]. However, contamination by the inorganic cations is often unavoidable because the solubility of the inorganic salts in the ionic liquids is not negligible. Thus some alternative methods have been used for preparing these ionic liquids [10, 11]. Some water-stable and hydrophobic ionic liquids, represented by $N(CF_3SO_2)_2^-$ ionic liquids, can be prepared easily by making interact the equimolar amounts of an organic halide and $LiN(CF_3SO_2)_2$ in water. The $N(CF_3SO_2)_2^-$ ionic liquid phase is separated from the aqueous phase containing lithium halide. A trace of lithium salts in the $N(CF_3SO_2)_2^-$ ionic liquid can be removed by extracting the ionic liquid into dichloromethane [8]. The removal of water from these water-stable ionic liquids can be done simply by vacuum drying at elevated temperature.

It is usually difficult to remove protons from the ionic liquids that undergo hydrolysis. In the chloroaluminate ionic liquids, the evacuation coupled by the treatment with phosgene can be proposed for removing protons [12].

3.3 WORKING ELECTRODES

Platinum, glasslike carbon, and tungsten are often used as inert working electrodes for the fundamental electrochemical studies in the ionic liquids. For such transient electrochemical techniques as cyclic voltammetry, chronoamperometry, and chronopotentiometry, it is safer to use the working electrode with a small active area. This is because most of the ionic liquids will have low conductivity, and this often causes the ohmic drop in the measured potentials by the current flowing between the working and counter electrode. Microelectrodes may be useful for the electrochemical measurements in the case of handling low conductive media.

3.4 REFERENCE ELECTRODES

The potential of an electrode gives essential information about the phenomena being observed at the electrode. The electrode potential determines both direction and rate of the electrode reaction. Therefore it is necessary to measure or control the electrode potential against a reference electrode, of which the potential is stable and reproducible during a series of measurements. In aqueous solutions there are several such well-established reference electrodes, namely Ag/AgCl, Hg/Hg_2Cl_2,

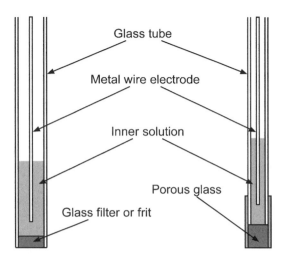

Figure 3.2 Examples of the reference electrodes for ionic liquids.

Hg/HgO, and Hg/HgSO$_4$. These reference electrodes may be usable also for the ionic liquids with some liquid–liquid junctions: however, contamination by water into the ionic liquids is unavoidable and not desirable in most cases. The reference electrodes based on aprotic solvents (e.g., Ag/Ag(I) in CH$_3$CN) can be used also in the ionic liquids, though there is still a possibility of contamination by the solvents. Thus it is always better to prepare the reference electrodes based on the ionic liquids.

Figure 3.2 shows the typical reference electrodes used for the ionic liquids as well as for nonaqueous solutions. The reference electrode is basically composed of a metal electrode immersed in an inner solution. The supporting electrolyte of the inner solution should be identical to that of a test solution in order to minimize the contamination and the junction potential. The inner solution contains the electroactive species determining the potential of the reference electrode. Consequently the inner solution should be separated from the test solution by a glass filter, or frit, in order to avoid the mixing of these solutions by keeping the electrochemical junction. Porous glass is also used for this purpose. The potential of the reference electrode generally depends on both electroactive species and supporting electrolyte. Whereas it is difficult to compare the potentials measured in different ionic liquids, even if the same redox couple is employed for the reference electrodes, the redox potential of ferrocenium (Fc$^+$)/ferrocene (Fc) is often used as a common potential standard for nonaqueous systems, since its redox potential is regarded as independent of solvents. Thus the potential of the reference electrode should be given against the redox potential of the Fc/Fc$^+$, also in ionic liquids. In addition the rest potential of a metal electrode immersed in the supporting electrolyte without any redox couple (sometimes called as a quasi- or pseudoreference electrode) is less reliable and should not be used for electrochemical measurements unless its potential is proved to be stable and calibrated properly with a true

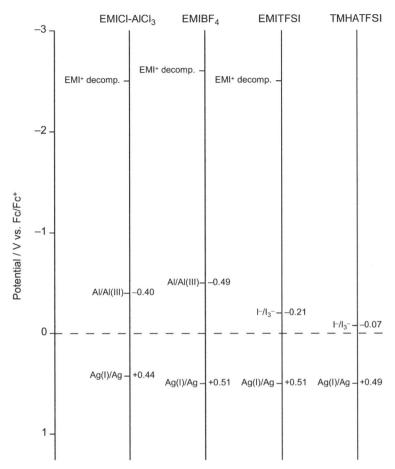

Figure 3.3 *Potentials of the selected redox couples in some ionic liquids.*

reference electrode. The low conductivity of the ionic liquids often causes the ohmic drop in the potential measurements. In order to reduce the ohmic drop, the tip of the reference electrode should be placed at the working electrode, if necessary, using a Luggin-Haber capillary.

Figure 3.3 shows the potentials of the selected redox couples in some ionic liquids against Fc/Fc^+. The cathodic decomposition potential of 1-ethyl-3-methyl-imidazolium, EMI^+, is located at about -2.5 V in relation to Fc/Fc^+ in EMICl–$AlCl_3$, $EMIBF_4$, and $EMIN(CF_3SO_2)_2$. It seems that the difference in the potentials of Ag(I)/Ag [10, 13, 14] in these ionic liquids is not so significant, whereas the potential of I^-/I_3^- is likely to be affected by the kinds of the organic cations [8, 15, 16].

Various redox couples have been used for the reference electrodes in the electrochemical studies in the ionic liquids. The following are some examples of the reference electrodes, which are actually used in some ionic liquids.

Chloroaluminate Systems. In case of chloroaluminate ionic liquids, the potential of an aluminum electrode immersed in the acidic ionic liquids which contain $AlCl_3$ at more than 50 mol% is usually assumed as the potential standard.

The electrode reaction is considered as follows:

$$4Al_2Cl_7^- + 3e^- = Al + 7AlCl_4^- \tag{3.1}$$

The dimeric chloroaluminate anion, $Al_2Cl_7^-$, is present under acidic condition. The molar fraction of $AlCl_3$ in the acidic ionic liquid, which is used as the reference electrode, is often assigned a 66.7 or 60.0 mol%. It should be noted that the potential of this reference electrode is dependent on the mole fraction (activity) of $AlCl_3$, as is evident in equation (3.1). This reference electrode is used not only for the acidic ionic liquids but also for basic and neutral ones in which no redox reaction between Al and monomeric chloroaluminate anion, $AlCl_4^-$, is available at near room-temperature. In addition this Al/Al(III) reference electrode is also used for some nonchloroaluminate ionic liquids. The potential of the Al/Al(III) electrode ($[AlCl_3] = 60.0$ mol%) is reported as $-0.406 \sim -0.400$ V against Fc/Fc^+ in the $EMICl$-$AlCl_3$ ionic liquid [9].

Chlorozincate Systems. The reference electrode similar to Al/Al(III) is used in chlorozincate ionic liquids. The electrodeposition of Zn is reported to be possible in the acidic chlorozincate melt containing $ZnCl_2$ at more than 33.3 mol% at near room-temperature [17]. The reference electrode is composed of a Zn electrode immersed in the acidic melt that contains $ZnCl_2$, usually at 50 mol%. Although the details of the electrode reaction are not yet known, it is thought that such chlorozincate species as $ZnCl_3^-$, $Zn_2Cl_7^{3-}$, and $(ZnCl_2)_n$, behave as electroactive species. The potential of this Zn/Zn(II) electrode has not been measured against Fc/Fc^+.

Other Nonchloroaluminate Systems. In case of the ionic liquids having no reversible electroactive species, a certain redox couple must be selected for the reference electrode. The Al/Al(III) electrode is sometimes used in nonchloroaluminate ionic liquids: however, the Al/Al(III) electrode is not recommended for those who have no experience of preparing chloroaluminate ionic liquids. It is possible to prepare the reference electrode composed of a platinum electrode immersed in the ionic liquid which contains both Fc^+ and Fc if a soluble ferrocenium salt is available. It should be noted, however, that the volatility of Fc can lead to a change in the concentration of Fc (i.e., the electrode potential) at elevated temperatures.

The redox reaction between silver and monovalent silver species, Ag(I), is one of the typical redox couples used for the reference electrodes in aqueous and nonaqueous media. The Ag/Ag(I) electrode is composed of a Ag electrode immersed in an ionic liquid that contains Ag(I) at 0.01 or 0.1 mol dm^{-3}. Ag(I) is introduced into the ionic liquid by dissolving silver salts having the same anion as the ionic liquid. For example, $AgBF_4$ is reported to be soluble in $EMIBF_4$ [10], and it is probably soluble in other BF_4^- ionic liquids. The potential of Ag(I)/Ag ($[Ag(I)] = 0.2$ mol dm^{-3} is estimated as $+0.51$ V against Fc/Fc^+ in $EMIBF_4$ at room-temperature

[9, 10]. In case of $N(CF_3SO_2)_2^-$ ionic liquids, it is possible to prepare $AgN(CF_3SO_2)_2$ by the reaction of excess Ag_2O with $HN(CF_3SO_2)_2$ aqueous solution, followed by filtration, evaporation of water, and drying as

$$Ag_2O + 2HN(CF_3SO_2)_2 \rightarrow 2AgN(CF_3SO_2)_2 + H_2O \qquad (3.2)$$

The $HN(CF_3SO_2)_2$ aqueous solution can be prepared by passing a $LiN(CF_3SO_2)_2$ aqueous solution through a protonated cation-exchange resin. $AgN(CF_3SO_2)_2$ is soluble in the $N(CF_3SO_2)_2^-$ ionic liquids. The formal potential of $Ag(I)/Ag$ is reported as $+ 0.49 \sim + 0.51$ V against Fc/Fc^+ at room-temperature [14]. Of course, the anion of the silver salt does not need to be the same as the ionic liquid. For example, $Ag(I)$ can be intriduced into the $N(CF_3SO_2)_2^-$ ionic liquids by dissolving $AgCF_3SO_3$.

The redox reaction between iodides, I^- and I_3^-, is also used as the reference electrodes in nonchloroaluminate ionic liquids especially for the $N(CF_3SO_2)_2^-$ ionic liquids. Platinum wire is used for the electrode and immersed in an ionic liquid that contains iodine and an iodide salt having the same cation as the ionic liquid. The concentration ratio of the iodine salt to iodide is often specified as $1 : 4$, namely $[I_3^-] : [I^-] = 1 : 3$. The potential of this I^-/I_3^- electrode is reported as -0.16 V in $TMHAN(CF_3SO_2)_2$ at $50°C$ [15] ($TMHA^+$ = trimethyl-*n*-hexylammonium) and -0.21 V in $EMIN(CF_3SO_2)_2$ at room-temperature [16] against Fc/Fc^+.

REFERENCES

1. A. J. Bard, L. R. Faulkener, eds., *Electrochemical Methods Fundamentals and Applications*, 2nd ed., Wiley, 2001.

2. M. T. Carter, C. L. Hussey, S. K. D. Strubinger, R. A. Osteryoung, *Inorg. Chem.*, **30**, 1149 (1991).

3. I. M. AlNashef, M. L. Leonard, M. C. Kittle, M. A. Matthews, J. W. Weidner, Electrochem. *Solid State Lett.*, **4**, D16 (2001).

4. I. M. AlNashef, M. L. Leonard, M. A. Matthews, J. W. Weidner, *Ind. Eng. Chem. Res.*, **41**, 4475 (2002).

5. M. C. Buzzeo, O. V. Klymenko, J. D. Wadhawan, C. Hardacre, K. R. Seddon, R. G. Compton, *J. Pjys. Chem. A*, **107**, 8872 (2003).

6. Y. Katayama, H. Onodera, M. Yamagata, T. Miura, *J. Electrochem. Soc.*, **151**, A59 (2004).

7. J. S. Wilkes, M. J. Zaworotko, *J. Chem. Soc., Chem. Commun.*, 965 (1992).

8. P. Bonhôte, A-P. Dias, N. Papageorgiou, K. Kalyanasundaram, M. Grätzel, *Inorg. Chem.*, **35**, 1168 (1996).

9. J. Fuller, R. T. Carlin, R. A. Osteryoung, *J. Electrochem. Soc.*, **144**, 3881 (1997).

10. Y. Katayama, S. Dan, T. Miura, T. Kishi, *J. Electrochem. Soc.*, **148**, C102 (2001).

11. J. G. Huddleston, A. E. Visser, W. M. Reichert, H. D. Willauer, G. A. Broker, R. D. Rogers, *Green Chemistry*, **3**, 156 (2001).

12. M. A. M. Noël, P. C. Trulove, R. A. Osteryoung, *Anal. Chem.*, **63**, 2892 (1991).

13. Q. Zhu, C. L. Hussey, G. R. Stafford, *J. Electrochem. Soc.*, **148**, C88 (2001).

14. Y. Katayama, M. Yukumoto, M. Yamagata, T. Miura, T. Kishi, in *Proc. 6th Int. Symp. Molten Salt Chemistry and Technology*, C. Nianyi and Q. Zhiyu, eds., Shanghai University Press 2001, p. 190.

15. K. Murase, K. Nitta, T. Hirato, Y. Awakura, *J. Appl. Electrochem.*, **31**, 1089 (2001).

16. R. Kawano, M. Watanabe, *Chem. Commun.*, **2003**, 330 (2003).

17. Y.-F. Lin, I.-W. Sun, *Electrochim. Acta*, **44**, 2771 (1999).

Chapter 4

Electrochemical Windows of Room-Temperature Ionic Liquids

Hajime Matsumoto

4.1 INTRODUCTION

Room-temperature ionic liquids (denoted RTILs) have been studied as novel electrolytes for a half-century since the discovery of the chloroaluminate systems. Recently another system consisting of fluoroanions such as BF_4^- and PF_6^-, which have good stability in air, has also been extensively investigated. In both systems the nonvolatile, noncombustible, and heat resistance nature of RTILs, which cannot be obtained with conventional solvents, is observed for possible applications in lithium batteries, capacitors, solar cells, and fuel cells. The nonvolatility should contribute to the long-term durability of these devices. The noncombustibility of a safe electrolyte is especially desired for the lithium battery [1]. RTILs have been also studied as an electrodeposition bath [2].

In considering RTILs as an electrolyte for these applications, it is important to know the electrochemical stability of the RTILs toward a particular electrode. For this purpose the electrochemical potential range, starting from the point where no electrochemical reaction is observed, has been estimated using the electrochemical method. "The electrochemical window" (denoted EW) of RTILs is a term commonly used to indicate both the potential range and the potential difference; it is

Electrochemical Aspects of Ionic Liquids Edited by Hiroyuki Ohno
ISBN 0-471-64851-5 Copyright © 2005 John Wiley & Sons, Inc.

calculated by subtracting the reduction potential from the oxidation potential. In the history of the development of RTILs, improvement of the EW RTILs was one of the important landmarks, especially since this meant an improvement in the cathodic limiting potential of the RTILs [2–7].

Cyclic voltammetry and linear sweep voltammetry are usually used for the EW estimation of the RTILs, as in conventional electrolytes. These methods are easy to use despite the complication often encountered in comparing the resulting voltammogram with other reported data and even for the same RTILs. Except for technical problems such as "IR drop," this appear to be due to the difference in the measurement conditions, including differences between the working and/or reference electrode, differences in potential scanning rate and/or cut-off current density, that determine the cathodic and anodic limiting potentials, and the like. Further the quality of the RTILs affects the resulting voltammogram. For example, the presence of a small amount of water in the RTILs cannot be neglected even at 100 ppm because the electrochemical measurements are generally very sensitive to impurities. Despite these difficulties comparisons of EW data among the various RTILs provides important information to prepare electrochemically stable RTILs.

In this chapter, the electrochemical windows of RTILs are reviewed and compared for different RTILs. The effects of the measurement conditions on the resulting voltammograms are an important part of the discussion. The use of "ferrocene" as an internal potential reference in the RTILs is also introduced.

4.2   EFFECTS OF THE MEASUREMENT CONDITIONS ON THE VOLTAMMETRY IN THE RTILS

4.2.1   Evaluation of the Reduction or Oxidation Potential of the RTILs by Voltammetry

To estimate the EW of RTILs, one obtains the reduction and oxidation potentials of the RTILs toward a certain reference electrode (RE) as in conventional electrolyte solutions. Similar problems arise with the techniques used for finding the EW in the RTILs as in estimations of organic solvent systems. For example, the different reference electrodes make it difficult to compare obtained potential data. These issues will be discussed below (see Section 4.2.2).

There is additionally the important problem involved in choosing the reduction or oxidation potential of the electrolyte solutions from either cyclic voltammetry (CV) or linear sweep voltammetry (LSV). Since the oxidation or reduction reaction of cations or anions contained in the RTILs are electrochemically irreversible in general [8–10], the corresponding reduction or oxidation potential cannot be specifically obtained, unlike the case of the redox potential for an electrochemically reversible system. Figure 4.1 shows the typically observed voltammogram (LSV) for RTILs. Note that both the reduction and oxidation current monotonically increase with the potential sweep in the cathodic and anodic directions, respectively. Since no peak is observed even at a high current density (10 mA cm^{-2}), a certain

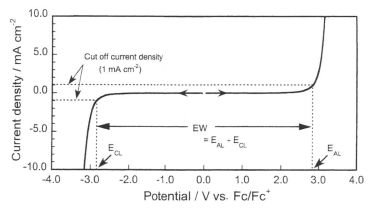

Figure 4.1 *Linear sweep voltammogram of RTILs as typically observed. Two separate voltammograms are indicated here.*

current density, which is called the cut-off current, must be selected to evaluate the reduction or oxidation potential. In this figure the reduction or the oxidation potential—in other words, the cathodic limiting or anodic limiting potential, which are denoted as E_{CL} and E_{AL}, respectively—is evaluated when the reduction or oxidation current reaches $1.0 \, mA \, cm^{-2}$. This means that both E_{CL} and E_{AL} will change because of the arbitrarily selected cut-off current density. In many studies, the cut-off current density is selected as between 0.1 and $1.0 \, mA \, cm^{-2}$. In studies of the application in capacitors, however, the cut-off current density has been selected to be below $0.1 \, mA/cm^2$ [11–13]. This may be due to the dependence of the performance of devices on the decomposition potential of the electrolyte. The comparison of different RTILs indicates that the current scale of a voltammogram reflects the current density or surface area of the working electrode.

RTILs differ from conventional electrolytes by their viscosity. Even relatively low viscous RTILs possess 20 to nearly 200 mPas at 25°C, which is two or three orders of magnitude greater than that of conventional solvents except for the $F(HF)_{2.3}^{-}$ systems [14]. The conductivity of RTILs is not so much low ($0.1 \sim 20 \, mS \, cm^{-1}$ at 25°C) due to its high concentration ($3 \sim 6 \, mol \, dm^{-3}$), however, it seemed to be necessary to considerer the compensation of the "IR-drop" or "ohmic drop" especially in cases of highly viscous RTILs. However, this problem can be circumvented by selecting an appropriate size of working electrode to reduce the specific current. Table 4.1 shows rough estimates of the IR-drop for a cut-off current of $1.0 \, mA \, cm^{-2}$ through a working electrode disk of various diameters. Note from the rough estimates that the microelectrode appears to be more suitable for use for the EW even in highly viscous RTILs because the IR-drop is almost negligible ($<100 \, \mu V$). An additional merit to the use of a microelectrode is that the mass transport property is improved with the change in the diffusion mechanism [15]. The same situation is also achieved by the enhancement of the mass transport by means of the forced convection with a rotating disk electrode (RDE) system [16]. Both

TABLE 4.1   Calculated IR-drop at 1.0 mA cm^{-2} Flow Through a Disk Working Electrode

Conductivity (mS cm^{-1})	Viscosity^a (mPas)	Resistance^b (kΩ)	iR-drop^c		
			$d = 32$ mm	$d = 1.6$ mm	$d = 1.0$ cm
0.1	500 ~1000	10	80 μV	0.2 V	7.85 V
1	100~500	1	8 μV	20 mV	785 mV
5.0	40~100	0.2	1.6 μV	4 mV	157 mV
10.0	20~40	0.1	0.8 μV	2 mV	78.5 mV

^{a}Viscosity is a typical value actually observed.
^{b}Roughly estimated by multiplying the distance between the working electrode and the reference electrode with the conductivity of RTILs, assuming that the distance is 1.0 cm. It is noted that the estimated value change with the distance and both electrode must be closed as much as possible.
^{c}d = diameter.

methods seem more suitable to a highly viscous RTIL ($>$500 mPa s at 25°C) with low conductivity. There is a brief report on the microelectrode [17–20] or the RDE [18] applied to the estimation of EW of RTILs. On the other hand, in the case of RTILs with a moderate viscosity ($<$100 mPa s at 20°C), it is not necessary to excessively consider IR-drop even when working disk electrodes with a relatively small size (d $<$ 5 mm) are used as shown in Table 4.1. The roughly estimated IR-drop in such cases might not be as large (lower than tens of milli-volts).

4.2.2   Reference Electrode

The configuration of the reference electrode (RE) obtained in RTILs so far is listed in Table 4.2. Note that the configuration of the RE used in RTILs is almost the same as that in an organic solvent. In all cases, the reference electrode contains a redox system that has a stable potential. For the chloroaluminate system, the chloroaluminate anion ($Al_2Cl_7^-$) itself is involved in the potential determining reaction. To avoid contaminating the target RTILs from the redox species in the RE, the content of the reference electrode is separated by a filtering material such as glass frit.

Figure 4.2 shows the EW data for EMI–BF$_4$ obtained from voltammograms of different reference electrodes [31, 33, 34]. Note that although the range of the EW is almost the same (4.4 V), the cathodic and anodic limiting potentials are different, even in the same RTIL. One way around this problem is to use one redox compound as the internal reference. For an organic solvent system, the IUPAC recommends the use of a redox potential of ferrocene (dicyclopentadienyliron(II)) as the reference potential regardless of the kind of solvent molecule [35]. For RTILs, there are a few reports referring to the redox potential of ferrocene in RTILs. Table 4.2 shows that the redox potential of ferrocene in RTILs is identical to the content of the RE [21, 24, 30]. Though the validity of ferrocene has not yet been sufficiently RTILs, there are a few reports of voltammograms with a ferrocene internal reference [36–38]. This discussion will show that the reported reduction and oxidation

TABLE 4.2 Reference Electrode (RE) Used in RTILs

Number	RTILs in RE	Redox Material	Electrode Material	Potential Determing Reaction	Redox Potential of Ferrocene[a]	Reference
Chloroaluminate system						
1	BPC + AlCl$_3$ ($N = 0.67$)	Al$_2$Cl$_7^-$	Al wire	Al + 7AlCl$_4^-$ ↔ 4Al$_2$Cl$_7^-$ + 3e$^-$	0.24	21
2	BPC + AlCl$_3$ ($N = 0.60$)	Al$_2$Cl$_7^-$	Al wire	Al + 7AlCl$_4^-$ ↔ 4Al$_2$Cl$_7^-$ + 3e$^-$	0.42[b]	22
3	EMIC + AlCl$_3$ ($N = 0.67$)	Al$_2$Cl$_7^-$	Al wire	Al + 7AlCl$_4^-$ ↔ 4Al$_2$Cl$_7^-$ + 3e$^-$	0.27[b]	23
4	EMIC + AlCl$_3$ ($N = 0.60$)	Al$_2$Cl$_7^-$	Al wire	Al + 7AlCl$_4^-$ ↔ 4Al$_2$Cl$_7^-$ + 3e$^-$	0.40	24
5	EMIC + AlCl$_3$ ($N = 0.44$)	Al$_2$Cl$_7^-$	Al wire	Al + 7AlCl$_4^-$ ↔ 4Al$_2$Cl$_7^-$ + 3e$^-$	—	4
6	DMABPC + AlCl$_3$ ($N = 0.67$)	Al$_2$Cl$_7^-$	Al wire	Al + 7AlCl$_4^-$ ↔ 4Al$_2$Cl$_7^-$ + 3e$^-$	0.29[c]	5
7	DMEEMAC + AlCl$_3$ ($N = 0.40$)	Al$_2$Cl$_7^-$	Al wire	Al + 7AlCl$_4^-$ ↔ 4Al$_2$Cl$_7^-$ + 3e$^-$	—	25
8	DMPIC+ AlCl$_3$($N = 0.67$)	Al$_2$Cl$_7^-$	Al wire	Al + 7AlCl$_4^-$ ↔ 4Al$_2$Cl$_7^-$ + 3e$^-$	—	26
9	DMPIC+ AlCl$_3$($N = 0.60$)	Al$_2$Cl$_7^-$	Al wire	Al + 7AlCl$_4^-$ ↔ 4Al$_2$Cl$_7^-$ + 3e$^-$	—	27
10	PMIC+ AlCl$_3$($N = 0.60$)	Al$_2$Cl$_7^-$	Al wire	Al + 7AlCl$_4^-$ ↔ 4Al$_2$Cl$_7^-$ + 3e$^-$	—	28
Nonchloroaluminate system						
11	EMI–CF$_3$SO$_3$	[TMAI] = 60 mM, [I$_2$] = 12 mM	Pt wire	3I$^-$ ↔ I$_3^-$ + 2e$^-$	—	29
12	EMI–CH$_3$SO$_3$	[TMAI] = 60 mM, [I$_2$] = 12 mM	Pt wire	3I$^-$ ↔ I$_3^-$ + 2e$^-$	—	29
13	EMI–TFSI	[TPAI] = 60 mM, [I$_2$] = 15 mM	Pt wire	3I$^-$ ↔ I$_3^-$ + 2e$^-$	0.195	30
14	EMI–BF$_4$	AgBF$_4$ (saturated)	Ag wire	Ag ↔ Ag$^+$ + e$^-$	—	31
15	EMI–TaF$_6$	AgTaF$_6$ (saturated)	Ag wire	Ag ↔ Ag$^+$ + e$^-$	—	33

Note: **BPC**: *N*-butylpyridiniumchloride, **EMIC**: 1-ethyl-3-methylimidazolium-chloride; **DMABPC**: *p*-dimethylamino-*N*-butylpyridiniumchloride; **DMEEMAC**: dimethylethylethoxymethylammonium-cholride; **DMPIC**: 1,2-dimethyl3-propylimidazoliumchloride; **PMIC**: 1-propyl-3-methylimidazoliumchloride; **TMAI**: tetramethylammoniumiodide; **TPAI**: tetrapropyl-ammoniumiodide; **TFSI**: bis(trifluoromethylsulfonyl) imide.

[a]Redox potential of ferrocene in the same RTILs that contained in RE.

[b]These value are not obtained experimentally. Calculated with considering the reported potential difference depending on the composition of chloraualuminate. These data are shown in the reference in the table.

[c]The potential difference between #1 and #5 is 0.05 V.

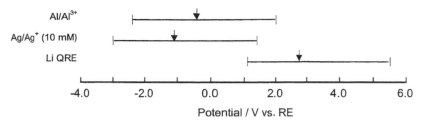

Figure 4.2 *Electrochemical windows of EMI–BF$_4$ estimated with various reference electrodes. The arrow shows the position of a certain redox material such as ferrocene (the position is virtually indicated).*

potential of RTILs can be corrected by the use of the redox potential of ferrocene, if available. A comparison of E_{Cl} and E_{AL} for various RTILs corrected by the use of ferrocene will be presented later.

When the RTIL contained in the RE is different from the target RTIL, an experimental error can be generated because of the junction potential between these RTILs. However, the use of an internal reference such as ferrocene can remove the junction potential. This is because the redox potential measured with the RE contains the same junction potential.

A quasi-reference electrode (QRE) is also used for voltammetry in both conventional electrolyte solutions and RTILs. Pt [18–19, 38], Li [39], and Ag [12, 34, 40, 47] metals are the most commonly used QRE in reports on the EW of RTILs. The merit of the QRE is that the junction potential, as stated above, is negligible because these metals are directly immersed in the target RTILs. Taking the easiness and convenience of a QRE into consideration, it should be noted that the potential determining reaction is unclear when only a metal is used as the QRE. Furthermore, there is the possibility that some by-products generated during the voltammetry can affect the potential of the QRE. To avoid this problem, the relatively high concentration of the metal cation (10 $\sim$ 100 mM) dissolved in a target RTIL or an electrolytic current is kept as low as possible (nA $\sim$ μA). The addition of a small amount of Ag$^+$ has been shown to shift the redox potential of Ag/Ag$^+$ over a few hundred mV compared with that of Ag QRE. Therefore an internal reference such as ferrocene should be used as in conventional organic electrolytes.

4.2.3 Effect of Residual Water on Voltammetry

The major impurities contained in RTILs are water and oxygen, even in highly pure samples. This is because these molecules are easily dissolved into the RTILs from air. Since these molecules are electrochemically active, the removal of these molecules is essential before any voltammetric measurement. The effect of water on the EW of RTILs has been studied in RTILs containing the imidazolium cation and fluoroanions (1-butyl-3-methylimidazolium tetrafluoroborate or hexafluorophosphate). Both the E_{CL} and E_{AL} are reported to have shifted toward positive and negative potentials, respectively, with the addition of more than 3 wt% of water, and this has been shown to cause a decrease in the wide EW of over 2.0 V [19]. However, there is a study of the effects of a very small amount of water (10.5 mM, ca. 100 ppm) on the voltammogram of an RTIL containing N-butylpyridinium (BP) and chloroaluminate anion [48]. For glassy carbon (GC) and tungsten (W) working electrodes, no significant effects were observed regardless of the addition of water. For a platinum working electrode, a reduction current of HCl, which is generated from the reaction between water and the chloroaluminate anion, was observed at a potential much more positive than that of the cathodic limiting potential (E_{CL}) of this system. These studies indicate the importance of removing water from the RTILs.

Figure 4.3(a) shows a linear sweep voltammogram (LSV) with platinum as the working electrode in TMPA–TFSI (trimethylpropylammonium-bis(trifluoromethyl)-imide) containing various amounts of water. The amount of water in the RTIL was

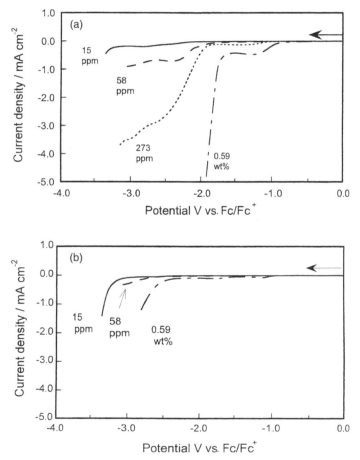

Figure 4.3 Linear sweep voltammogram of trimethylpropylammonium-bis(trifluoromethylsulfo-nyl)imide [TMPA-TFSI] containing various amounts of water. T = 25° C; working electrodes: (a) Pt (0.02 cm²) and (b) GC (0.00785 cm²). The scan rate is 50 mV s⁻¹.

controlled by vacuum drying of an air-saturated sample [49]. Before sample drying (water content was 0.59 wt%), the peak for the reduction of oxygen was observed at -1.2 V vs. Fc/Fc$^+$, and the reduction wave for water was observed at below -1.5 V vs. Fc/Fc$^+$. Both reduction currents decreased with the decreasing amount in the TMPA–TFSI. This fact indicates that the reduction wave derived from oxygen and water both overlapping the reduction wave for TMPA–TFSI, which appeared after sufficient dehydration of the sample (15 ppm). Even with the presence of 58 ppm water, the reduction in current was observed, and the current may mislead the cathodic limiting potential of the sample by about a few hundred mV toward the positive potential. This phenomenon indicates the importance of dehydration in samples below 50 ppm. A glove box filled with an inert gas such as Ar or N$_2$ is essential to avoid contamination by water and oxygen during the electrochemical measurements. As stated above, GC and W are not sensitive to residual amounts of water in the chloroaluminate system [48]. For a nonchloroaluminate system the glassy carbon (GC) electrode is insensitive to water, as shown in Figure 4.3(b). Fortunately, because of RTILs' nonvolatile nature the water can be easily removed by vacuum drying, unlike the case of conventional organic solvents. For example, after vacuum drying (10^{-4} Torr) at 100°C, the water content in a hydrophobic RTIL such as EMI–TFSI will be less than 20 ppm. In general, the time needed to decrease the water content below 20 ppm will depend on the sample volume (e.g., 1 mL, 30 min) [49].

4.2.4 Effect of Scan Rate on the Voltammogram

The scan rate is an important parameter for potential sweep methods such as CV or LSV. The current is proportional to the square root of the scan rate in all electrochemical systems—irreversible, reversible, and quasi-reversible systems. Figure 4.4 shows the LSV for EMI–TFSI using a glassy carbon (GC) [49]. Note in this figure

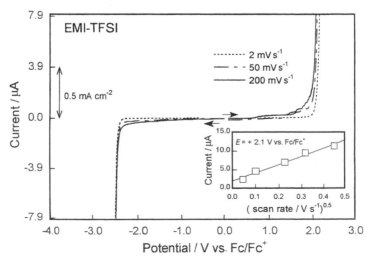

Figure 4.4 *Linear sweep voltammogram of EMI–TFSI at various scan rates. Inset shows the relationships between the current density at +2.1 V vs. Fc/Fc$^+$ and square root of current density. T = 25° C; working electrode: GC (0.00785 cm^2).*

that in the relationships between the oxidation current at a certain potential and the square root of the current, the oxidation current density obeys the theoretical prediction. This indicates that the limiting potential estimated at the same current density can be shifted in the negative or positive direction by changing the scan rate. Therefore, in the case of mutual comparison among various reported voltammgrams of RTILs taken at different scan rates, the cut-off current density in each case should be changed with the relationship between square root of scan rate and a current density.

4.3   MUTUAL COMPARISONS OF ELECTROCHEMICAL WINDOWS OF RTILS

One of the aims of this chapter is to show how the electrochemical windows compare for the various RTILs reported so far. To reduce the difference in the measurement conditions, a cathodic and anodic limiting potential is tabulated from the reported voltammogram, whose current scale can be indicated by a current density. The cut-off current density is changed by the scan rate based on the relationship between the electrolytic current and the scan rate. The basis for the selection of the cut-off current is $0.5 \, \text{mA} \, \text{cm}^{-2}$ at $50 \, \text{mV} \, \text{s}^{-1}$. In a case where the maximum current density observed in a voltammogram is lower than the cut-off current density, the limiting potential is taken at the maximum current. However, when the current scale is much greater than the cut-off current density, the limiting potential is taken at a cut-off current density that is as low as possible.

To show the possible effect of the internal potential reference, the redox potential of ferrocene in the RTILs is used here for the correction of the potential data estimated for the various reference electrodes.

4.3.1   Chloroaluminate Systems

Chloroaluminate systems have been intensively studied over the past few decades mainly in electrochemical research fields. Many of the electrochemical properties of RTILs have been sorted out using the chloroaluminate systems [50]. Chloroaluminate anions form RTILs with various organics, including not only EMI and BP but also triazolium [51] and aliphatic onium cations [7, 52–54]. Other unique RTILs similar to the chloroaluminate systems such as the chlorogallate ($GaCl_4^-$) [55], chloroborate (BCl_4^-) [56], and bromoaluminate ($AlBr_4^-$) systems [57] have been reported.

The most significantly different feature of this system, compared with other nonchloroaluminate systems, is that the composition changed with the mole ratio between the organic chloride and $AlCl_3$. Also the corresponding EW changed because of the change in the potential determining reaction. Usually the mole fraction of $AlCl_3$ is denoted as N ($=[AlCl_3]/([AlCl_3]+[RCl])$). The chloroaluminate systems are categorized by the N value as follows: $N > 0.5$, $N = 0.5$, and $N < 0.5$, which are called acidic, neutral, and basic melts, respectively. The cathodic or anodic limiting potential estimated from the reported voltammogram is shown in Table 4.3. Often aluminum wire immersed in an acidic melt ($N = 0.60$ or

TABLE 4.3 Electrochemical Windows of RTIL Based on Chloroaluminate

Cation	N^a	WE^b	S^c (mm^2)	RE^d	v^e (mV s^{-1})	I^f (mAcm^{-2})	E_{CL}^g V vs. RE	(V vs. Fc/Fc$^+$)	E_{AL}^h V vs. RE	(V vs. Fc/Fc$^+$)	EW Range^i (V)	Reference
DMPI	0.60	Pt	20	8	20	0.32	0.0		2.5		2.5	26
DMPI	0.52	W	7.85	9	100	0.20	0.1		2.3		2.2	27
DMPI	0.50	GC	70	8	20	0.32	−2.5		2.3		4.7	26
DMPI	<0.50	Pt	20	8	20	0.32	−2.4		1.0		3.5	26
DMPI	0.45	W	7.85	9	100	0.40	−2.3		—		—	27
PMI	0.55	W	7.85	10	100	0.70	—		2.7		—	28
PMI	0.45	W	7.85	10	100	0.50	−2.2		—		—	28
EMI	0.60	W	78.4	3	100	0.70	−0.2	−0.4	2.3	2.1	2.5	58
	0.55	W	78.4	4	50	0.50	−0.5	−0.9	2.5	2.1	3.0	6
	0.55	W	28.3	4	20	0.32	0.0	−0.4	2.6	2.2	2.6	59
	0.50	GC	70	8	20	0.32	−2.2		2.4		4.6	26
	0.50	W	78.4	4	50	0.50	−1.9	−2.3	2.5	2.1	4.4	6
	0.50	W	28.3	4	20	0.32	−1.9	−2.3	2.5	2.1	4.4	59
	0.47	W	28.3	4	20	0.32	−2.0	−2.4	1.0	0.6	3.0	59
	0.47	W	78.4	3	100	0.70	−2.2	−2.4	0.8	0.5	3.0	58
	0.44	W-micro	$d = 25$ mm	4	2000	50	−2.2	−2.6	1.2	0.8	3.4	17
	0.40	W	78.4	4	50	0.50	−2.0	−2.4	0.9	0.5	2.9	6
DMABP	0.67	GC	—	6	200	0.14	−0.5	−0.8	1.7	1.4	2.1	5
	0.49	GC	—	6	200	0.14	−1.7	−2.0	0.9	0.6	2.6	5
BP	0.67	GC	—	1	100	0.01	−0.1	−0.3	1.9	1.7	2.0	21
	0.60	W	78.4	1	100	0.70	−0.1	−0.4	2.4	2.1	2.5	58
	0.50	W	78.4	4	50	0.50	−1.1	−1.5	2.5	2.1	3.6	6
	0.50	GC	—	1	100	0.01	−1.1	−1.3	1.4	1.2	2.5	21
	0.47	W	78.4	1	100	0.70	−1.2	−1.4	0.7	0.5	1.9	58

0.44	W	77	1	50	0.20	−1.1	−1.3	0.9	0.7	2.0	48
0.44	GC	71	1	50	0.20	−1.1	−1.3	0.8	0.5	1.8	48
0.44	Pt	49	1	50	0.20	−1.2	−1.4	0.7	0.5	1.9	48

Note: BP: *N*-butylpyridinium; **EMI**: 1-ethyl-3-methylimidazolium; **DMABP**: *p*-dimethylamino-*N*-butylpyridinium; **DMPI**: 1,2-dimethyl3-propylimidazolium; **PMI**: 1-propyl-3-methylimidazolium.

[a]Mole fraction of $AlCl_3$.
[b]Working electrode.
[c]Area of working electrode.
[d]Reference electrode (see Table 4.2).
[e]Scan rate.
[f]Cut-off current.
[g]Cathodic limiting potential.
[h]Anodic limitng potential.
[i]Range $= E_{AL} - E_{CL}$.

0.67) is used as the RE. Since the potential determining reaction is aluminum deposition from the acidic melt on an Al wire, the potential scale in the voltammogram is often denoted as V vs. Al/Al^{3+}. The potential differences between these two REs is not very large (130 ∼ 180 mV), and it can be obtained from the reported potentiometric titration curve, if available [22–24, 26]. Figure 4.5 shows the EWs of various chloroaluminate systems, excluding the potential differences in the RE. From this figure it can be confirmed that the EW of the chloroaluminate systems depends on the composition of the melt. The ions involved in the potential determining reaction are specified in Table 4.4. However, the anodic limiting potential is not always determined by the oxidation of AlCl$_4^-$, as shown in the *p*-dimethylamino-*N*-butylpyridnium (DMABP) system. The anodic limiting potential of this system is caused by the oxidation of the DMABP cation [5].

Figure 4.6 shows the EWs of chloroaluminate systems with the correction of the potential differences derived from the melt composition and the kind of cation using the redox potential of ferrocene. This figure clearly indicates the effect of using the ferrocene internal reference. The cathodic limiting or anodic limiting potential is almost the same when the same cation is contained in RTILs, and is independent of the measurement conditions such as the difference of melt composition, not considered in Figure 4.5. The potential variation seen in Figure 4.5 even for the same RTILs is almost eliminated.

4.3.2 Nonchloroaluminate System

The nonchloroaluminate system has been intensively investigated, since fluoro-anions are also attractive candidates for RTILs [29, 60, 61]. The significant difference from the chloroaluminate system is its stability in air. The number of anions, which form RTILs with various cations, demonstrate an upward trend [40–45, 47, 62, 63].

The EW data for the nonchloroaluminate systems reported so far are summarized in Table 4.5. Problems arising from the difference in the measurement conditions become apparent from the data obtained in the same RTILs such as EMI–BF$_4$ and EMI–TFSI. The potential data obviously differ based on the type of RE. Furthermore the range of the EW decreases with the decreasing current density, which is used to estimate both the cathodic and anodic limiting potentials as stated above (Section 4.2.1). For these reasons comparison of the EW data can contain relatively large errors aside from the effects of the measurement conditions such as the scan rate and current density used for the determination of the limiting potential.

As was stated above, ferrocene is considered to be a potential internal reference for chloroaluminate systems. There are only a few studies that mention the redox potential of ferrocene in nonchloroaluminate systems. Table 4.5 gives the potential data corrected with the redox potential of ferrocene in each RTIL, if available. Figure 4.7 shows the EW of the RTILs. Shown in the figure are the EWs of both chloroaluminate (solid line) and nonchloroaluminate (dotted line) systems. It is interesting to note that both the cathodic and anodic limiting potentials of the RTILs based on the same cation are the same whether the system is chloroaluminate

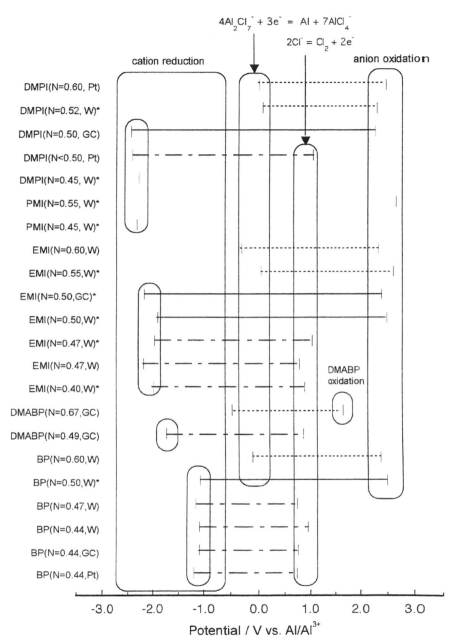

Figure 4.5 *Electrochemical window of chloroaluminate system. The asterisk indicates the composition of the melt in RE of* N = 0.60. *Another case is* N = 0.67. *The potential difference between* N = 0.60 *and* 0.67 *(130 ∼ 180 mV) is not considered. Working electrodes: GC (glassy carbon); W (tungsten); Pt (platinum).*

TABLE 4.4 Ionic Species Associated with a Cathodic or Anodic Limiting Potential of Chloroaluminate

Composition (N)	Anions in Melts	Cathodic Limiting Potential	Anodic Limiting Potential
$N > 0.5$ acidic	$Al_2Cl_7^-$, $AlCl_4^-$	$Al_2Cl_7^-$	$AlCl_4^-$
$N = 0.5$ neutral	$AlCl_4^-$	Cation	$AlCl_4^-$
$N < 0.5$ basic	$AlCl_{l4}^-$, Cl^-	Cation	Cl^-

or nonchloroaluminate. This fact suggests ferrocene to be a valid internal reference for comparing the EWs of different RTILs.

In comparisons using the ferrocene internal reference, the origin of the cathodic and anodic limiting potentials of the RTILs can be described as follows:

1. The cathodic limiting potential of the RTILs is basically determined by the reduction of the cations.

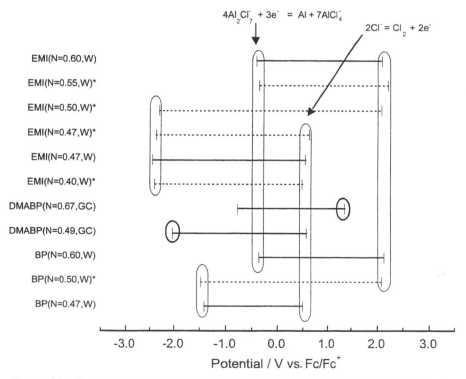

Figure 4.6 Electrochemical windows of chloroaluminate system with a correction by the ferrocene internal reference. Original data were measured with the different REs. Solid: N = 0.67; broken: N = 0.60. Working electrode: GC (glassy carbon); W (tungsten); Pt (platinum).

TABLE 4.5 Electrochemical Windows of RTIL Based on Nonchloroaliminate System

Cation	Anion	WE[a]	S^b (mm²)	RE[c]	Scan Rate (mV s⁻¹)	Cut-off Current (mAcm⁻²)	E_{CL}^d V vs. RE	E_{CL}^d V vs. Fc/Fc⁺	E_{AL}^e V vs. RE	E_{AL}^e V vs. Fc/Fc⁺	EW range[f] (V)	Reference
DMEI	TFSI	Pt	0.78	13	50	10	−2.6		2.3		4.9	30
DMPI	TFSI	GC	70	Li QRE	20	1.0	0.2		5.4		5.2	39
DMPI	TFSI	GC	77	Ag QRE	20	0.02	−1.9		2.3		4.2	12
DMPI	Methide	GC	70	Li QRE	20	1.0	0.3		5.7		5.4	39
DMPI	Methide	GC	70	Li QRE	20	0.32	0.5		5.4		4.9	39
DMFP	BF$_4$	GC	0.785	Ag QRE	50	0.5	−1.4		2.6		4.0	66
EMP	BF$_4$	Pt	1.77	Li QRE	50	0.5	1.2		5.5		4.4	67
EMI	CF$_3$SO$_3$	Pt	—	11	50	1.0	−2.2		2.1		4.3	29
EMI	CF$_3$SO$_3$	Pt	0.78	13	50	10	−2.0		2.2		4.2	30
EMI	CH$_3$SO$_3$	Pt	—	12	50	1.0	−2.0		2.1		4.1	29
EMI	BF$_4$	GC	70	6	100	0.7	−2.1	−2.5	—	—	4.4	33
EMI	BF$_4$	Pt	20	6	100	0.7	−2.0	−2.4	2.4	2.0	4.4	33
EMI	BF$_4$	Pt	0.785	13	50	5	−3.0		1.4		4.4	31
EMI	BF$_4$	Pt	—	Ag QRE	5	1.0	−2.0	(1.1)[g]	2.4	(5.5)[g]	4.4	46
EMI	BF$_4$	GC	0.785	Ag QRE	50	0.5		(0.8)[g]	—	(5.4)[g]	4.6	34
EMI	TFSI	Pt	0.78	13	50	10	−2.1	−2.3	2.5	2.3	4.6	30
EMI	TFSI	GC	77	Ag QRE	20	0.02	−2.0		2.1		4.1	12
EMI	TFSI	GC	0.785	13	50	0.5	−2.3	−2.4	2.2	2.1	4.5	64
EMI	TFSA	Pt	1.77	Ag QRE	100	0.1	−1.8		2.0		3.8	47
EMI	CF$_3$COO	Pt	0.78	13	50	10	−2.1		1.4		3.5	30
EMI	F(HF)$_{2.3}$	GC	—	Pt QRE	10	0.5		−1.9		2.1	3.9	38
EMI	TSAC	GC	0.785	12	50	0.5	−2.1	−2.3	2.0	1.9	4.1	36
EMI	TaF$_6$	Pt	—	15	—	0.5	−2.6		1.9		4.5	32
EMI	BETI	GC	77	Ag QRE	20	0.02	−2.0		2.1		4.1	12
EMI	DCA	Pt	1.77	Ag QRE	100	0.1	−1.5		1.5		3.1	47
DMI	F(HF)$_{2.3}$	GC	—	Pt QRE	10	0.5		−1.5		1.9	3.4	38
PrMI	F(HF)$_{2.3}$	GC	—	Pt QRE	10	0.5		−1.9		2.1	4.0	38
PrMI	BF$_4$	GC	0.785	Ag QRE	50	0.5		(0.9)[g]	—	(5.5)[g]	4.6	34
BMI	PF$_6$	Pt micro	0.00049	Ag QRE	100	0.5	−2.0		1.7		3.7	20
BMI	F(HF)$_{2.3}$	GC	—	Pt QRE	10	0.5		−1.9		2.1	4.1	38
BMI	BF$_4$	GC	0.785	Ag QRE	50	0.5		(0.73)[g]		(5.4)[g]	4.7	34

TABLE 4.5 (Continued)

Cation	Anion	WE[a]	S[b] (mm²)	RE[c]	Scan Rate (mV s⁻¹)	Cut-off Current (mAcm⁻²)	E_{CL}^{d} V vs. RE	E_{CL}^{d} V vs. Fc/Fc⁺	E_{AL}^{d} V vs. RE	E_{AL}^{d} V vs. Fc/Fc⁺	EW range[f] (V)	Reference
BMI	BF₄	Pt micro	0.002	Pt QRE	100	0.02	−1.8		2.6		4.4	19
PeMI	F(HF)₂.₃	GC	—	Pt QRE	10	0.5		−2.1		2.1	4.2	38
HMI	F(HF)₂.₃	GC		Pt QRE	10	0.5		−2.4		2.1	4.5	38
TBMA	TFSI	Pt micro	0.00049	Ag QRE	100	0.5	−3.5		1.7		5.2	20
P13	TFSA	Pt	1.77	Ag QRE	100	0.5	−1.9		2.8		4.7	47
P13	PTS	Pt	1.77	Ag QRE	100	0.1	−1.3		1.6		2.9	47
PP13	TFSI	GC	0.785	13	50	0.5	−3.2	−3.3	2.7	2.5	5.8	37
TMPA	TFSI	GC	0.785	13	50	0.5	−3.0	−3.1	2.6	2.5	5.6	64
TMMMA	TFSI	GC	0.785	13	50	0.5	−2.7		2.5		5.2	64
TMPhA	TFSI	GC	0.785	13	50	0.5	−2.3	−2.4	2.3	2.2	4.6	64
TeEA	TSAC	GC	0.785	13	50	0.5	−2.4	−2.6	2.2	2.1	4.6	36
TES	TFSI	GC	0.785	13	50	0.5	−2.1		2.7		4.8	65
TBS	TFSI	GC	0.785	13	50	0.5	−2.3		2.6		4.9	65
BP	TFSI	GC	0.785	13	50	0.5	−1.3	−1.4	2.5	2.4	3.9	64
DMABP	TFSI	GC	0.785	13	50	0.5	−2.0	−2.1	1.5	1.4	3.5	49

Cations: DMEI: 1,2-dimethyl-3-ethylimidazolium; DMPI: 1,2-dimethyl-3-propylimidazolium; EMI: 1-ethyl-3-methylimidazolium; DMFP: 1,2-dimethyl-4-fluoropyrazolium; EMP: 1-ethyl-2-methylpyrazolium; DMI: 1,3-dimethylimidazolium; PrMI: 1,propyl-3-methylimidazolium; BMI: 1-butyl-3-methylimidazolium; PeMI: 1-pentyl-3-methylimidazolium; HMI: 1-hexyl-3-methylimidazolium; TBMA: tributylmethylammonium; P13: *N*-methyl-*N'*-propyl-pyrrolidinium; TMMMA: trimethylmethoxymethylammonium; TMPhA: trimethylphenylammonium; TeEA: tetraethylammonium; TES: triethylsulfonium; TBS: tributylsulfonium; BP: *N*-butylpyridinium; DMABP: *p*-dimethylamino-*N*-butylpyridinium. Anions: TFSI: *bis*(trifluoromethylsulfonyl)imide, Methide: *tris*(trifluoromethylsulfonyl)methane; TFSA:*bis*(trifluoromethylsulfonyl)amide (=TFSI); TSAC: 2,2,2-trifluoro-*N*-(trifluoromethylsulfonyl)amide; BETI: bis(perfluoroethylsulfonyl)imide; DCA: dicyanamide; PTS: *p*-toluensulfonate.

[a]Working electrode.
[b]Area of working electrode.
[c]Reference electrode (see Table 4.1).
[d]Cathodic limiting potential.
[e]Cut-off current.
[f]Anodic limiting potential.
[g]Range $= E_{AL} - E_{CL}$.
[h]These potential data (V vs. Li/Li⁺) are recorrected with the potential of Li/Li⁺ in each RTILs.

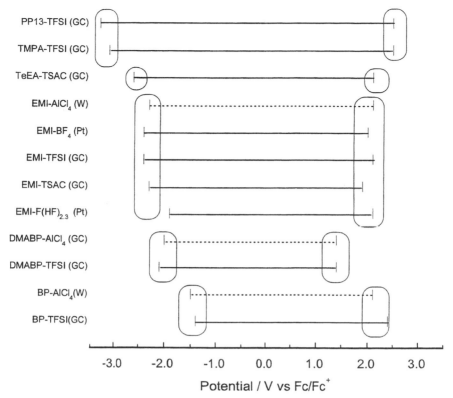

Figure 4.7 *Electrochemical windows for RTILs using the ferrocene internal reference. Solid line: nonchloroaluminate system; dotted line: chloroaluminate system. Working electrode: GC (glassy carbon); W (tungsten); Pt (platinum).*

2. However, if the cathodic stability of the anion is poor, the cathodic limiting potential is determined by the anion reduction and not the cation reduction as seen for the TSAC anion. In comparing the cathodic and anodic limiting potentials of PP13 or TMPA–TFSI with that of TeEA–TSAC, the TSAC is seen to determine not only the cathodic limiting potential, but also anodic limiting potential.

3. The anodic limiting potential of RTILs is basically determined by the oxidation of anions. However, the anodic limiting potential of EMI and DMABP is determined not by the oxidation of the anions, but by the cation, since the anodic limiting potential of the EMI and DMABP system is independent of the anions.

4. The EW of aliphatic quaternary ammonium systems based on fluoroanions has the widest EW compared with that of aromatic systems. This result can be explained by the presence of quaternary ammonium salts, which are usually used as a supporting electrolyte in the conventional electrolyte and which have wide electrochemical windows.

4.4 CONCLUSION

This chapter described the effects of voltammetric measurement conditions on the electrochemical window. With the use of ferrocene as an internal reference, a comparison between the electrochemical windows studied so far could be carried out to a greater degree (± 0.1 V).

Other possible comparisons of EW data that have not been described in the text are as follows:

1. The effect of a surface passivation layer simultaneously generated with the decomposition of the RTILs, with respect to the effect of the electrode material on the cathodic or anodic limiting potential.
2. A comparison the EW data of the RTILs with that of conventional solvent systems with respect to the effect of the interaction between cations and anions on the cathodic or anodic limiting potential.

These investigatiions can help elucidate the origin of the electrochemical windows of RTILs. The purity of the RTILs must always be a major consideration when discussing the electrochemical window of RTILs.

REFERENCES

1. A. Webber, G. E. Blomgren, in *Advances in Lithium-Ion Batteries*, W. van Schalkwijk, B. Scrosati, eds., Kluwer Academic, 2002, p. 185.
2. F. H. Hurley, T. P. Wier Jr., *J. Electrochem. Soc.*, **98**, 203 (1951).
3. H. L. Chum, V. R. Koch, L. L. Miller, R. A. Osteryoung, *J. Am. Chem. Soc.*, **97**, 3264 (1975).
4. J. S. Wikes, J. A. Levisky, R. A. Wilson, C. L. Hussey, *Inorg. Chem.*, **21**, 1263 (1982).
5. G. T. Cheek, R. A. Osteryoung, *Inorg. Chem.*, **21**, 3581 (1982).
6. M. Lipsztajin, R. A. Osteryoung, *J. Electrochem. Soc.*, **130**, 1968 (1983).
7. G. E. Blomgren, S. D. Jones, in *Molten Salts*, R. J. Gale, G. Blomgren, H. Kojima, eds., Proc. Vol. 92-16, Electrochemical Society, 1992, p. 379.
8. M. Lipsztajn, R. A. Osteryoung, *Electrochim. Acta*, **29**, 1349 (1984).
9. J. Xie, T. L. Riechel, *J. Electrochem. Soc.*, **145**, 2660 (1998).
10. L. Xiao, K. E. Johnson, *J. Electrochem. Soc.*, **150**, E307(2003).
11. M. Ue, M. Takeda, M. Takehara, S. Mori, *J. Electrochem. Soc.*, **144**, 2684(1997).
12. A. B. McEwen, H. L. Ngo, K. LeCompte, J. L. Goldman, *J. Electrochem. Soc.*, **146**, 1687(1999).
13. K. Xu, M. S. Ding, T. R. Jow, *J. Electrochem. Soc.*, **148**, A267(2001).
14. R. Hagiwara, Y. Ito, *J. Fluorine. Chem.*, **99**, 1, (1999)
15. J. Heinze, K. Borgwarth, in *Electrochemical Microsystem Technologies*, J. W. Schultze, T. Osaka, M. Datta, eds., Taylor and Francis, 2002, p. 32.
16. A. J. Bard, L. R. Faulkner, *Electrochemical Methods*, Wiley, 1980, p. 280.

17. R. T. Carlin, R. A. Osteryoung, *J. Electroanal. Chem.*, **252**, 81 (1988).

18. P. A. Z. Suarez, V. M. Selbach, J. E. L. Dullius, S. Einloft, C. M. S. Piatnicki, D. S. Azambuja, R. F. de Souza, J. Dupont, *Electrochim. Acta*, **42**, 2533 (1997).

19. U. Schroder, J. D. Wadhawan, R. G. Compton, F. Marken, P. A. Z. Suarez, C. S. Consorti, R. F. de Souza, J. Dupont, *New J. Chem.*, **24**, 1009 (2000).

20. B. M. Quinn, Z. Ding, R. Moulton, A. J. Bard, *Langmuir*, **18**, 1734 (2002).

21. J. Robinson, R. A. Osteryoung, *J. Am. Chem. Soc.*, **101**, 323 (1979).

22. R. J. Gale, R. A. Osteryoung, *Inorg. Chem.*, **18**, 1603 (1979).

23. C. L. Hussey, T. B. Scheffler, J. S. Wilkes, A. A. Fannin Jr., *J. Electrochem. Soc.*, **133**, 1389 (1986).

24. C. Scordilis-Kelley, J. Fuller, R. T. Carlin, J. S. Wilkes, *J. Electrochem. Soc.*, **139**, 694 (1992).

25. J. R. Stuff, S. W. Lander, J. W. Rovang, J. S. Wilkes, *J. Electrochem. Soc.*, **137**, 1492 (1990).

26. P. R. Gifford, J. B. Palmisano, *J. Electrochem. Soc.*, **134**, 610 (1987).

27. G. E. Gray, J. Winnick, P. A. Kohl, *J. Electrochem. Soc.*, **143**, 2262 (1996).

28. G. E. Gray, J. Winnick, P. A. Kohl, *J. Electrochem. Soc.*, **143**, 3820 (1996).

29. E. I. Cooper, E. J. M. O'Sullivan, in *Molten Salts*, R. J. Gale, G. Blomgren, H. Kojima, eds., Proc. Vol. 92-16, Electrochemical Society, 1992, p. 386.

30. P. Bonhôte, A. Dias, N. Papageorgiou, K. Kalyanasundaram, M. Grätzel, *Inorg. Chem.*, **35**, 1168 (1996).

31. Y. Katayama, S. Dan, T. Miura, T. Kishi, *J. Electrochem. Soc.*, **148**, C102 (2001).

32. K. Matsumoto, R. Hagiwara, Y. Ito, *J. Fluorine. Chem.*, **115**, 133 (2002).

33. J. Fuller, R. T. Carlin, R. A. Osteryoung, *J. Electrochem. Soc.*, **144**, 3881 (1997).

34. T. Nishida, Y. Tashiro, M. Yamamoto, *J. Fluorine. Chem.*, **120**, 135 (2003).

35. A. K. Covington, *Pure Appl. Chem.*, **57**, 531 (1985).

36. H. Matsumoto, H. Kageyama, Y. Miyazaki, in *Molten Salts XIII*, P. C. Trulove, H. C. DeLong, R. A. Mantz, G. R. Stafford, M. Matsunaga, eds., Proc. Vol. 2002-19, Electrochemical Society, 2002, p. 1057.

37. H. Sakaebe, H. Matsumoto, *Electrochem. Commun.*, 594 (2003).

38. R. Hagiwara, K. Matsumoto, Y. Nakamori, T. Tsuda, Y. Ito, H. Matsumoto, K. Momota, *J. Electrochem. Soc.*, **150**, D195 (2003).

39. V. R. Koch, L. A. Dominey, C. Nanjundiah, M. J. Ondrechen, *J. Electrochem. Soc.*, **143**, 798 (1996).

40. J. Sun, M. Forsyth, D. R. MacFarlane, *J. Phys. Chem. B*, **102**, 8858 (1998).

41. D. R. MacFarlane, P. Meakin, J. Sun, N. Amini, M. Forsyth, *J. Phys. Chem. B*, **103**, 4164 (1999).

42. D. R. MacFarlane, J. Golding, S. Forsyth, M. Forsyth, G. B. Deacon, *Chem. Commun.*, 1430 (2001).

43. D. R. MacFarlane, S. Forsyth, J. Golding, G. B. Deacon, *Green Chem.*, **4**, 444 (2002).

44. S. Forsyth, D. R. MacFarlane, *J. Mater. Chem.*, **13**, 2451 (2003).

45. J. M. Pringle, J. Golding, K. Baranyai, C. M. Forsyth, G. B. Decon, J. L. Scott, D. R. MacFarlane, *New. J. Chem.*, **27**, 1504 (2003).

46. H. Nakagawa, S. Izuchi, K. Kuwana, T. Nukuda, Y. Aihara, *J. Electrochem. Soc.*, **150**, A695 (2003).

47. J. N. Barisci, G. G. Wallace, D. R. MacFarlane, R. H. Baughman, *Electrochem. Commun.*, **6**, 22 (2004).

48. S. Sahami, R. A. Osteryoung, *Anal. Chem.*, **55**, 1970 (1983).

49. H. Matsumoto, unpublished results.

50. J. S. Wilkes, *Green Chem.*, **4**, 73 (2002).

51. B. Vestergaard, N. J. Bjerrum, I. Petrushina, H. A. Hjuler, R. W. Berg, M. Begtrup, *J. Electrochem. Soc.*, **140**, 3108 (1993).

52. S. D. Jones, G. E. Blomgren, *J. Electrochem. Soc.*, **136**, 424 (1989).

53. A. P. Abbott, C. A. Eardley, N. R. S. Farley, G. A. Griffith, A. Pratt, *J. Appl. Electrochem.*, **31**, 1345 (2001).

54. S. D. Jones, G. E. Blomgren, in *Molten Salts*, C. L. Hussey, S. N. Flengas, J. S. Wilkes, Y. Ito, eds., Proc. Vol. 90-17, The Electrochemical Society, 1990, p. 273.

55. S. P. Wicelinski, R. J. Gale, J. S. Wilkes, *J. Electrochem. Soc.*, **134**, 263 (1986).

56. S. D. Williams, J. P. Schoebrechts, J. C. Selkirk, G. Mamantov, *J. Am. Chem. Soc.*, **109**, 2218 (1987).

57. J. A. Boon, J. S. Wilkes, J. A. Lanning, *J. Electrochem. Soc.*, **138**, 465 (1991).

58. Z. J. Karpinski, R. A. Osteryoung, *Inorg. Chem.*, **23**, 1491 (1984).

59. T. J. Melton, J. Joyce, J. T. Maloy, J. A. Boon, J. S. Wilkes, *J. Electrochem. Soc.*, **137**, 3865 (1990).

60. E. I. Cooper, C. A. Angell, *Solid State Ionics*, **18**, 570 (1986).

61. R. Hagiwara, Y. Ito, *J. Fluorine Chem.*, **105**, 221, (2000), and references there in.

62. H. Matsumoto, H. Kageyama, Y. Miyazaki, *Chem. Commun.*, 1726 (2002).

63. Z. B. Zhou, M. Takeda, M. Ue, *J. Fluorine Chem.*, **123**, 127 (2003).

64. H. Matsumoto, M. Yanagida, K. Tanimoto, T. Kojima, Y. Tamiya, Y. Miyazaki, in *Molten Salts XII*, P. C. Trulove, H. C. DeLong, G. R. Stafford, S. Deki, eds., Proc. Vol. 99-41, Electrochemical Society, 2000, p. 186.

65. H. Matsumoto, T. Matsuda, Y. Miyazaki, *Chem. Lett.*, 1430 (2000).

66. J. Caja, T. D. J. Dunstan, D. M. Ryan, V. Katovic, in *Molten Salts XII*, P. C. Trulove, H. C. DeLong, G. R. Stafford, S. Deki, eds., Proc. Vol. 99-41, Electrochemical Society, 2000, p. 150.

67. J. Caja, T. D. J. Dunstan, and V. Katovic, in *Molten Salts XIII*, P. C. Trulove, H. C. DeLong, R. A. Mantz, G. R. Stafford, M. Matsunaga, eds., Proc. Vol. 2002-19, Electrochemical Society, 2002, p. 1014.

Chapter *5*

Diffusion in Ionic Liquids and Correlation with Ionic Transport Behavior

**Md. Abu Bin Hasan Susan, Akihiro Noda,
and Masayoshi Watanabe**

5.1 DIFFUSION AND DIFFUSIVITY—FUNDAMENTAL ASPECTS

Diffusion, which occurrs in essentially all matter, is one of the most ubiquitous phenomena in nature. It is the process of transport of materials driven by an external force field and the gradients of pressure, temperature, and concentration. It is the net transport of material that occurs within a single phase in the absence of mixing either by mechanical means or by convection. The rates of different technical as well as many physical, chemical, and biological processes are directly influenced by diffusive mass transfer, and also the efficiency and quality of processes are governed by diffusion [1].

The intrinsic nature of particles to perform a perpetual, irregular movement is referred to as Brownian motion, which provides the basic mechanism for diffusion [1]. Because the random fluctuations in the positions of particles in space are often translational, the kinetics of these processes can be considered comparable to the decay of a concentration gradient by translational Brownian motion. Likewise, as the orientation of any molecule undergoes similar random fluctuations in space,

Electrochemical Aspects of Ionic Liquids   Edited by Hiroyuki Ohno
ISBN 0-471-64851-5   Copyright © 2005 John Wiley & Sons, Inc.

rotational motion also occurs. In the presence of an external field, the extent of rotational order produced depends on a balance between the magnitude and direction of the applied field and the intensity of the rotational motion, which in turn depends on the thermal energy of the system. Removal of the field causes the rotational order [2]. The rotational and translation motion can be separated, and it may be assumed that the rotational motion occurs through a series of very small displacement steps, each of which occurs very rapidly [2]. The discussion in this chapter is limited to translational diffusion in isothermal, isobaric systems with no external force field gradients, and the focus is on diffusion studies under only the concentration gradient.

To get a sense of the physics behind diffusion, diffusion fluxes and diffusion potentials need to be defined in terms of driving forces. Fick's first law is used to correlate the diffusion flux density with the concentration gradient of the diffusants, which yields the diffusion coefficient as the factor of proportionality. For a system with two components and one-dimensional diffusion in the z-direction, the particle flux $\bar{j}_i$ is related to the gradient of concentration, $\delta c_i/\delta z$, of these particles as

$$\bar{j}_i = -D_i \frac{\delta c_i}{\delta z} \qquad (i = 1, 2) \tag{5.1}$$

D_i is called diffusion coeffcient or diffusivity.

Combining equation (5.1) with the law of matter conservation,

$$\frac{\delta c_i}{\delta t} + \frac{\delta[D_i(\delta c_i/\delta z)]}{\delta z} = 0 \tag{5.2}$$

results in Fick's second law:

$$\frac{\delta c_i}{\delta t} = D_i \frac{\delta^2 c_i}{\delta z^2} \tag{5.3}$$

This partial differential equation simplifies if D_i is constant and can be solved for particular initial and boundary conditions. This allows determining D_i from measurements of the concentration distribution as a function of position and time.

5.2   THE PROCESSES OF DIFFUSION IN LIQUIDS

Molecules in liquid, as compared to gases, are densely packed and are strongly affected by force fields of neighboring molecules, and the values of diffusivity for liquids are much smaller than those for low-pressure gases. The mass transfer by diffusion, however, depends on the concentration gradient, and it does not have to be low, since for a large concentration gradient the transfer rates can be high [3]. In a liquid system, translational diffusion is the most fundamental form of transport and is responsible for all chemical reactions, since the reacting species must collide before they can react. Translational diffusion can take different form, and for a systematic description, at least two elementary categories need to be considered: self-diffusion and mutual diffusion [2–5].

In a liquid that is in thermodynamic equilibrium, the individual particles move with random translational motion, and they have an equal opportunity of taking up any point in the total space occupied by the liquid. The motion, characterized as a random walk of the particles, is self-diffusion. If it becomes possible to label a particle without otherwise changing its properties, and to follow its motion through the unlabeled molecules, a self-diffusion coefficient can be described in terms of concentration gradient and be defined as in equation (5.3) [2]. A particle under such conditions can be tagged, and its trajectory can be followed over a long time. In a medium without long-range correlations, the squared displacement of the species from its original position will eventually grow linearly with time. The molecular displacements of a single particle then quantify self-diffusion. To reduce statistical errors, the self-diffusivity, D, of a species is defined by averaging Einstein's equation over a large enough number of molecules, N [1, 4]:

$$D = \left(\frac{1}{6N}\right) \sum_{k=1}^{N} \lim_{t \to \infty} \frac{1}{t} \left\langle |\vec{r}_k(t) - \vec{r}_k(0)|^2 \right\rangle \tag{5.4}$$

where $\vec{r}_k(t)$ and $\vec{r}_k(0)$ are the position of the k-th molecule of species i at time t and 0. Since species diffuse independently of each other, only one species is considered and the index i is generally dropped in the notations. However, self-diffusion is not limited to one-component systems. The random walk of particles of each component in any composition of a mixture containing multiple components can be observed.

Self-diffusivity is a measure of the translational mobility of individual molecules (or ions) driven by an internal kinetic energy. It therefore provides molecular description of matter, liquids in this case. Self-diffusion can be combined with other data such as viscosities, electrical conductivities, and densities for evaluating and improving solvodynamic models, such as the Stokes-Einstein type, and providing a deep insight into molecular transport in liquid systems [4]. The study of temperature dependencies of the self-diffusion coefficient, as well as those of other physical properties, allows to calculate activation energies of the system. The temperature dependencies help to propose diffusion models and to confirm the validity of experimental and computational data on single and multicomponent systems of electrolytes and nonelectrolytes. The temperature dependencies in most cases follow Arrhenius behavior, with some obeying the type of the Vogel-Fulcher-Tamman equation (VFT) [6].

If a liquid system containing at least two components is not in thermodynamic equilibrium due to concentration inhomogeneities, transport of matter occurs. This process is widely called mutual diffusion and it denotes the macroscopically perceptible relative motion of the individual particles due to concentration gradients.

The process is also known as chemical diffusion, interdiffusion, transport diffusion, and in the case of systems with two components, binary diffusion.

Whereas mutual diffusion characterizes a system with a single diffusion coefficient, self-diffusion gives different diffusion coefficients for all the particles in the system. Self-diffusion thereby provides a more detailed description of the single chemical species. This is the molecular point of view [7], which makes the self-diffusion more significant than that of the mutual diffusion. In contrast, in practice, mutual diffusion, which involves the transport of matter in many physical and chemical processes, is far more important than self-diffusion. Moreover mutual diffusion is cooperative by nature, and its theoretical description is complicated by nonequilibrium statistical mechanics. Not surprisingly, the theoretical basis of mutual diffusion is more complex than that of self-diffusion [8]. In addition, by definition, the measurements of mutual diffusion require mixtures of liquids, while self-diffusion measurements are determinable in pure liquids.

5.3 SELF-DIFFUSION AND IONIC TRANSPORT IN IONIC LIQUIDS

Regardless of the stochastic nature of its elementary steps, diffusion follows well-defined dependencies [1]. The wealth and beauty of these dependencies is particularly impressive when diffusion occurs in concentrated electrolyte solutions like ionic liquids. Owing to their structural variability and importance, ionic liquids represent a particularly attractive system for the study of self-diffusion and ionic transport behavior.

Ionic liquids, as the name implies, are comprised entirely of ions. These are receiving an upsurge of interest for potential use in multidisciplinary areas due to, *inter alia*, their immeasurably low vapor pressure, high ionic conductivity, non-flammability, and greater thermal and electrochemical stability [9–30]. Research to date includes numerous attempts to prepare ionic liquids based on new anions and cations with a view toward realizing unique physicochemical properties. It has already been observed that certain combinations of organic cations, such as imidazolium and pyridinium derivatives, and bulky and soft anions, such as PF_6^-, BF_4^-, $CF_3SO_3^-$, and $(CF_3SO_2)_2 N^-$, form ionic liquids at near ambient temperature [14–16,18, 22–23]. However, the fundamentals of the ionic transport properties in ionic liquids still need to be systematically explored. Conceptual understanding of ionic transport behavior in ionic liquids is not only complicated by the various physical conditions under which the diffusion phenomena may appear, it is also complicated by the fact that the spatial temporal ranges over which both diffusion phenomena are perceived by the different experimental techniques can dramatically vary.

Impedance spectroscopy, meaning the measurement of complex resistivities with ac current methods, is an important tool to study diffusion and to correlate it with ionic transport behavior. The diffusion coefficient, D_σ, obtained from conductivity measurements (*vide infra*) is related to the self-diffusion coefficient, D

(NMR measurements) by $D = H_R \cdot D_\sigma$. Here H_R is the Haven ratio, which can give information on the ionicity of the ionic liquids [22]. The alternate way is to compare the molar conductivity calculated from the self-diffusion coefficients (Λ_{NMR}) of ionic liquids with the experimental molar conductivity calculated from the ionic conductivity and the density (Λ). The ratio of the molar conductivities (Λ/Λ_{NMR}), the other form of H_R, is the indicative of the percentage of ions, which can contribute to ionic conduction [22]. Thus the fraction of the ion pairs or aggregates may be estimated from the Haven ratio as $1 - \Lambda/\Lambda_{NMR}$.

In this chapter we deal with the ionic diffusion coefficient, the ionic association, and the interaction between ions in ionic liquids. We employ self-diffusivity to interpret the ionic transport behavior and aim at evaluating the ionicity and the ionic states of the ionic liquids.

5.4 PULSED-GRADIENT SPIN-ECHO (PGSE) NMR FOR MEASUREMENTS OF SELF-DIFFUSION COEFFICIENTS

The pulsed-gradient spin-echo NMR (PGSE–NMR) method is noninvasive and provides a convenient means to independently measure the self-diffusion coefficient of each ionic species in the system of an ionic liquid, provided that the components contain NMR-sensitive nuclei. In general, the method involves the attenuation of a spin-echo signal resulting from the dephasing of the nuclear spins that comes from the combination of the translational motion of the spins under the imposition of spatially well-defined gradient pulses for the measurement of the displacement of the observed spins [31]. The PGSE experiment utilizes a spin-echo experiment coupled with two gradient pulses. By the subsequent application of these pulses, the magnetic field is made inhomogeneous over two short time intervals. The gradient pulses create a linear gradient magnetic field that varies the precessions of the nuclei in the field. The NMR signal becomes sensitive to the differences in the locations of the nuclei (and hence of the respective atoms and molecules) between the two field gradient pulses. These differences are revealed by analyzing the attenuation of the NMR signal as a function of the intensity of the gradient pulses. Self-diffusion causes an attenuation of the spin-echo amplitude due to incomplete refocusing. Differences in molecular positions at two subsequent intervals, however, represent nothing else than their displacements.

Figure 5.1 represents schematic illustration of the simplest pulse sequence for measuring self-diffusion [4, 31]. This is a simple Hahn spin-echo pulse sequence ($90°–\tau–180°–\tau–$acq) with a rectangular gradient pulse inserted into each τ delay. The time interval between the leading edges of the gradient pulses is denoted by Δ. After a 90° radio frequency (RF) pulse, the macroscopic magnetization is rotated from the z-axis into the x-y plane. When a gradient pulse of duration δ and magnitude g is applied, the spins dephase. The spin precession is reversed after application of a 180° RF pulse after a time τ. A second gradient pulse of equal duration δ and magnitude g follows to tag the spins in the same way. If the positions of the spins remain unchanged in the sample, the effects of the gradient pulses

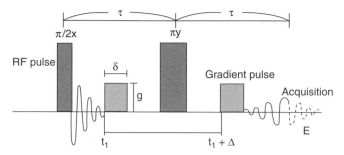

Figure 5.1 *Pulsed-gradient spin-echo NMR sequence for the measurement of self-diffusion coefficient.*

compensate each other, and all spins refocus. On the other hand, if the self-diffusion causes the spins to move and the effects of the gradient pulses do not compensate, the echo amplitude is reduced. The decrease of the amplitude with the applied gradient is proportional to the movement of the spins. The echo signal attenuation, E, is related to the experimental parameters by [32]

$$\ln(E) = \ln\left(\frac{S}{S_{g=0}}\right) = -\gamma^2 g^2 D \delta^2 \left(\Delta - \frac{\delta}{3}\right) \tag{5.5}$$

where S is the spin-echo signal intensity and γ is the gyromagnetic ratio. The PGSE measurement involves the collection of spin-echo amplitudes as the gradient pulse strengths are varied. The value of D can be easily determined from the slope of a plot of $\ln(S/S_{g=0})$ versus $-\gamma^2 g^2 D \delta^2 (\Delta - \delta/3)$

Owing to the large γ value, protons offer the best measuring conditions with respect to both the minimum concentration of the diffusants and their minimum displacements [1]. However, PGSE–NMR diffusion studies can be performed with other nuclei including ^{13}C, ^{15}N, and ^{19}F. The lower limit of observable displacements is of the order of 100 nm. With an observation time of a few 100 ms as a typical upper value, this corresponds to minimum diffusivities of about 10^{-10} cm^{2}s^{-1}.

5.5 IONIC LIQUIDS FOR DIFFUSION STUDIES AND IONIC TRANSPORT BEHAVIOR

To reveal ionic transport behavior through self-diffusion studies [22], we choose four different ionic liquids: 1-ethyl-3-methylimidazolium tetrafluoroborate (EMIBF$_4$), 1-ethyl-3-methylimidazolium *bis*(trifluoromethylsulfonyl)imide (EMITFSI), 1-butylpyridinium tetrafluoroborate (BPBF$_4$), 1-butylpyridinium, and *bis*(trifluoromethylsulfonyl)imide (BPTFSI) (Scheme 5.1). The ionic liquids are comprised of two cations and anions, and they allow us to visualize the dependence of ionic transport behavior on structural variety from the comparative self-diffusion behavior.

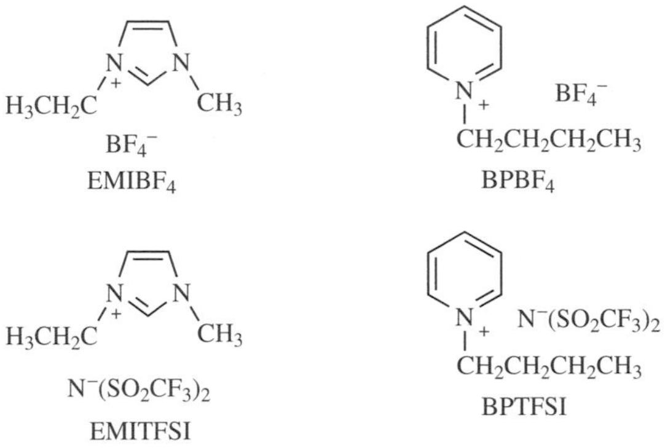

Scheme 5.1 *Molecular structures of the ionic liquids, EMIBF$_4$, EMITFSI, BPBF$_4$, and BPTFSI.*

TABLE 5.1 Thermal Properties for EMIBF$_4$, EMITFSI, BPBF$_4$, and BPTFSI

	T_g/K^a	T_c/K^a	T_m/K^b	T_d/K^b
EMIBF$_4$	181	193	279	664
EMITFSI	187	218	257	717
BPBF$_4$	202	251	272	615
BPTFSI	—	224	299	677

^{a}Onset temperatures of a heat capacity change (T_g), an exotherm peak (T_c) and an endotherm peak (T_m) during heating scan from 123 K by using differential scanning calorimetry.
^{b}Temperature of 10% weight loss during heating scan from room temperature by using thermo-gravimetry.

The thermal properties of the ionic liquids, as determined by differential scanning calorimetry (DSC) and thermogravimetry/differential thermal analysis (TG–DTA), are summarized in Table 5.1. The ionic liquids investigated in this study are all liquid at ambient temperature and therefore are room-temperature ionic liquids (RTILs). All of the RTILs have wide liquid temperature ranges together with high thermal stability. In particular, EMITFSI [18, 21] and BPTFSI exhibit higher thermal stability than EMIBF$_4$ and BPBF$_4$. During cooling from 373 to 123 K at the rate of 10 Kmin^{-1}, only a heat capacity change corresponding to the glass transition can be observed for EMIBF$_4$ and BPBF$_4$. This indicates that the crystallization rates of both of the ionic liquids are very slow, and that the super-cooled liquids are fairly stable [19–20]. For instance, EMIBF$_4$ does not crystallize at 213 K for more than 5 hours, and BPBF$_4$ does not crystallize at 243 K for more than 7 hours. In contrast, an exothermic peak based on the crystallization and a heat capacity change assigned to the glass transition have been observed for EMITFSI and BPTFSI during the cooling scans, indicating relatively fast crystallization rates of these ionic liquids. In the following heating scans, the thermograms show glass transition temperature (T_g), crystallization temperature (T_c), and melting temperature (T_m), successively, which is a common feature for most of the ionic liquids [21].

TABLE 5.2   Density and Molar Concentration for EMIBF$_4$, EMITFSI, BPBF$_4$, and BPTFSI^a

T/K	EMIBF$_4$ ρ/gcm^{-3}	EMITFSI ρ/gcm^{-3}	BPBF$_4$ ρ/gcm^{-3}	BPTFSI ρ/gcm^{-3}
313	1.266	1.502	1.208	1.436
308	1.271	1.507	1.212	1.440
303	1.275	1.512	1.216	1.444
298	1.279	1.518	1.220	1.449
293	1.283	1.523	1.224	1.453

T/K	EMIBF$_4$ M/moldm^{-3}	EMITFSI M/moldm^{-3}	BPBF$_4$ M/moldm^{-3}	BPTFSI M/moldm^{-3}
313	6.39	3.84	5.42	3.45
308	6.42	3.85	5.43	3.46
303	6.44	3.86	5.45	3.47
298	6.46	3.88	5.47	3.48
293	6.48	3.89	5.49	3.49

^{a}Molecular weight/gmol^{-1}; EMIBF$_4$ 197.97; EMITFSI 391.32; BPBF$_4$ 223.01; BPTFSI 416.36.

The ionic liquids are dense systems, with the density exhibiting well-defined temperature dependencies. Table 5.2 shows the density of the ionic liquids in a range of temperatures. The density decreases linearly with the temperature increase, regardless of the ionic liquid structures. Based on the density data and the molecular weight, molar concentration of each ionic liquid has also been calculated and listed in Table 5.2. Molar concentrations of the ionic liquids are high, and they change with structures and temperature.

5.6   SELF-DIFFUSION COEFFICIENT AND ITS CORRELATION WITH VISCOSITY

The value of self-diffusion coefficient (D) for the ionic liquids has been determined from the slope of a plot of signal attenuation, $\ln(S/S_{g=0})$ versus $-\gamma^2 g^2 D\delta^2(\Delta-\delta/3)$. In these measurements, ^{19}F and ^{1}H are used for the detection of the anions and the cations, respectively. Figure 5.2 gives the results for ^{1}H PGSE–NMR echo signal for EMIBF$_4$ in terms of the relationship between left-hand term and right-hand term in equation (5.5). Although the linearity of the plots (Figure 5.2) is consistent with the measured species in free diffusion in the time scale of Δ (50 ms), the measured species, that is, each ion, interact with each other. The presence of strong coulombic attractive force between the anions and the cations suggests that the ions associate to form ion-pairs and ion-aggregates. However, coulombic repulsive force between the like charges also exists. The equilibrium between the attractive and repulsive interaction at each temperature, while maintaining the charge valance, relates to the chemical equilibrium for association

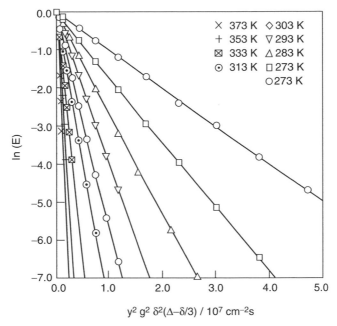

Figure 5.2 *Plots of signal attenuation versus* $\gamma^2 g^2 \delta^2 (\Delta - \delta/3)$ *for* ^1H *PGSE–NMR echo signals of EMIBF$_4$. In this experiment g is $5.78T_m^{-1}$, δ ranges from 0 to 3.5 ms, and Δ is 50 ms.*

and dissociation of the components in the ionic liquids. Since NMR measurement can detect a nucleus (i.e., ^1H or ^{19}F) but cannot distinguish between the ion and their associated species, the PGSE–NMR diffusion coefficients obtained in this experiment are an average self-diffusion coefficient of the ions and associated ions. In addition, since the signal splitting between the ions and their aggregates cannot be observed, the rate of exchange constants between the ions and ion-aggregates are shorter than NMR time scale.

Figure 5.3 depicts the Arrhenius plots of the apparent self-diffusion coefficient of the cation (D_{cation}) and anion (D_{anion}) for EMIBF$_4$ and EMITFSI (Figure 5.3a) and for BPBF$_4$ and BPTFSI (Figure 5.3b). The Arrhenius plots of the summation ($D_{cation} + D_{anion}$) of the cationic and anionic diffusion coefficients are also shown in Figure 5.4. The fact that the temperature dependency of each set of the self-diffusion coefficients shows convex curved profiles implies that the ionic liquids of interest to us deviate from ideal Arrhenius behavior. Each result of the self-diffusion coefficient has therefore been fitted with VFT equation [6].

$$D = D_0 \exp\left(\frac{-B}{T - T_0}\right) \qquad (5.6)$$

where D_0 (cm^{2}s^{-1}), B (K), and T_0 (K) are constants. The solid lines in Figures 5.3a,b, and 5.4 are the fitted profiles of the experimental points in equation (5.6). The

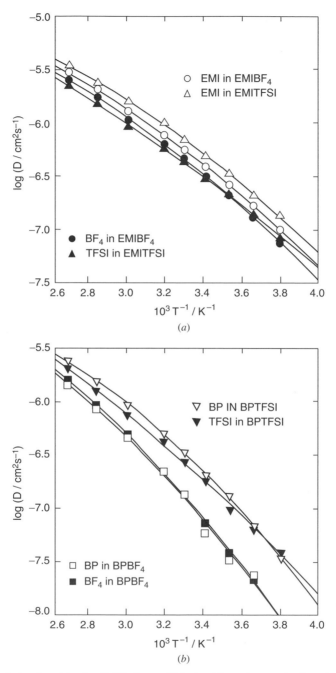

Figure 5.3   *Arrhenius plots of self-diffusion coefficients of the anions and cations for (a) EMIBF$_4$ and EMITFSI and (b) BPBF$_4$ and BPTFSI.*

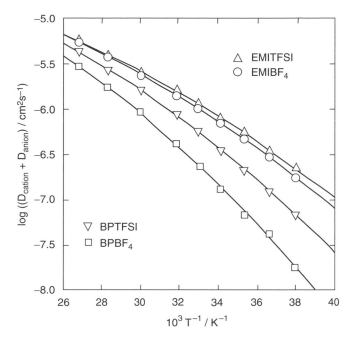

Figure 5.4 *Arrhenius plots of simple summation of cationic and anionic self-diffusion coefficients ($D_{cation} + D_{anion}$) for EMIBF$_4$, EMITFSI, BPBF$_4$, and BPTFSI.*

best-fit parameters of the VFT equation for these ionic liquids are summarized in Table 5.3.

Apparent cationic transference numbers ($D_{cation}/(D_{cation} + D_{anion})$) can be used for the comparison of the self-diffusion coefficients between the anion and the cation (Figure 5.5). The cationic diffusion coefficients for EMIBF$_4$ and BPBF$_4$ are similar to the anionic diffusion coefficient. When the anion is replaced by more bulky TFSI, the relative diffusivity of the cations to that of the anion becomes larger, and the B parameters of the VFT equation for the cation become smaller than those of TFSI anion (Table 5.3). The larger B values for BP cation compared to EMI indicates that larger activation energies (larger B values) are required for the ionic liquids with BP cation. The differences of the D_0 and T_0 parameters between the cation and the anion are small, and especially, T_0 parameters are around 1.3×10^2 K for all of the ionic liquids. These results reveal that the differences of the ionic self-diffusion coefficient at each temperature are dependent mainly on the B parameters, that is, the activation energies.

Viscosities of the ionic liquids at different temperature have been depicted in Figure 5.6 as Arrhenius plots. Resembling the self-diffusion behavior, the temperature dependence of viscosity follows the VFT equation:

$$\eta = \eta_0 \exp\left(\frac{B}{T - T_0}\right) \qquad (5.7)$$

TABLE 5.3   VFT Equation Parameters for Apparent Self-Diffusion Coefficient Data, $D = D_0 \exp[-B/(T - T_0)]$

	$D_0/10^{-4}$ cm^{2}s^{-1}	$B/10^2$ K	$T_0/10^2$ K	$R^2/10^{-1}$
EMIBF$_4$				
Cation	1.6	9.7	1.3	9.99
Anion	1.6	10	1.3	9.99
Cation + anion	3.3	9.9	1.3	9.99
EMITFSI				
Cation	1.7	9.4	1.3	9.99
Anion	1.7	11	1.2	9.99
Cation + anion	3.3	9.9	1.3	9.99
BPBF$_4$				
Cation	6.9	15	1.3	9.96
Anion	7.8	15	1.3	9.99
Cation + anion	15	15	1.3	9.97
BPTFSI				
Cation	2.5	11	1.4	9.99
Anion	5.3	14	1.2	9.98
Cation + anion	6.6	12	1.3	9.99

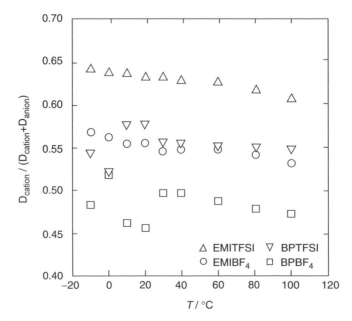

Figure 5.5   *Temperature dependences of apparent cationic transference numbers for EMIBF$_4$, EMITFSI, BPBF$_4$, and BPTFSI.*

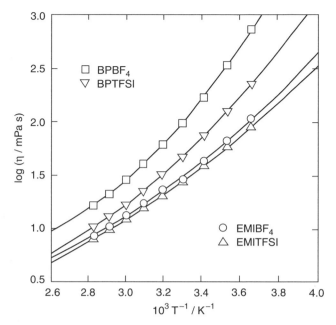

Figure 5.6 Arrhenius plots of viscosity for EMIBF$_4$, EMITFSI, BPBF$_4$, and BPTFSI.

TABLE 5.4 VFT Equation Parameters for Viscosity Data,
$\eta = \eta_0 \exp[B/(T - T_0)]$

	$\eta_0/10^{-1}$ mPas	$B/10^2$ K	$T_0/10^2$ K	$R^2/10^{-1}$
EMIBF$_4$	2.0	7.5	1.5	9.99
EMITFSI	1.5	8.4	1.4	9.99
BPBF$_4$	2.3	7.2	1.8	9.99
BPTFSI	1.2	8.5	1.6	9.99

where η_0 (mPa s), B (K), and T_0 (K) are constants. Solid lines in Figure 5.6 are the fitted profile with the best-fit VFT parameters of viscosity listed in Table 5.4. The temperature dependency of viscosity for EMIBF$_4$ and EMITFSI is in good agreement with literature [18, 21].

The temperature dependencies of the viscosity (Figure 5.6) and the summation of the self-diffusion coefficient ($D_{cation} + D_{anion}$) (Figure 5.4) interestingly show the contrasted profiles with the indication of inverse relationship between viscosity and self-diffusion coefficient. This can be explained in terms of the Stokes-Einstein equation, which correlates the self-diffusion coefficient (D_{cation} and D_{anion}) with viscosity (η) by the following relationship:

$$D = \frac{kT}{c\pi\eta r} \tag{5.8}$$

where k is the Boltzmann's constant, T is absolute temperature, c is a constant (4 to 6), and r is effective hydrodynamic or Stokes radius. Figure 5.7 shows the

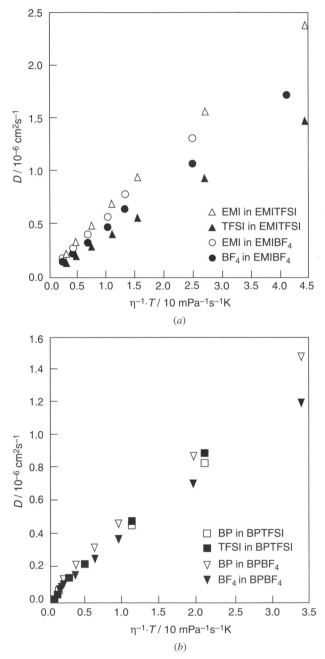

Figure 5.7 *Relationship between* $T\eta^{-1}$ *(T: absolute temperature, η: viscosity) and self-diffusion coefficients of anion and cation for (a) EMIBF$_4$ and EMITFSI and (b) BPBF$_4$, and BPTFSI.*

relationships between the self-diffusion coefficient of the anion (D_{anion}) and the cation (D_{cation}) and $T\eta^{-1}$ for EMIBF$_4$ and EMITFSI (Figure 5.7a), and for BPBF$_4$ and BPTFSI (Figure 5.7b). In all cases, the straight lines passing through the origin indicate that the ionic diffusivity in the ionic liquids obeys equation (5.8). However, it should be noticed that the slopes of the plots do not reflect the size of each ionic species [21]. As shown in Figure 5.7a, the EMI cations in EMITFSI and EMIBF$_4$ indicate similar and the largest slopes in the relationships, followed by the BF$_4$ anion and TFSI anion. If these are explained by equation (5.8), the hydrodynamic radius of each ion follows the order, TFSI > BF$_4$ > EMI, and does not reflect the calculated size of each ionic species [21. Figure 5.7b indicates that the BP cation, TFSI anion, and BF$_4$ anion have a similar hydrodynamic radius, regardless of the difference in the calculated ion size. Since the ionic liquids are highly concentrated electrolyte solutions (Table 5.2), the diffusion of an ion in the liquids may be dependent on that of other species. The influence of coulombic interaction between ionic species (i.e., attractive and repulsive interaction) and the equilibrium between dissociated ions and associated ions also seem to be important for the relation between the ionic size and the diffusivity. It has already been shown that the ionic charge has a strong effect on the transport properties. Hussey et al. [33] reported that the increase of overall negative anionic charge of metal complexes increases the relative hydrodynamic radii in chloroaluminate- or bromoaluminate-ionic liquids. It is interesting that the monomeric and polymeric anions exhibit almost identical hydrodynamic radii if their overall negative charge is the same. However, the ionic liquids in our study consist of only singly charged anions and cations. The apparent cationic transference number shows that the cation diffuses a little faster than the anion, with the exception for BPBF$_4$. In addition the results of the Stokes-Einstein relationships (Figure 5.7) show the cationic hydrodynamic radii to be relatively smaller than the anionic radii for both of the TFSI and BF$_4$ anions in our ionic liquids. If the calculated van der Waals radius is considered, BF$_4$ will have the smaller value compared to that of EMI and BP. Consequently the cation will diffuse more easily than the anion. Owing to their resonance ring structures, the de-localization of cationic charge as well as the planar structures seems to be important.

5.7 MOLAR CONDUCTIVITY AND ITS CORRELATION WITH DIFFUSION COEFFICIENT

The temperature dependence of molar conductivity, calculated from ionic conductivity determined from complex impedance measurements and molar concentrations, and the VFT fitting curves are shown Figure 5.8. The VFT equation for molar conductivity is

$$\Lambda = \Lambda_0 \exp\left(\frac{-B}{T - T_0}\right) \tag{5.9}$$

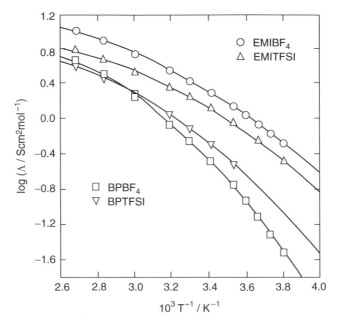

Figure 5.8 *Arrhenius plots of molar conductivity (Λ, obtained experimentally from conductivity and molar concentration) for EMIBF$_4$, EMITFSI, BPBF$_4$, and BPTFSI.*

TABLE 5.5 VFT Equation Parameters for Molar Conductivity Data

	$\Lambda_0/10$ Scm^{-2} mol^{-1}	$B/10^2$ K	$T_0/10^2$ K	$R^2/10^{-1}$
	$\Lambda = \Lambda_0 \exp[-B/(T - T_0)]$			
EMIBF$_4$	13	5.0	1.7	9.99
EMITFSI	4.9	3.9	1.9	9.99
BPBF$_4$	19	6.8	1.8	9.99
BPTFSI	5.0	4.3	2.0	9.99
	$\Lambda_0/10^2$ Scm^{-2} mol^{-1}	$B/10^2$ K	$T_0/10^2$ K	$R^2/10^{-1}$
	$\Lambda_{NMR} = \Lambda_0 \exp[-B/(T - T_0)]$			
EMIBF$_4$	4.5	7.6	1.4	9.99
EMITFSI	4.5	7.6	1.4	9.99
BPBF$_4$	20	13	1.4	9.97
BPTFSI	9.2	9.9	1.4	9.99

where Λ_0 (Scm^{-2}mol^{-1}), B (K), and T_0 (K) are constants. The best-fit parameters for VFT equation are tabulated in Table 5.5.

The Nernst-Einstein equation is applied to calculate molar conductivity (Λ_{NMR}) from the PGSE–NMR diffusion coefficients:

$$\Lambda_{NMR} = \frac{N_A e^2 (D_{cation} + D_{anion})}{kT} = \frac{F^2 (D_{cation} + D_{anion})}{RT} \qquad (5.10)$$

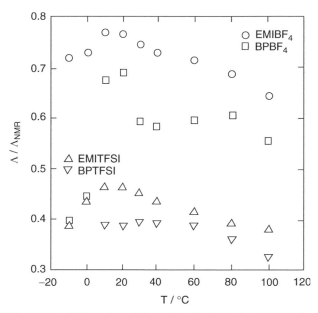

Figure 5.9 Molar conductivity ratios (Λ/Λ_{NMR}) plotted against temperature for EMIBF$_4$, EMITFSI, BPBF$_4$, and BPTFSI.

where N_A is the Avogadro's number, e is electric charge on each ion, F is the Faraday's constant, and R is the universal gas constant. It should be noted that equation (5.10) is derived without considering ionic association and is also derived on the assumption that each ion has the activity of unity. In the temperature range studied, the Λ values are found to be lower than the Λ_{NMR} values at each temperature. The Λ/Λ_{NMR} is thus plotted against temperature in Figure 5.9. The Λ/Λ_{NMR} ratios range from 0.6 to 0.8 for EMIBF$_4$ and BPBF$_4$, whereas the ratios range from 0.3 to 0.5 for EMITFSI and BPTFSI. Depending on the ionic liquid, the Λ/Λ_{NMR} differs, indicating the difference in the percentage of ions contributing to ionic conduction in the diffusion components. Thus the fraction of the ion pairs or aggregations $(1 - \Lambda/\Lambda_{NMR})$ also significantly differs. However, NMR cannot distinguish between the ions and ionaggregates, and their exchange rate is faster than the NMR time scale (*vide supra*). In this case the experimental results are average, and the percentage of ions or aggregations are also the "time-averaged ion" or the "time-averaged aggregations," which is the "time-averaged solvation," as suggested by Hussey et al. [33]. We recognize that the percentage of the ions can be defined as the existence probability, if NMR and conductivity measurements can be conducted on the same time scale.

A large value of Λ/Λ_{NMR} refers to a high ionicity of an ionic liquid and indicates that the ionic liquid is comprised mainly of charged cationic and anionic species, which can diffuse and contribute to the ionic conduction. The Λ/Λ_{NMR} ratios for EMIBF$_4$ and BPBF$_4$ are relatively high enough to consider these ionic liquids to consist mostly of the ions. This suggests that NMR diffusion coefficients in these cases reflect the individual ionic diffusivity and most of the ions contribute to the ionic conduction. In other words, the activity of each ion is close to unity. The situation might not be so straightforward and become a little complicated in EMITFSI and BPTFSI. The Λ/Λ_{NMR} ratios are evidently smaller than those for EMIBF$_4$ and BPBF$_4$. The summation of the self-diffusion coefficient $(D_{cation} + D_{anion})$ at each temperature (Figure 5.4) follows the order EMITFSI > EMIBF$_4$ > BPTFSI> BPBF$_4$, which reflects the magnitude of the viscosity (Figure 5.6). However, the high diffusivity for EMITFSI and BPTFSI is not reflected in their molar conductivity (Figure 5.8). This is the key factor for the small Λ/Λ_{NMR} ratios. The viscosity of the ionic liquids seems to be influenced by many molecular parameters, such as molecular weight and molecular shape of each ion, and coulombic interaction between the ionic species. A comparison of the viscosity of the ionic liquids with EMI cation and with BP cation shows that the larger cationic size and the larger cationic molecular weight can cause larger viscosity of the latter ionic liquids. Although the formula weight of TFSI anion is much larger than that of BF$_4$ anion, the viscosity of the ionic liquids with TFSI anion is lower than that of BF$_4$ anion. The consideration of anionic structures thus invalidates the effect of molecular weight on the viscosity. This may be attributed to the presence of ion-pairs and neutral ion-aggregates in the ionic liquids with TFSI anion, which dilutes the ionic concentration and reduces coulombic interaction between the ionic species. The reduced coulombic interaction may decrease the viscosity, which in turn enhances the ionic diffusion coefficient. However, the ion-pairs and neutral ion-aggregates cannot contribute to the ionic conduction, and thus the molar conductivity of the ionic liquids with TFSI anion becomes small. Certain molecular interaction between the cations and TFSI anion, for instance, hydrogen bonding interaction, might play an important role in understanding the low Λ/Λ_{NMR} ratios for the EMI ionic liquids with TFSI anion. However, there is no evidence of hydrogen bonds involving BP cation. Finally, the Λ/Λ_{NMR} has been found to have no significant influence of temperature. Although the data in Figure 5.9 are scattered, the Λ/Λ_{NMR} ratios do not vary much with temperature. This implies that the change in temperature cannot bring about significant change in the structures of the ionic liquids.

5.8   CONCLUSIONS

Self-diffusivity, cooperatively with ionic conductivity, provides a coherent account of ionicity of ionic liquids. The PGSE–NMR method has been found to be a convenient means to independently measure the self-diffusion coefficients of the anions and the cations in the ionic liquids. Temperature dependencies of the self-diffusion coefficient, viscosity and ionic conductivity for the ionic liquids, cannot be explained simply by Arrhenius equation; rather, they follow the VFT equation. There is a simple correlation of the summation of the cationic and the anionic diffusion coefficients for each ionic liquid with the inverse of the viscosity. The apparent cationic transference number in ionic liquids has also been found to have dependence on the size of the anion, although the size of each ion does not directly affect the ionic diffusion coefficient. The Λ/Λ_{NMR} ratio of the ionic liquids varies, depending on the structures of the ionic liquids and the extent of ionic association and/or ionic activity consequently changes.

ACKNOWLEDGMENTS

This research was supported in part by Grants-in-Aid for Scientific Research 14350452 and 16205024 from the Japanese Ministry of Education, Science, Sports, and Culture and by Technology Research Grant Program from the NEDO of Japan. M.A.B.H.S. also acknowledges a postdoctoral fellowship from JSPS. In addition the authors are thankful to Dr. Kikuko Hayamizu for her useful discussions and support in conducting NMR experiments.

REFERENCES

1. J. Kärger, *Diffusion under Confinement*, Sitzungsbericht der Sachsischen Akademie im Hirzel-Verlag, 2003.

2. H. J. V. Tyrrell, K. R. Harris, *Diffusion in Liquids*, Butterworths, 1984.

3. B. E. Poling, J. M. Prausnitz, J. P. O'Connell, *The Properties of Gases and Liquids* McGraw-Hill, 2001.

4. P. Wasserscheid, T. Welton, eds., *Ionic Liquids in Synthesis*, Wiley-VCH, 2003.

5. A. L. van Geet, A. W. Adamson, *J. Phys. Chem.*, **68**, 238 (1964).

6. H. Vogel, *Phys. Z.* **22**, 645 (1921); G. S. Fulcher, *J. Am. Ceram. Soc.*, **8**, 339 (1923).

7. H. Weingartner, in *Diffusion in Condensed Matter*, J. Karger, P. Heitjans, R. Harberlandt, eds., Vieweg, 1998.

8. J. Crank, *The Mathematics of Diffusion*, 2nd ed., Clarendon Press, 1975.

9. J. D. Holbrey, K. R. Seddon, *Clean Products Processes*, **1**, 223 (1999).

10. T. Welton, *Chem. Rev.*, **99**, 2071 (1999).

11. P. Wasserscheid, W. Keim, *Angew. Chem. Int. Ed.*, **39**, 3772 (2000).

12. J. Dupont, R. F. de Souza, P. A. Z. Suarez, *Chem. Rev.*, **102**, 3667 (2002).

13. Ohno, H., ed., *Ionic Liquids: The Front and Future of Material Development*, CMC Press, Tokyo, 2003 (in Japanese).

14. J. S. Wilkes, M. J. Zaworotko, *J. Chem. Soc., Chem. Commun.*, 965 (1992).

15. J. Fuller, R. T. Carlin, H. C. De Long, D. J. Haworth, *Chem. Soc., Chem. Commun.*, 299 (1994).

16. R. T. Carlin, H. C. De Long, J. Fuller, P. C. Trulove, *J. Electrochem. Soc.*, **141**, L73 (1994).

17. V. R. Koch, C. Nanjunduah, G. B. Appetecchi, B. Scrosati, *J. Electrochem. Soc.*, **142**, L116 (1995).

18. P. Bonhôte, A.-P. Dias, N. Papageorgiou, K. Kalyanasundaram, M. Grätzel, *Inorg. Chem.*, **35**, 1168 (1996).

19. J. Fuller, A. C. Breda, R. T. Carlin, *J. Electrochem. Soc.*, **144**, L67 (1997).

20. J. Fuller, A. C. Breda, R. T. Carlin, *J. Electroanal. Chem.*, **459**, 29 (1998).
21. A. B. McEwen, H. L. Ngo, K. LeCompte, J. L. Goldman, *J. Electrochem. Soc.*, **146**, 1687 (1999).
22. A. Noda, K. Hayamizu, M. Watanabe, *J. Phys. Chem. B*, **105**, 4603 (2000).
23. A. Noda, M. Watanabe, *Electrochim. Acta*, **45**, 1265 (2000).
24. H. Matsumoto, H. Kageyama, Y. Miyazaki, *Chem. Commun.*, 1726 (2002).
25. R. Kawano, M. Watanabe, *Chem. Commun.*, 330 (2003).
26. M. A. B. H. Susan, A. Noda, S. Mitsushima, M. Watanabe, *Chem. Commun.*, 938 (2003).
27. A. Noda, M. A. B. H. Susan, K. Kudo, S. Mitsushima, K. Hayamizu, M. Watanabe, *J. Phys. Chem. B*, **107**, 4024 (2003).
28. M. A. B. H. Susan, M. Yoo, H. Nakamoto, M. Watanabe, *Chem. Lett.*, **32**, 836 (2003).
29. W. Xu, E. I. Cooper, C. A. Angell, *J. Phys. Chem. B*, **107**, 6170 (2003).
30. M. A. B. H. Susan, H. Nakamoto, M. Yoo, M. Watanabe, *Trans. MRS-J*, **29**, 1043 (2004).
31. W. S. Price, *Concepts Magn. Reson.*, **9**, 299 (1997); **10**, 197 (1998).
32. E. O. Stejskal, *J. Chem. Phys.*, **43**, 3597 (1965).
33. C. L. Hussey, I.-W. Sun, S. K. D. Strubinger, P. A. Barnard, *J. Electrochem. Soc.*, **137**, 2515 (1990).

Chapter 6

Ionic Conductivity

Hiroyuki Ohno, Masahiro Yoshizawa, and Tomonobu Mizumo

Electrolyte solutions are essential for electrochemical devices. However, almost all the solvents for electrolyte solutions have a crucial drawback: they are volatile solvents. Recently ionic liquids (ILs), which are liquids composed only of ions, have received much attention as new electrolyte materials. These ILs have two attractive features of very high concentrations of ions and high mobilities of the component ions at room temperature, and many of these ILs show the ionic conductivity of over 10^{-2} S cm^{-1} at room temperature [1–3]. In this chapter we focus on the conductivity of ILs. Ionic conductivity is generally given by the equation

$$\sigma_i = \sum ne\mu \qquad (6.1)$$

where n is carrier ion number, e is electric charge, and μ is mobility of carrier ions.

The electrochemical properties of ILs studied are mainly their ionic conductivity, transference number, and potential window. These electrochemical properties differ in accord with factors such as moisture absorption, experimental atmosphere, and electrode species. Therefore the conductivity measurements must be made under consistent conditions. Often, electrochemical measurements and construction of electrochemical cells are performed in a glove box filled with an inert gas (usually argon).

Ionic conductivity is defined as the reciprocal of proper resistance (R_b). It is obtained by calculating

$$\sigma_i = \frac{l}{R_b S} \qquad (6.2)$$

Electrochemical Aspects of Ionic Liquids Edited by Hiroyuki Ohno
ISBN 0-471-64851-5 Copyright © 2005 John Wiley & Sons, Inc.

where l is the distance between the two electrodes and S is the mean area of the electrodes. Since it is almost impossible to measure actual surface area of the applied electrode, l/S, which is known as cell constant, is determined by measuring the conductivity of standard solutions like aqueous KCl. For ionic conductivity, S cm^{-1} or mS cm^{-1} are used as the unit of measure.

The R_b based on the sample cannot be calculated correctly, since the electric charge transfer resistance and the electric double layer in an electrode interface are also detected as a resistance, even if bias voltage is impressed to the measurement cell in order to measure the ionic conductivity. For the ionic conductivity measurement, a dc four-probe method, or the complex-impedance method, is used to separate sample bulk and electrode interface [4]. In particular, the complex-impedance method has the advantage that it can be performed with both nonblocking electrodes (the same element for carrier ion M^{n+} and metal M) and blocking electrodes (usually platinum and stainless steel were used where charge cannot be transferred between the electrode and carrier ions). The two-probe cell, where the sample is sandwiched between two polished and washed parallel flat electrodes, is used in the ionic conductivity measurement by complex-impedance method as shown in Figure 6.1.

The detected current is influenced considerably by any incomplete contact between sample and electrode, bubbles, grain boundary, and so on. So it is necessary to be careful in preparing the contact conditions for the sample and electrode interface in polymer systems. When the polymers are flexible films or soft paste, good contact is obtained by simply pressing. For hard or brittle polymer systems, it is difficult to get good contact by the pressing pressure. To improve the contact, it is necessary to melt the polymer material or to sputter electrode material on the

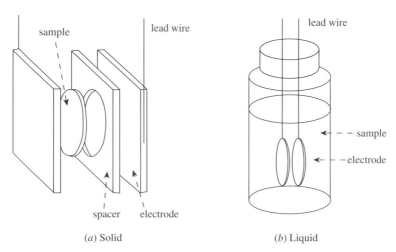

(a) Solid   (b) Liquid

Figure 6.1 *Schematic representation of cells used for ionic conductivity measurement by impedance method as an example. (a) Solid-state samples; (b) liquid-state samples.*

sample surface. These processes must be carried out with appropriate spacers to prevent deformation of the polymer sample.

When using metals, one must be aware of the strong reduction ability of metals as nonblocking electrodes. For instance, lithium metal must be handled in a completely dry atmosphere. It cannot therefore be used in an aqueous system or in samples containing active protons. In addition, since lithium nitride is produced by the reaction of lithium metal and nitrogen gas, argon gas must be used as the inert gas. Generally, highly activated metals are used on blocking electrodes to measure ionic conductivity.

The impedance analyzer is frequently used under an alternating current. The reliability of the observed impedance depends on the frequency applied. It decreases with increasing frequency. Measurement frequency range is generally from 10 Hz to 1 MHz, of course depending on the performance of the equipment used. Although about 10 mV amplitude voltage is desirable, it depends on sample, and the appropriate value is chosen according to each sample. When absolute value of impedance is large (generally around the order of $G\Omega$), sufficient current cannot be obtained under ordinary conditions. However, when the absolute value of the obtained impedance is small (generally under several Ω), the experimental error becomes large due to large current. These values can be adjusted by changing the sample's thickness.

Figure 6.2 shows a typical Nyquist plot (or Cole-Cole plot) and its equivalent circuits for measurements with nonblocking electrodes [5, 6]. In Figure 6.2b where the samples show fast ion transport and the distribution of relaxation process for ion transport is small, two arcs are obtained in the Nyquist plot. The arcs show the movements of the ion and dipole moments at their frequency ranges. The left-side arc (high-frequency range) is explained by both the resistance component for ion transfer and the capacity component for the dielectric constant. The right-side arc (low-frequency range) relates to the response at the electrode surface. The resistance component responds to the charge transfer and the capacity component to the electric double layer. In the smaller frequency range, the density distribution in the sample is due to the fully progressed redox reaction at the electrode surface. The impedance is based on the diffusion of ions. A straight line, declined 45 degrees, is depicted in the Nyquist plot (Warburg impedance).

When blocking electrodes are used, a straight line is constructed perpendicularly to the real axis because electrode surface has only a double-layer capacitor. In practice, the angle obtained is often less than the 90 degrees shown in Figure 6.3. This is due to the incomplete double-layer capacitor. When redox species contact the electrode surface, the resulting Nyquist plot should be similar to plots where electrodes are nonblocking.

When the temperature is changed, the Nyquist plot also changes, as shown in Figure 6.4a, due to the change in conductivity. Figure 6.4 shows the Nyquist plots for the ionic liquid prepared by the neutralization of 1-benzyl-2-methylimidazole and HTFSI. The temperature dependence of the ionic conductivity is generally depicted by an Arrhenius plot (Figure 6.4b).

(a)   1. Equivalent circuit for high-frequency range

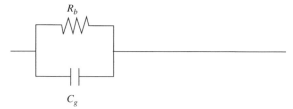

2. Equivalent circuit for low-frequency range

(b)

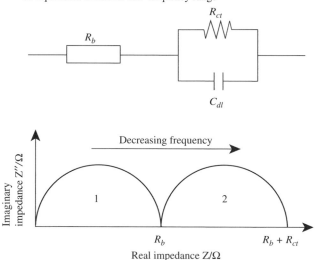

Figure 6.2 *(a) Schematic representation of equivalent circuit for an ion conductor put between a pair of nonblocking electrode, and (b) the corresponding Nyquist plot. The high-frequency equivalent circuit (a – 1) is composed of the bulk resistance of the sample (R_b) and geometrical capacitance (C_g), while the low-frequency equivalent circuit (a – 2) is composed of the resistance to charge transfer across the sample/electrode interface (R_{ct}) and double-layer capacitance (C_{dl}).*

The typical temperature dependence of the ionic conductivity has been described [5]. Since most of ILs including polymer systems show upper convex curvature in the Arrhenius plot, and not a straight line, the temperature dependence of the ionic conductivity is expressed by Vogel-Fulcher-Tamman (VFT) equation [7]:

$$\sigma_i(T) = \frac{A}{\sqrt{T}} \exp\left[\frac{-B}{(T-T_0)}\right] \tag{6.3}$$

where A and B are constants and T_0 is ideal glass transition temperature. Parameters A and B relate to carrier ion number and activation energy for the ion transport, respectively. Parameter T_0 is usually 30° to 50°C lower than that of T_g obtained by the DSC measurement. It is understood that the ions migrate in the amorphous

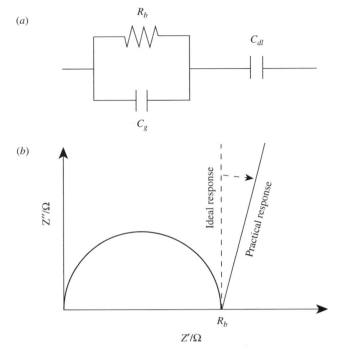

Figure 6.3 *(a) Schematic representation of equivalent circuit for an ion conductor put between a pair of blocking electrode, and (b) the corresponding Nyquist plot. Ideally the sample–electrode interface is composed only of the double-layer capacitance. However, the practical Nyquist plot that corresponds to this frequency region is not vertical to the real axis. The rate-limiting process of this plot is that the ion diffuses to form a double layer.*

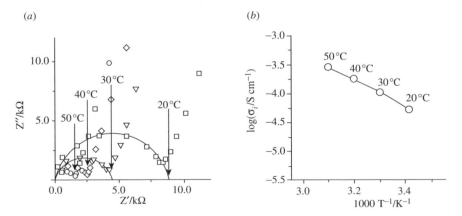

Figure 6.4 *Temperature dependence of the ionic conductivity for BzEImHTFSI. (a) Nyquist plots for various temperatures; (b) the corresponding Arrhenius plot.*

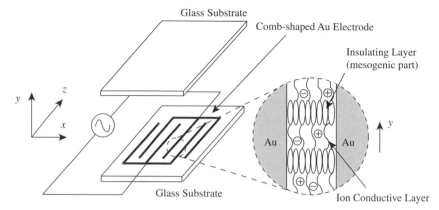

Figure 6.5 *Schematic representation of the comb-shaped Au electrode cell, and the domain formation of the liquid crystalline-type electrolyte in Au electrodes.*

phase and that their migration is influenced by viscosity. This is surmised the fact that most systems show straight lines in the VFT plots.

The hysteresis that may appear depends on the direction or the scan rate of temperature change. This tendency is based on the phase transition or slow diffusion process of sample. Remarkably such hysteresis can even appear where a sample shows crystallization within the measurement temperature range. Therefore the ionic conductivity measurement is usually performed at each temperature after reaching the constant value. On the other hand, ionic conductivity is measured at scan rate of $1°$ to $2°C \cdot min^{-1}$. Therefore, when hysteresis appears during the heating or cooling process, the relationship between the phase transition and the ionic conductivity can be used in the analysis at this scan rate with the DSC measurement. It is better to use a small cell design to avoid the temperature distribution in samples.

Finally, we consider the method for anisotropic ionic conductivity. Anisotropic ionic conductivity is a phenomenon that appears in a sample undergoing liquid crystalline phase or micro phase separation. In the nanotechnology field such samples are expected to yield important materials because they can raise the value of ion conduction. To measure the anisotropic ionic conductivity, one has to design a special cell. Figure 6.5 shows an example of a cell shaped like a comb [8]. In the comb-shaped cell, the ionic conductivity is measured between two teeth of the comb. As the liquid crystalline molecules orient perpendicularly on the glass substrate, the ionic conductivity is obtained along with the minor axis. The important task is to prepare the specific phase homogeneously in the measurement cell. The details will soon be presented in published papers.

REFERENCES

1. P. Bonhôte, A.-P. Dias, M. Armand, N. Papageorgiou, K. Kalyanasundaram, M. Grätzel, *Inorg. Chem.*, **35**, 1168 (1996).

2. R. Hagiwara, Y. Ito, *J. Fluorine Chem.*, **105**, 221 (2000).

3. P. Wasserscheid, T. Welton, eds., *Ionic Liquids in Synthesis*, Wiley-VCH, 2003.

4. P. M. S. Monk, in *Fundamentals of Electroanalytical Chemistry*, Wiley, 2001.

5. J. R. MacCallum, C. A. Vincent, eds., *Polymer Electrolyte Reviews 1 and 2*, Elsevier, 1987 and 1989.

6. P. R. Sφrensen, T. Jacobsen, *Electrochim. Acta*, **27**, 1671 (1982).

7. (a) H. Vogel, *Phys. Z.*, **22**, 645 (1921); (b) G. S. Fulcher, *J. Am. Ceram. Soc.*, **8**, 339 (1925); (c) G. Tamman, W. Hesse, *Z. Anorg. Allg. Chem.*, **156**, 245 (1926).

8. (a) K. Ito-Akita, N. Nishina, Y. Asai, H. Ohno, T. Ohtake, Y. Takamitsu, T. Kato, *Polym. Adv. Technol.*, **11**, 529 (2000); (b) M. Yoshizawa, T. Mukai, T. Ohtake, K. Kanie, T. Kato, H. Ohno, *Solid State Ionics*, **154–155**, 779 (2002); (c) M. Yoshio, T. Mukai, K. Kanie, M. Yoshizawa, H. Ohno, T. Kato, *Adv. Mater.*, **14**, 351 (2002).

Chapter 7

Optical Waveguide Spectroscopy

Hiroyuki Ohno and Kyoko Fujita

7.1 INTRODUCTION

Spectroelectrochemistry has proved to be a powerful technique for the study of redox processes. In the transmission experiment spectroelectrochemistry is used to provide optical data that is complementary in nature to standard current or potential measurements [1]. Spectroelectrochemistry was originally conceived as a means to monitor, in the visible spectral region solution electrochemical processes through absorbance changes occurring within the diffusion layer thickness adjacent to optically transparent electrodes. The power of spectroelectrochemical techniques in monitoring electrochemical processes was immediately apparent because the data obtained on the redox processes was possible in the absence of background current due to nonfaradaic charging and/or electroactive impurities. Further it was found that through the choice of an appropriate wavelength, spectroelectrochemistry can add a dimension of selectivity to an electrochemical measurement. In part because of these features, conventional transmission spectrochemistry in the visible wavelength regime is now widely used to characterize spectroscopically the redox process of diverse materials, ranging from electron transfer proteins [2], to electropolymers [3], to electrochromic organic and inorganic thin films [4], and to a host of organic radical-producing processes [1]. However, the standard transmission method of spectroelectrochemistry suffers from poor sensitivity.

Electrochemical Aspects of Ionic Liquids   Edited by Hiroyuki Ohno
ISBN 0-471-64851-5   Copyright © 2005 John Wiley & Sons, Inc.

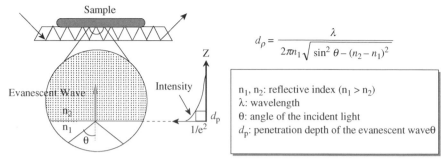

Figure 7.1 Characteristics of the evanescent wave.

However, a total internal reflection technique can be used to enhance the sensitivity at the interface [5, 6]. This is because the total internal reflection of light at an interface generates an evanescent wave whose intensity falls exponentially with distance from the waveguide surface (Figure 7.1). The characteristic distance defined by this decay, corresponding to the penetration depth from the waveguide surface [7, 8], is related to the wavelength of the incident light, and for visible light it is several hundred nanometers. The evanescent wave is used to observe UV-visible absorption [9, 10], or IR [11], and fluorescence [12, 13] spectra of molecules located within the penetration depth. This technique is useful for analyzing the effects of temperature [14], pH [15], and gas molecules [16] on the internally reflected spectral density.

Optical waveguide (OWG) spectroscopy is a total internal reflection technique that is used to detect visible light absorption spectra [17–20]. Like ordinary visible spectroscopy, it gives information about both the structure and the electronic state of the molecules on a waveguide [17, 20]. It also provides qualitative information about the molecules on the waveguide [19]. As the incident light propagates by repeating total reflection in the waveguide, the evanescent wave generated at every reflection is absorbed by samples on the waveguide. The accumulation of absorbed light allows the absorption spectra of the samples to provide excellent signal to noise ratios. We have been using this technique to study surface-adsorbed molecules, so we can confirm that OWG spectroscopy provides reliable quantitative information about the molecules on the waveguide [19]. A combination of OWG spectroscopy with electrochemical methods can give even better information [21–25]. OWG spectroelectrochemistry is useful technique for sensitive and simple analysis of redox reaction of molecules dissolved in a small amount of solution [22] or adsorbed on opaque electrodes [23, 24]. Figure 7.2 shows the spectral change due to the redox reaction of Methylene blue. This OWG spectral change was repeatedly observed in situ, and it corresponded to the potential switching, since the electrochemical cell was constructed on the waveguide. The redox reaction of the molecules in the ionic liquid was also easily analyzed by constructing an electrochemical cell system on the waveguide composed by working, reference, and counterelectrodes [25].

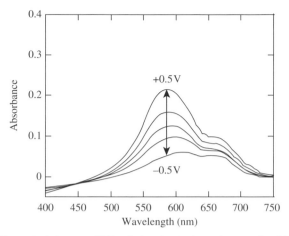

Figure 7.2 *OWG spectral change of MB solution due to the redox reaction (W.E.: carbon; C.E.: Pt; R.E.: Ag).*

7.2 ANALYSIS OF REDOX REACTIONS IN IONIC LIQUID [26]

OWG spectroscopy has proved to be a highly sensitive detection method for molecules. The visible absorption spectrum has been studied with the OWG spectrophotometer (SIS-50, System Instruments Inc.) [19]. The light source used was a 150 W Xenon lamp (Hamamatsu Photonics). The beam was modulated by a mechanical chopper and introduced into an optical fiber through a microscope object lens; the beam was focused on the edge of the waveguide. After propagating through the waveguide, the transmitted light was collected by the second micro-lens, and directed through the optical fiber to CCD spectrometer (Figure 7.3).

An important task is to select a suitable waveguide for the OWG analysis in ionic liquid. This is because most molten salts have high refractive indexes. The waveguide should have a higher refractive index than the refractive indexes of the ionic

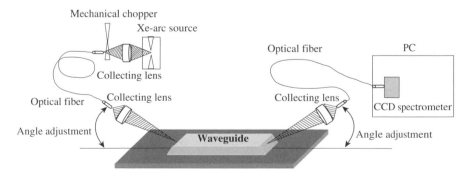

Figure 7.3 *Schematic view of the OWG spectroscopy system.*

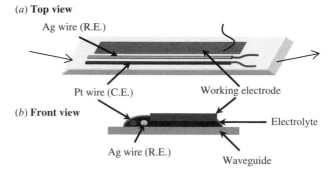

Figure 7.4   Electrochemical cell system on the waveguide.

liquids, which are generally around 1.4 to 1.5 [27, 28]. Infact the OWG analysis in ionic liquid cannot be carried out with a quartz glass waveguide. For this purpose, an optical glass plate with $n = 1.75$ is used [29]. The waveguide plate was prepared by both edges being cut at 60°, and then the plate was polished accurately enough for optical use. The angle of incident light was directed at 60° to the surface, and then the average number of reflections in this OWG estimated to be 7 times per centimeter. In typical ILs the OWG plate enables spectral measurements between wavelengths 350 to 600 nm apart.

The electrochemical redox reaction of molecules dissolved in ILs is detected as changes in the absorption spectra using the electrochemical cell system constructed on an optical glass plate (Figure 7.4) [29]. This cell system consists of a working electrode together with Pt and Ag wires as the counter- and reference electrodes, respectively. The sample ionic liquid is directly injected into this electrochemical cell system, and the corresponding spectra are measured by the applied potential. Only μl range analysis is available, and it is suitable for expensive and highly valuable samples such as ILs.

REFERENCES

1. A. Bard, ed., *Electroanalytical Chemistry*, **7**, Dekker, 1974.
2. M. Collinson, E. F. Bowden, *Anal. Chem.*, **64**, 1470 (1992).
3. T. A. Skotheim, R. L. Elsenbaumer, J. R. Rsynolds, eds., *Handbook of Conducting Polymers*, Dekker, 1998.
4. A. Ferencz, N. R. Armstrong, G. Wegner, *Macromolecules*, **27**, 1517 (1994).
5. N. J. Harrick, ed., *Internal Reflection Spectroscopy*, Wily, 1967.
6. W. N. Hansen, *J. Opt. Soc. Amer.*, **58**, 380 (1968).
7. R. M. Dickson, D. J. Norris, Y. Tzeng, W. E. Moerner, *Science*, **274**, 966 (1996).
8. (a) X. Xu, E. Yeung, *Science*, **275**, 1106 (1997). (b) X. Xu, E. Yeung, *Science*, **281**, 1650 (1998).

9. S. B. Mended, L. Li, J. J. Burke, *Langmuir*, **12**, 3374 (1996).

10. K. Kato, A. Takatsu, N. Matsuda, *Chem. Lett.*, 31 (1999).

11. A. Bentaleb, A.Abele, Y. Haikei, P. Scheef, J. C. Voegel, *Langmuir*, **14**, 6493 (1998).

12. T. R. E. Simpson, D. J. Revell, M. J. Cook, D.A. Russell, *Langmuir*, **13**, 460 (1997).

13. A. N. Asanov, W. W. Wilson, P. B. Oldhan, *Anal. Chem.*, **70**, 1156 (1998).

14. L. M. Johnson, F. J. Leonberger, G. W. Pratt, *Appl. Phys. Lett.*, **41**, 134 (1982).

15. L. Yang, and S. S. Saavedra, *Anal. Chem.*, **67**, 1307 (1995).

16. P. J. Skrdla, S. S. Saavedra, N. R. Armstrong, S. B. Mendes, N. Peyghambarian, *Anal. Chem.*, **71**, 1332 (1999).

17. N. Matsuda, A. Takatsu, K. Kato, Z. Shigesato, *Chem. Lett.*, 125 (1998).

18. J. T. Bradshaw, S. B. Mendes, S. S. Saavedra, *Anal. Chem.*, **74**, 1751 (2002).

19. H. Ohno, S. Yoneyama, F. Nakamura, K. Fukuda, M. Hara, M. Shimomura, *Langmuir*, **18**, 1661 (2002).

20. M. Mitsuishi, T. Tanuma, J. Matsui, J. Chen, T. Miyashita., *Langmuir*, **17**, 7449 (2001).

21. (a) K. Kato, A. Takatsu, N. Matsuda, R. Azumi, M. Matsumoto, *Chem. Lett.*, 437 (1995).
    (b) N. Matsuda, A. Takatsu, K. Kato, *Chem. Lett.*, 105 (1996).

22. W. J. Doherty, C. L. Donley, N. R. Armstrong, S. S. Saavedra, *Appl. Spectrosc.*, **56**, 920 (2002).

23. H. Ohno, K. Fukuda F. Kurusu, *Chem. Lett.*, 76 (2000).

24. K. Fujita, C. Suzuki, H. Ohno, *Electrochem. Comm.*, **5**, 47 (2003).

25. K. Fukuda, H. Ohno, *Electroanalysis*, **14**, 605 (2002).

26. H. Ohno, C. Suzuki, K. Fukumoto, M. Yoshizawa, K. Fujita, *Chem. Lett.*, 450 (2003).

27. P. Bônhote, A.P. Dias, N. Papageprgiou, K. Kalyanasundaram, M. Grätzel, *Inorg. Chem.*, **35**, 1168 (1996).

28. J. G. Huddleston, A. E. Visser, W. M. Reichart, H. D. Willauer, G. A. Broker, R. D. Rogers, *Green Chem.*, **3**, 156 (2001).

29. K. Fujita, H. Ohno, *Polym. Adv. Technol.*, **14**, 486 (2003).

Chapter 8

Electrolytic Reactions

Toshio Fuchigami

Solvents have the important function of dissolving substrates in solutions and controlling desired reactions. Since many organic compounds do not dissolve in aqueous solutions, organic solvents have often been employed for organic reactions. Similarly to organic reactions, organic solvents have been very often used in organic electrolytic reactions. However, water is one of the ideal electrolytic solvents. Water does have some limitations such as poor solubility of organic substrates and its competitive electrolytic reactions with substrates. Aprotic organic solvents are useful for avoiding such competitive reactions and also for dissolving both organic substrates and supporting electrolytes, but most organic solvents are flammable and not always safe from health and environmental perspectives. So far in attempts to solve these problems, SPE (solid polymer electrolyte) electrolysis and gas-phase electrolysis is being developed using electrocatalysts. In the case of inorganic electrolysis, solvent-free electrolysis, such as electrorefining processes using molten salts, has been already commercialized. In sharp contrast, there has yet been no established solvent-free organic electrode process.

In recent years room-temperature molten salts, namely ionic liquids, have proved to be a new class of promising solvents because of their good electroconductivity, nonflammability, thermal stability, nonvolatility, and reusability [1–3]. In addition, if a combination of cation and anion is appropriately made, aprotic media having a wide electrochemical window can be obtained. Therefore, when ionic liquids are used as electrolytic media, organic electrolytic reactions, particularly electroorganic synthesis should be possible without any organic solvents. Although ionic liquids have already been shown to be promising media for batteries, fuel

Electrochemical Aspects of Ionic Liquids   Edited by Hiroyuki Ohno
ISBN 0-471-64851-5   Copyright © 2005 John Wiley & Sons, Inc.

cells, photovoltaic cells, and electroplating processes, there have still been a limited number of papers dealing with organic electrolytic reactions and electrosynthesis [4, 5].

This chapter covers the applications of ionic liquids to organic electrolytic reactions, particularly electroorganic synthesis.

8.1 CELL DESIGN [6]

A proper choice of cell design is important for an electrolytic reaction. Organic electrolytic reactions are achieved on a laboratory scale by using an undivided cell. The simplest cell design is shown in Figure 8.1, but a cylindrical cell, shown in Figure 8.2, is the recommended design when anhydrous conditions or electrolysis under an inert gas atmosphere like a nitrogen gas is required. In the inert gas atmosphere the solvent and substrate are injected with a syringe into the cell through a rubber septum.

When a species reduced at a cathode is oxidized at an anode, and vice versa, a two-compartment cell, namely a divided cell like an H-type cell (Figure 8.3) with a sintered glass diaphragm or an ion-exchange membrane should be used in order to prevent mixing of both anodic and cathodic solutions. To decrease the cell voltage (voltage between an anode and cathode), the distance between both the electrodes should be kept as small as possible.

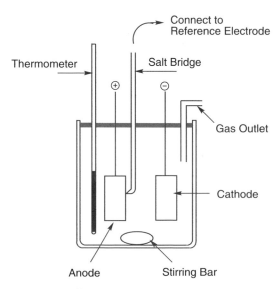

Figure 8.1 *Beaker-type cell.*

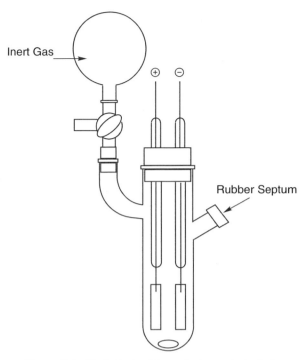

Figure 8.2   *Undivided cell for unhydrous electrolysis.*

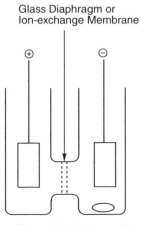

Figure 8.3   *H-type cell.*

8.2 ELECTROLYTIC METHOD

8.2.1 Constant Current Electrolysis and Constant Potential Electrolysis [6]

It is recommended that organic electrosynthesis be carried out at a constant current at first, since the setup and operation are simple. Then the product selectivity and yield can be improved by changing current density and the amount of electricity passed [current (A) × time (s) = electricity (C)]. However, the electrode potential changes with the consumption of the starting substrate (more positive in case of oxidation or more negative in case of reduction). Therefore the product selectivity and current efficiency sometimes decrease, particularly at the late stage of electrolysis.

Nevertheless, a constant potential electrolysis is suitable for achievement of high selectivity and clarification of the reaction mechanisms. Moreover, by the electrolysis, one can learn the number of electrons involved in the electrolytic reaction. As shown in Figure 8.1, a salt bridge terminated either by a Luggin capillary or a plug of porous Vycor glass is placed closed to the working electrode, and an appropriate constant potential relative to a reference electrode such as SCE (aqueous saturated calomel electrode) is applied using a potentiostat. The amount of electricity passed is measured by a coulombmeter.

8.2.2 Indirect Electrolysis Using Mediators [7]

Indirect electrolysis using mediators is an alternative powerful electrolytic method. This method is suitable for the following cases:

1. Electrolysis of hardly oxidizable and/or hardly reducible substrates
2. Precise control of the desired reaction
3. Avoidance of electrode passivation (nonconducting polymer formation on the electrode)

Halide ions, multiple valence metal ions (transition metal complexes), viologen, triarylamines, nitroxyl radicals, and fused arenes like naphthalene are used as mediators for this method. Mediators very often act as an electrocatalyst.

8.2.3 Electropolymerization [8]

Electropolymerization is highly useful for the preparation of π-conjugated electroconducting polymers. Potential scanning polymerization is often employed in addition to ordinary constant current and constant potential electropolymerizations. Since the polymer film is formed on the electrode surface, one can easily obtain valuable information about the propagation process from CV curves during the potential scanning.

8.3 SELECTIVE ANODIC FLUORINATION

Anodic partial fluorination, namely selective anodic fluorination of organic compounds, has much importance in the development of new types of pharmaceuticals, agrochemicals, and functional materials. However, the selective fluorination is not straightforward, and it very often requires hazardous reagents. Still selective anodic fluorination seems to be an ideal method because it can be carried out under mild conditions without any hazardous and/or reagents [9, 10]. The fluorination is usually conducted in aprotic solvents containing HF salt ionic liquids such as $Et_3N \cdot 3HF$ and $Et_4NF \cdot 3HF$. Since the discharge potential of fluoride ions is extremely high ($>+2.9$ V vs. SCE at Pt anode in MeCN), the fluorination proceeds via a (radical) cation intermediate as shown in Scheme 8.1.

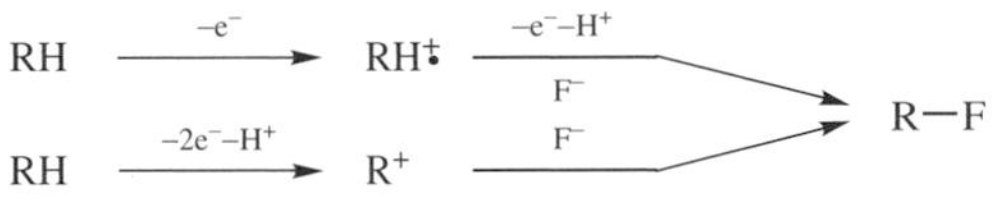

Scheme 8.1

When organic solvents are used for anodic fluorination, anode passivation (the formation of a nonconducting polymer film on the anode surface that suppresses faradaic current) takes place very often, which results in low efficiency. Moreover, depending on the substrates the use of acetonitrile can yield an acetoamidation by-product. To prevent acetoamidation and anode passivation, Meurs and his co-workers used an $Et_3N \cdot 3HF$ ionic liquid as both a solvent and supporting electrolyte (also a fluorine source) for the anodic fluorination of benzenes, naphthalene, olefins, furan, benzofuran, and phenanthroline. They obtained corresponding partially fluorinated products in less than 50% yields (Scheme 8.2) [11].

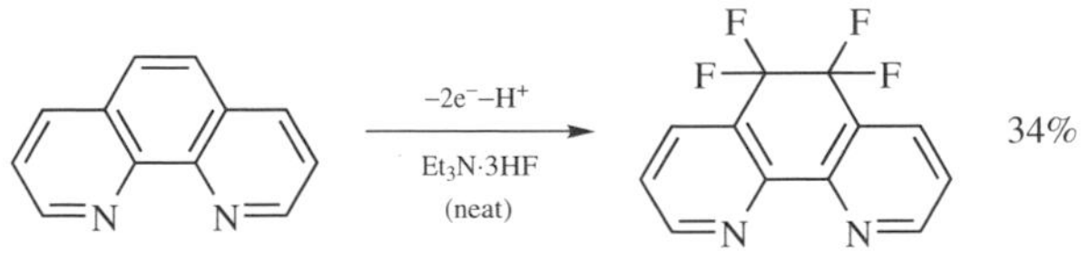

Scheme 8.2

Fuchigami's group and Laurent's group have independently achieved anodic α-fluorination of α-phenylthioacetate (1) in $Et_3N \cdot 3HF/MeCN$ [12, 13]. Since anode passivation takes place during the anodic fluorination, pulse electrolysis is necessary to avoid the anode passivation. Anodic difluorination of **1** was unsuccessful in the same electrolytic solution, while anodic fluorination of α-monofluorinated acetate (**2**) gave the α, α-difluorinated product **3** in moderate yield after passing a large excess amount of electricity (Scheme 8.3) [14].

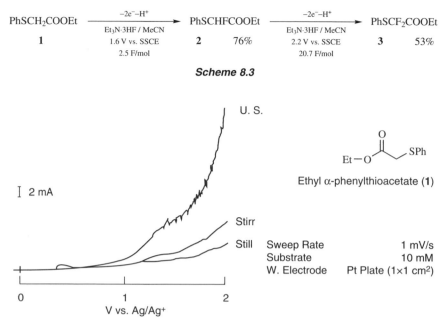

$$PhSCH_2COOEt \xrightarrow[\substack{Et_3N\cdot 3HF / MeCN \\ 1.6\ V\ vs.\ SSCE \\ 2.5\ F/mol}]{-2e^- -H^+} PhSCHFCOOEt \xrightarrow[\substack{Et_3N\cdot 3HF / MeCN \\ 2.2\ V\ vs.\ SSCE \\ 20.7\ F/mol}]{-2e^- -H^+} PhSCF_2COOEt$$

1 **2** 76% **3** 53%

Scheme 8.3

U. S.

Ethyl α-phenylthioacetate (**1**)

| 2 mA

Stirr

Still

Sweep Rate	1 mV/s
Substrate	10 mM
W. Electrode	Pt Plate (1×1 cm²)

0   1   2

V vs. Ag/Ag⁺

Figure 8.4 *Sonovoltammogram of ethyl α-phenylthioacetate (**1**) in ionic liquid, Et₃N·3HF.*

To overcome above-mentioned problems, anodic fluorination of **1** in neat $Et_3N\cdot 3HF$ without a solvent was investigated under ultrasonication [15]. As shown in Figure 8.4, a limiting diffusion oxidation current of **1** increased markedly under ultrasonication compared with that under mechanical stirring. The visocity of the ionic fluoride salt is higher than that of ordinary organic solvents, therefore, the mass transport of substrate **1** to the anode surface from the bulk of the fluoride salt is slower than that in organic solvents. The enhanced oxidation current is attributed to the markedly promoted mass transport of **1**. Namely ultrasonication to the ionic liquid $Et_3N\cdot 3HF$ generates cavitation, which makes micro-jet stream promoting mass transport of **1** to the anode surface from the bulk of the ionic liquid. The result indicates that ionic liquid $Et_3N\cdot 3HF$ behaves just like ordinary solvents under ultrasonication.

Anodic fluorination of α-phenylthioacetate **1** in ionic liquid $Et_3N\cdot 3HF$ proceeds smoothly without anode passivation under ultrasonication to provide the α-mono-fluorinated product **2** in high yield and with high current efficiency. Notably, anodic difluorination of **1** can be also achieved in the same ionic liquid under ultrasonication as shown in Scheme 8.4 [15]. Thus the current efficincy for the anodic difluorination of **1** was greatly improved as compared with that without ultrasonication. As described above, $Et_3N\cdot 3HF$ ionic liquid is highly useful for selective anodic fluorination.

However, $Et_3N\cdot 3HF$ is rather easily oxidized at around 2 V vs. Ag/Ag⁺. Therefore this ionic salt is not suitable for the fluorination of substrates having high oxidation potentials.

$$C_6H_4SCH_2COOEt \xrightarrow[\text{Et}_3\text{N} \cdot 3\text{HF}]{-2e, -H^+} C_6H_4SCHFCOOEt + C_6H_4SCF_2COOEt$$

$\quad$ **1** $\qquad\qquad\qquad\qquad\qquad\qquad$ **2** $\qquad\qquad$ **3**

Undersonication:	2F/mol	85%	4%
Non-sonication:	2F/mol	33%	24%
Undersonication:	6F/mol	0%	63%
Non-sonication:	6F/mol	9%	23%

Scheme 8.4

Nevertheless, Momota and his coworkers developed a new series of liquid fluoride salts, $R_4NF \cdot nHF(R=Me, Et,$ and n-Pr, $n > 3.5)$, that are useful in partial fluorination [16]. These electrolytes are non viscous liquids that have high electro-conductivity and anodic stability (Figure 8.5). As a result anodic partial fluorination of arenes such as benzene [16], mono-, di-, and trifluorobenzenes [17], chloroben-zene [18], bromobenzenen [19], toluene [20, 21], and quinolines [22] was success-fully carried out at high current densities by employing these liquid fluoride salts in the absence of solvent with good to high current efficiencies (66–90%) (Schemes 8.5 and 8.6). Nonconducting polymer films are often formed on anode

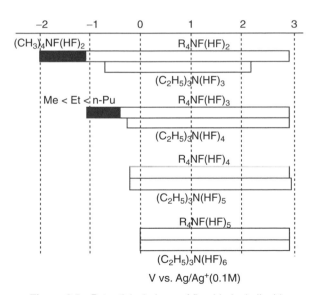

V vs. Ag/Ag$^+$(0.1M)

Figure 8.5 Potential windows of fluoride ionic liquids.

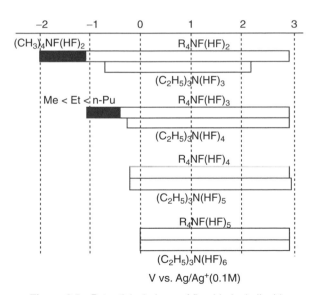

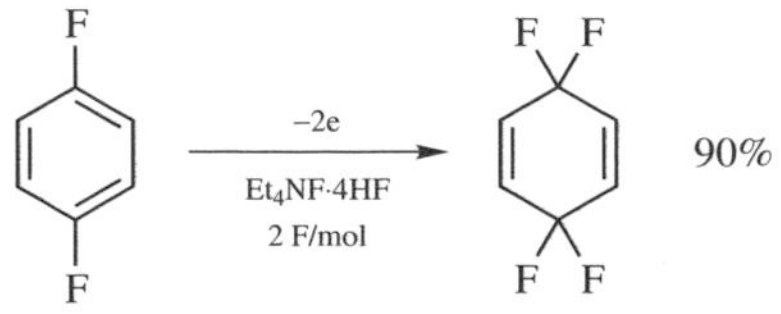

90%

Scheme 8.5

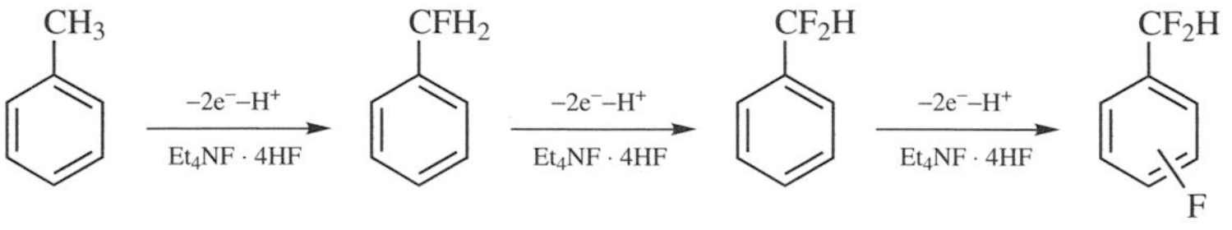

Scheme 8.6

surfaces when ordinary fluoride salts like $Et_3N \cdot 3HF$ are used. However, such film formation is reduced considerably when the new supporting fluoride salts are used.

Yoneda and Fukuhara also investigated the electrochemical stability (potential window) of ionic liquid $Et_3N \cdot nHF$ ($n = 3$–5) by cyclic voltammetry [23]. Their anodic stability increases with an increase of HF content (n) in the fluoride salts while the cathodic stability shows the reverse tendency. Thus $Et_3N \cdot 5HF$ is stable up to $+3$ V vs. Ag/Ag^+, but it is readily reduced at about -0.2 V to form hydrogen gas. The potential window of $Et_4NF \cdot 4HF$ is almost same as that of $Et_3N \cdot 5HF$, as shown in Figure 8.5. Thus the selective anodic fluorination of cyclic ketones and cyclic unsaturated esters in $Et_3N \cdot 5HF$ was successfully carried out to provide ring-opening and ring-expansion fluorinated products, respectively, as shown in Schemes 8.7 and 8.8 [23, 24].

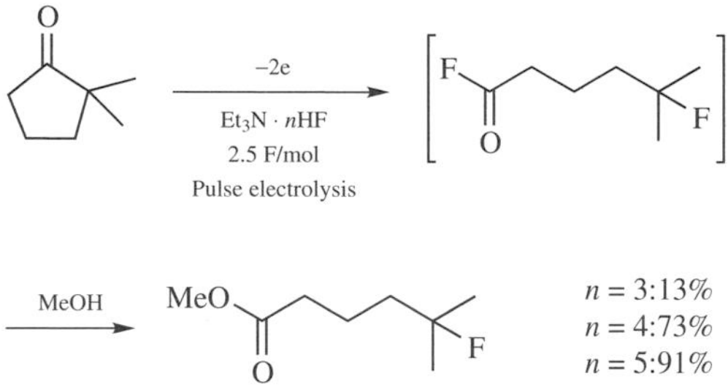

$n = 3$:13%
$n = 4$:73%
$n = 5$:91%

Scheme 8.7

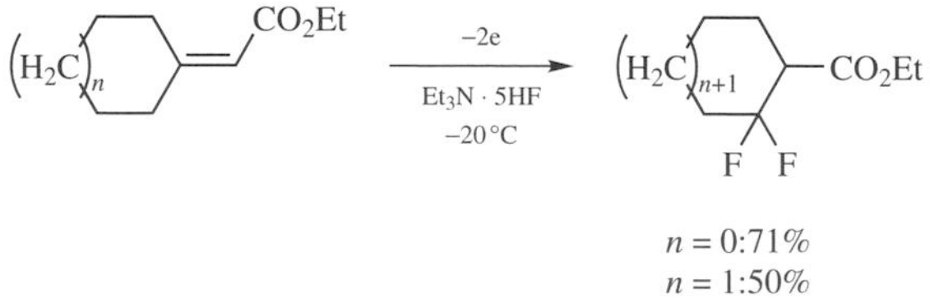

$n = 0$:71%
$n = 1$:50%

Scheme 8.8

As mentioned above, liquid fluoride salts like $Et_3N \cdot nHF$ ($n = 3$–5) and $Et_4NF \cdot 4HF$ have proved to be highly useful as the electrolytic media and fluoride ion source for selective anodic fluorination. However, this solvent-free method has an atom economy problem because of the use of an excessive amount of liquid salts in place of a solvent.

Fuchigami and his coworkers found that the fluorination of cyclic ethers like tetrahydrofuran, 1,4-dioxane, and 1,3-dioxolane was achieved by anodic oxidation of a mixture of a large amount of liquid cyclic ethers and a small amount of $Et_4NF \cdot 4HF$ (only 1.5–1.7 equiv. of F^- to the ether) at a high current density (150 mA cm^{-2}) (Scheme 8.9) [25]. In this method, the ether substrate was selectively oxidized

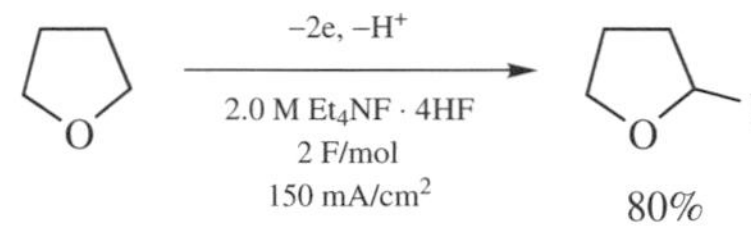

$$-2e, -H^+$$
2.0 M $Et_4NF \cdot 4HF$
2 F/mol
150 mA/cm^2
80%

Scheme 8.9

to provide the corresponding monofluorinated product in good yield and with good current efficiency. In sharp contrast, the use of organic solvents or a large amount of $Et_4NF \cdot 4HF$ resulted in no formation or low yield (ca. 10%) of the desired fluorinated product. Lactone and cyclic carbonates are also similarly fluorinated to provide the monofluorinated products in good yields as shown in Scheme 8.10 [25]. It

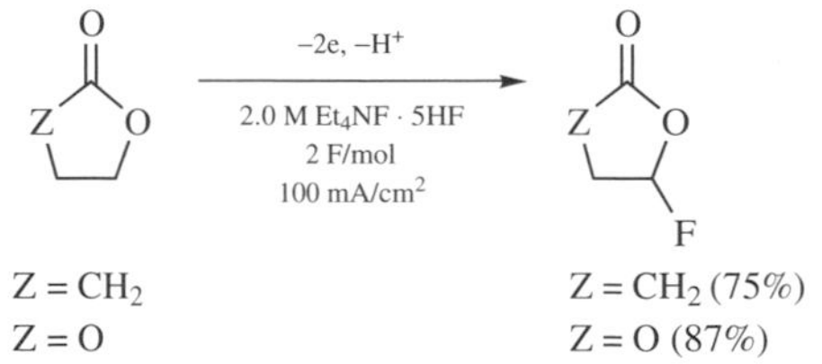

$$-2e, -H^+$$
2.0 M $Et_4NF \cdot 5HF$
2 F/mol
100 mA/cm^2

Z = CH$_2$ Z = CH$_2$ (75%)
Z = O   Z = O (87%)

Scheme 8.10

is notable that the fluorinated tetrahydrofuran is easily isolated, simply by distillation of the electrolytic ionic liquid fluoride salt after the electrolysis, with the remaining fluoride salt available for use in anodic fluorination. Thus completely solvent-free electroorganic synthesis was demonstrated by using liquid fluoride salt, as shown in Figure 8.6.

Moreover Fuchigami's group found that a combination of $Et_4NF \cdot nHF$ ($n = 4, 5$) and imidazolium ionic liquids was highly effective for anodic fluorination of phthalides [26]. As shown in Table 8.1, conventional electrolysis in organic solvents resulted in low yields. Since the oxidation potential of phthalide is extremely high (2.81 V vs. SCE), the oxidation of solvents takes place preferentially or predominantly (in DME). Even under solvent-free conditions (in liquid fluoride salts solely), the yield is also low due to simultaneous oxidation of the fluoride salts during the electrolysis. In sharp contrast, when ionic liquid, 1-ethyl-3-methylimidazolium trifluoromethanesulofonate ([EMIM][OTf]) is used, the yield increased markedly. The combination of [EMIM][OTf] and $Et_3N \cdot 5HF$ is the best choice to

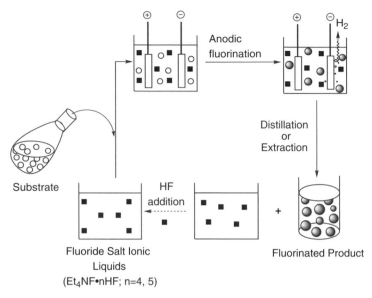

Figure 8.6   *Procedure of solvent-free electroorganic synthesis.*

provide excellent yield. On the other hand, the use of 1-ethyl-3-methylimidazolium fluoride·2.3HF ([EMIMF][2.3HF]) gave unsatisfactory yield (Scheme 8.11).

From these results it appears that the double ionic liquid system consisting of [EMIM][OTf] and a liquid fluoride salt like $Et_3N \cdot 5HF$ enhances not only the nucleophilicity of F^- but also the electrophilicity of a cationic intermediate of phthalide, as shown in Scheme 8.12. After generation of the cationic intermediate from phthalide, the cation should have TfO^- as the counter anion to form the

TABLE 8.1   Anodic Fluorination of Phthalide under Various Conditions

Run	Reaction Media	Supporting Electrolyte	Charge passed F/mol	Yield/%[a]
1	MeCN	$Et_3N \cdot 5HF$	6	16
2	CH_2Cl_2	$Et_3N \cdot 5HF$	8	23[b]
3	DME	$Et_4NF \cdot 4HF$	8	0
4	—	$Et_3N \cdot 3HF$	4	17
5	—	$Et_3N \cdot 5HF$	8	16[b]
6	[EMIM][OTf]	$Et_3N \cdot 3HF$	6	78
7	[EMIM][OTf]	$Et_4NF \cdot 4HF$	4.7	70
8	[EMIM][OTf]	$Et_3N \cdot 5HF$	4	90

[a] Determined by ^{19}F-NMR.
[b] Reaction was complicated.

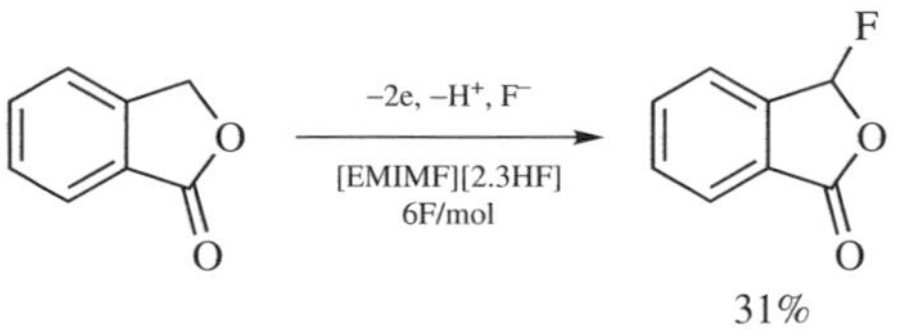

Scheme 8.11

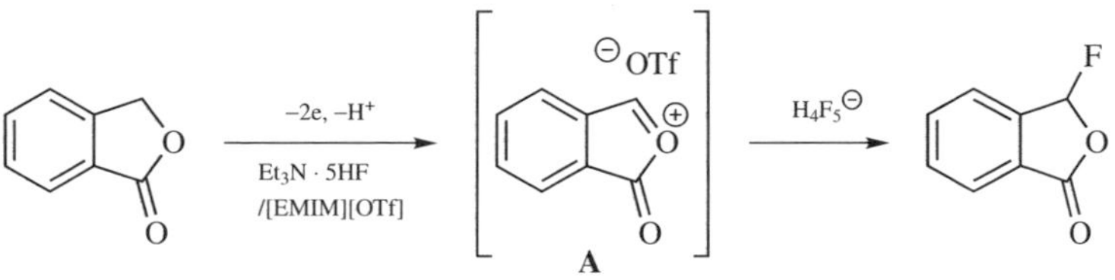

Activated cation

Scheme 8.12

activated cation **A**, which readily react with F^- to provide the fluorinated phthalide in good yield.

Fuchigami and his coworkers also found that imidazolium ionic liquid and dichloromethane show similar solvent effects on anodic fluorodesulfurization. As shown in Table 8.2, anodic fluorodesulfurization of 3-phenylthiophthalide (**4**) takes place exclusively in [EMIM][OTf] and CH_2Cl_2 to afford **5**, while α-fluorination

TABLE 8.2 Solvent Effect on Anodic Fluorination of 3-Phenylthiophthalide (4)

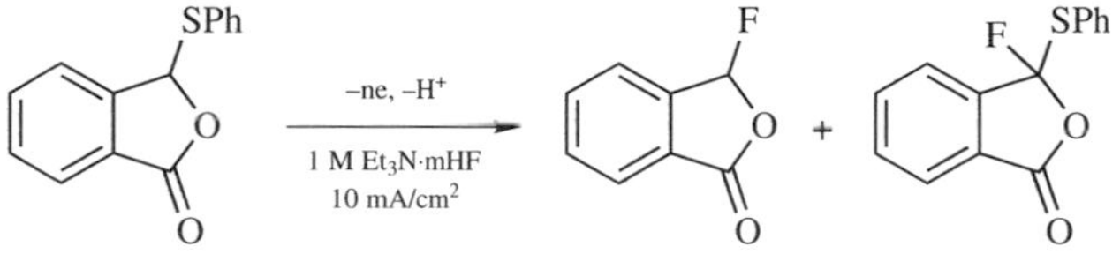

Run	Reaction Media	Supporting Electrolyte	Charge Passed F/mol	Yield %[a]	
				5	**6**
1	MeCN	$Et_3N \cdot 3HF$	4	44	—
2	CH_2Cl_2	$Et_3N \cdot 3HF$	4	86	—
3	THF	$Et_3N \cdot 3HF$	4	6	22
4	DME	$Et_3N \cdot 3HF$	3	9	72
5	[emim][OTf]	$Et_3N \cdot 3HF$	4	44	—
6	[emim][OTf]	$Et_3N \cdot 5HF$	4	83	Trace
7	[emim][BF_4]	$Et_3N \cdot 5HF$	3.5	65	—

[a]Determined by ^{19}F-NMR.

proceeds predominantly in DME to provide **6**. This suggests that [EMIM][OTf] destabilizes the radical cation intermediate **B** anodically generated from **4** similarly to CH_2Cl_2 as shown in Scheme 8.13. This anodic fluorodesulfurization can be performed by the reuse of the imidazolium and fluoride ionic liquids.

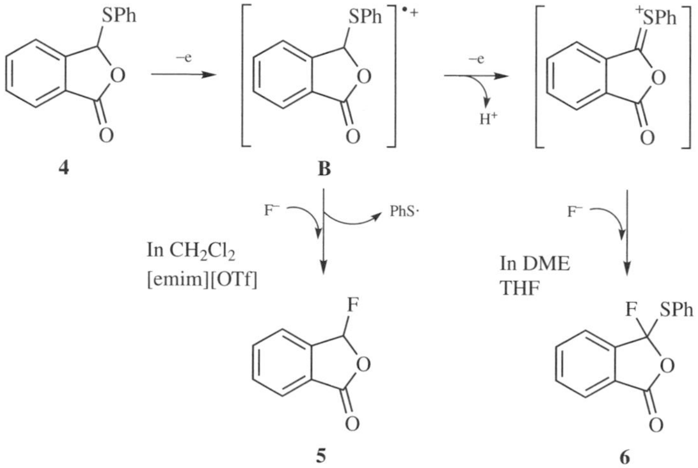

Scheme 8.13

Very recently it has been shown that anodically fluorinated product selectivity is greatly affected by fluoride ionic liquids, as shown in Scheme 8.14 [27]. The electrolysis of a phenylthioglycoside derivative **7** in $Et_3N \cdot 4HF$ provides fluorodesulfurization product **8** exclusively, while that in $Et_3N \cdot 3HF$ affords α-fluorinated product **9**

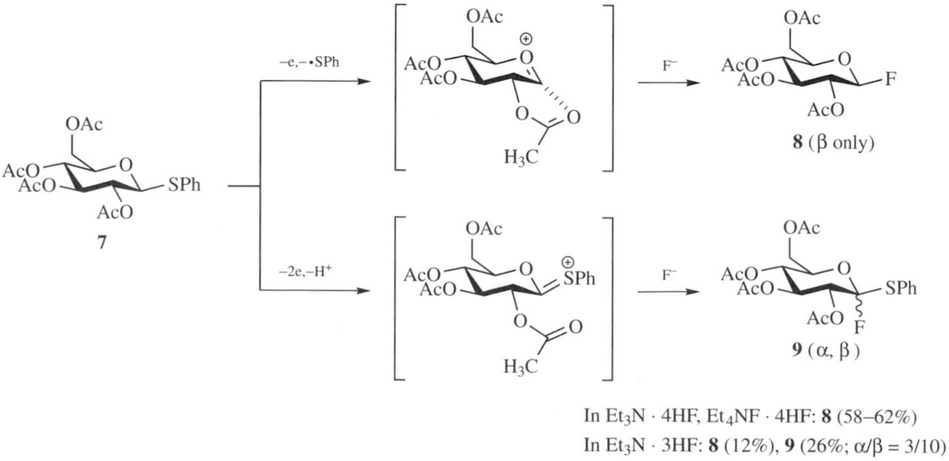

In $Et_3N \cdot 4HF$, $Et_4NF \cdot 4HF$: **8** (58–62%)
In $Et_3N \cdot 3HF$: **8** (12%), **9** (26%; α/β = 3/10)

Scheme 8.14

along with **8**. Since $Et_3N \cdot 3HF$ contains a considerable amount of free Et_3N, Et_3N seems to accelerate deprotonation of the radical cation intermediate resulting in the formation of α-fluorinated product **9**.

8.4 ELECTROCHEMICAL POLYMERIZATION

Electrochemical synthesis of electroconducting polymers such as polyarene [28–31], polypyrrole [32–34], polythiophene [35], and polyaniline [36, 37] has been carried out in moisture sensitive chloroaluminate ionic liquids. However, the polymer films are decomposed rapidly by the corrosive products like HCl generated by hydrolysis of the ionic liquids. In addition the treatment of the chloroalminate ionic liquids requires a special equipment such as glove box.

Nevertheless, Mattes's group and Fuchigami's group achieved independently electrooxidative polymerization of pyrrole, thiophene, and aniline in different moisture stable imidazolium ionic liquids [38–40]. Mattes's group used 1-butyl-3-methylimidazolium tetrafluoroborate and hexafluorophosphate ([BMIM][BF$_4$] and [BMIM][PF$_6$]) for electropolymerization, and they found that π-conjugated polymers thus obtained are highly stable and that they can be electrochemically cycles in the ionic liquids up to million cycles. In addition, since the polymers have cycle switching speeds as fast as 100 ms, they can be used as electrochemical actuators, electrochromic windows, and numeric displays.

Fuchigami and his coworkers employed [EMIM][OTf] for the electropolymerization. They found that the polymerization of pyrrole in the ionic liquid proceeds much faster than that in conventional media like aqueous and acetonitrile solutions containing 0.1 M [EMIM][OTf] as a supporting electrolyte, as shown in Figure 8.7. It is known that in the radical–radical coupling, further oxidation of oligomers and polymer deposition in the electrooxidative polymerization are favorably affected because the reaction products are accumulated in the vicinity of the electrode surface under slow diffusion conditions, and consequently the polymerization rate is increased. It is reasonable to assume that the polymerization rate in [EMIM][OTf] is higher than that in the conventional media, since neat [EMIM][OTf] (viscosity: 42.7 cP) has a higher viscosity than the others. Electropolymerization of thiophene is also accelerated in this ionic liquid, whereas that of aniline is decelerated.

It is notable that the morphology and some physical properties of polypyrrole and polythiophene films are greatly affected by the use of the ionic liquid [EMIM][OTf] as the electrolytic medium. As shown in Figure 8.8, grains are observed in the polypyrrole films prepared in aqueous and acetonitrile solutions, although the sizes of grain differ (Figure 8.8*a*, *b*). In sharp contrast, the surface of the polypyrrole film prepared in neat [EMIM][OTf] is so smooth that no grains are observed in the photograph (Figure 8.8*c*).

As summarized in Table 8.3, both the electrochemical capacity and electroconductivity are markedly increased when the polypyrrole and polythiophene films are prepared in the ionic liquid. This may be attributable to the extremely high concentration of anions as the dopants, which results in a much higher doping level.

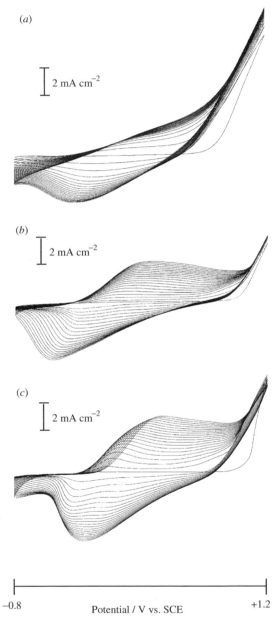

(a)

2 mA cm^{-2}

(b)

2 mA cm^{-2}

(c)

2 mA cm^{-2}

−0.8   Potential / V vs. SCE   +1.2

Figure 8.7 Cyclic voltammograms in the course of polymerization of 0.1 M pyrrole for 20 potential scan cycles in (a) 0.1 M [EMIM][OTf] + H$_2$O, (b) 0.1 M [EMIM][OTf] + MeCN, and (c) neat [EMIM][OTf].

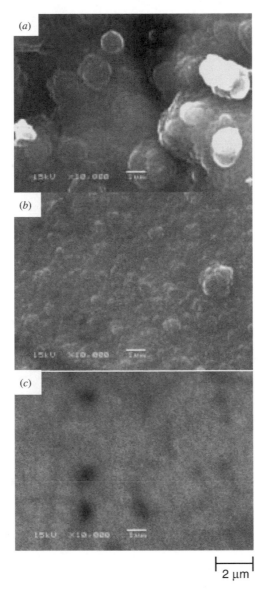

Figure 8.8 *SEM photographs of polypyrrole films (a) polymerized in 0.1 M [EMIM][OTf] + H₂O, (b) polymerized in 0.1 M [EMIM][OTf] + MeCN, and (c) polymerized in neat [EMIM][OTf].*

As described above, the polymer films prepared in the ionic liquid have a higher electrochemical density and highly regulated morphological structures. Therefore their possible utilizations are as high-performance electrochemical capacitors, ion-sieving films, ion-selective electrodes, matrices for hosting catalyst particles, and so on. Figure 8.9 shows SEM photographs of surfaces of palladium-polypyrrole

TABLE 8.3 Properties of Polypyrrole and Polythiophene Films Prepared in Various Media

Polymer	Media	Roughness Factor[a]/ Dimensional	Electro-chemical Capacity/C cm^{-3}	Electro-conductivity/ S cm^{-1}	Doping Level (%)[b]
Polypyrrole	H_2O	3.4	77	1.4×10^{-7}	22
Polypyrrole	CH_3CN	0.48	190	1.1×10^{-6}	29
Polypyrrole	$EMICF_3SO_3$	0.29	250	7.2×10^{-2}	42
Polythiophene	CH_3CN	8.6	9	4.1×10^{-8}	—
Polythiophene	$EMICF_3SO_3$	3.3	45	1.9×10^{-5}	—

[a]Roughness factor is standard deviation of the film thickness.
[b]The doping level was determined by elemental analysis.

composite films. The polypyrrole matrix of the composite films (*a*) and (*b*) was prepared in aqueous electrolyte and neat [EMIM][OTf], respectively, and the pale-gray particles observed in the photographs are attributed to the presence of palladium. It is evident that aggregated palladium clusters of a few micrometers are deposited

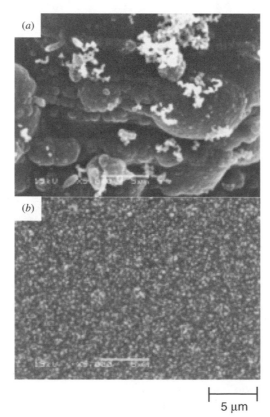

5 μm

Figure 8.9 *SEM photographs of polypyrrole films (a) polymerized in 0.1 M [EMIM] [OTf] + MeCN and(b) polymerized in neat [EMIM][OTf].*

locally on grains of polypyrrole prepared in the aqueous solution, while submicron clusters are highly dispersed on the whole surface of the polypyrrole matrix prepared in [EMIM][OTf]. This may be attributed to homogeneous polarization of the matrix in the palladium deposition process, since a polypyrrole matrix prepared in [EMIM][OTf] has a very smooth surface. In addition such a composite film can be used for a number of catalytic and electrocatalytic reactions.

The utility of the ionic liquid as a recyclable medium for the polymerization was also demonstrated. More than 90% of [EMIM][OTf] after the polymerization was easily recovered simply by extracting the remaining pyrrole monomer with chloroform. The recoved [EMIM][OTf] could be reused five times without significant loss of reactivity for the polymerization.

8.5 ELECTROCHEMICAL FIXATION OF CO₂

A quaternary ammonium salt $Et_4N^+ \cdot p\text{-}TsO^-$ can be changed to a melt at 120°C. More than 30 years ago, Weinberg successfully used this melt for electrochemical synthesis of an α-amino acid derivative from an imine and CO_2 using a mercury cathode, as shown in Scheme 8.15 [41]. The reaction proceeds via the radical anionic intermediate of the imine.

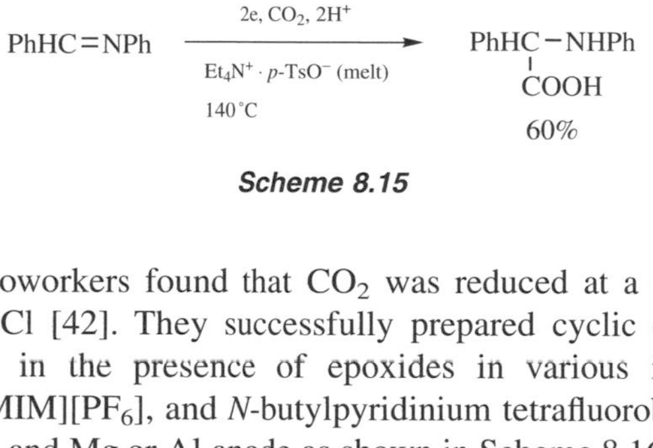

$$PhHC=NPh \xrightarrow[\substack{Et_4N^+ \cdot p\text{-}TsO^- \text{ (melt)} \\ 140°C}]{2e, CO_2, 2H^+} \begin{array}{c} PhHC-NHPh \\ | \\ COOH \\ 60\% \end{array}$$

Scheme 8.15

Deng and his coworkers found that CO_2 was reduced at a copper cathode at −2.4 V vs. Ag/AgCl [42]. They successfully prepared cyclic carbonates by the reduction of CO_2 in the presence of epoxides in various ionic liquids like [EMIM][BF₄], [BMIM][PF₆], and *N*-butylpyridinium tetrafluoroborate [BPy][BF₄] using a Cu cathode and Mg or Al anode as shown in Scheme 8.16. This electrolysis is nonfaradaic reaction in which a small amount of electricity engaged in the electroreduction of CO_2 generates catalytic species responsible for the addition of CO_2 to epoxides to form cyclic carbonates.

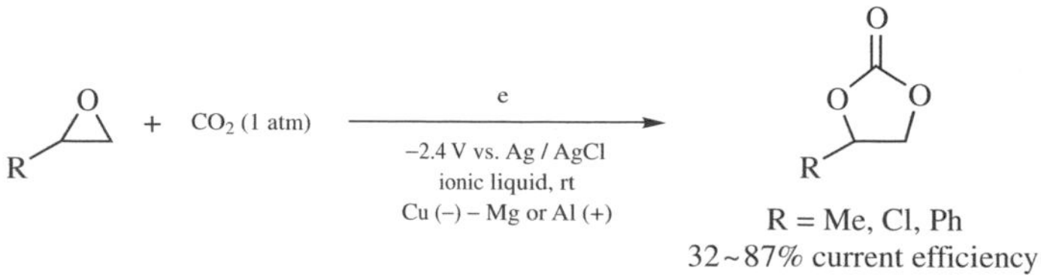

R = Me, Cl, Ph
32~87% current efficiency

Scheme 8.16

8.6 ELECTROREDUCTIVE COUPLING USING METAL COMPLEX CATALYSTS

Peters and Sweeny carried out cyclic voltammetry of nickel(II) salen at a glassy carbon electrode in [BMIM][BF$_4$]. They found that nickel(II) salen exhibits one-electron, quasi-reversible reduction to nickel(I) salen, and that the latter species serves as a catalyst for cleavage of carbon-halogen bonds in iodoethane and 1,1,2-trichlorotrifluoroethane (Freon^R 113) [43]. They also showed the diffusion coefficient for nickel(II) salen in the ionic liquid at room temperature is more than 500 times smaller than that $(1.0 \times 10^{-5} \, \text{cm}^2 \, \text{s}^{-1})$ in a typical organic solvent–electrolyte system such as DMF containg 0.1 M Et$_4$N·BF$_4$. This is due to the high viscosity of the ionic liquid.

Jouikov and his coworkers reported the electrocatalytic homo-coupling of PhBr and PhCH$_2$Br using NiCl(bpy) complex in [BMIM][NTf$_2$] as shown in Scheme 8.17 [44].

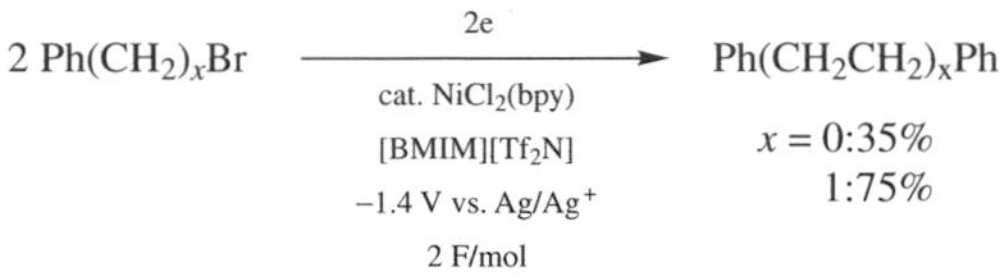

$$2 \, \text{Ph(CH}_2)_x\text{Br} \xrightarrow[\substack{\text{cat. NiCl}_2\text{(bpy)} \\ \text{[BMIM][Tf}_2\text{N]} \\ -1.4 \text{ V vs. Ag/Ag}^+ \\ 2 \text{ F/mol}}]{2e} \text{Ph(CH}_2\text{CH}_2)_x\text{Ph}$$

$$x = 0:35\%$$
$$1:75\%$$

Scheme 8.17

Barhdadi and his coworkers reported that direct electroreductive coupling of organic halides in [Octyl-MIM][BF$_4$] was unsuccessful because of its high viscosity. However, they successfully carried out the reaction in the ionic liquid containing DMF (10% v/v) using a Ni cathode and Mg or Al anode [45]. They performed homo-coupling of PhBr using NiBr$_2$(bpy) (5% vs. PhBr) in the same electrolytic solution. Then they obtained 100% current efficiency and 82% yield, using Ni, SUS, and Fe sacrificial anodes. They also achieved Ni-catalyzed electroreductive coupling of aryl halides and activated olefins, as shown in Scheme 8.18. One nitrogen atom bearing a free electron-pair of imidazolium cation seems to interact with the electrogenerated low-valent Ni-species, preventing precipitation of nickel metal.

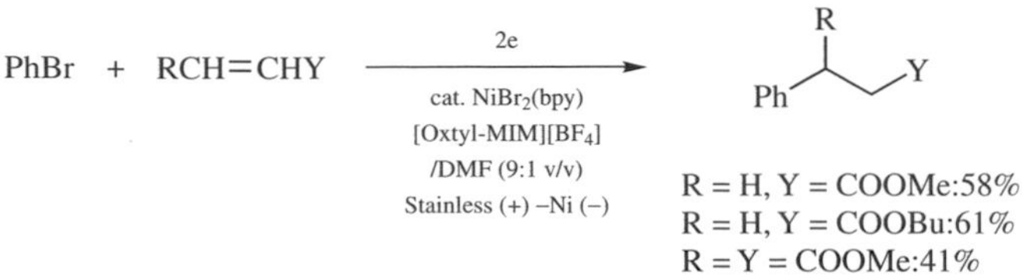

$$\text{PhBr} + \text{RCH=CHY} \xrightarrow[\substack{\text{cat. NiBr}_2\text{(bpy)} \\ \text{[Oxtyl-MIM][BF}_4\text{]} \\ \text{/DMF (9:1 v/v)} \\ \text{Stainless (+) –Ni (–)}}]{2e}$$

R = H, Y = COOMe:58%
R = H, Y = COOBu:61%
R = Y = COOMe:41%

Scheme 8.18

8.7 OTHER DEVELOPMENTS

Lu and his coworkers reported the electroreduction of benzoylformic acid at a glassy carbon cathode and Pt anode in ionic liquid [EMIM][Br] at 80°C giving mandelic acid in high yield and with moderate current efficiency as illustrated in Scheme 8.19 [46].

$$\text{PhCOCOOH} \xrightarrow[\substack{\text{[EMIM][Br]} \\ -1.3\,\text{V vs. SCE} \\ \text{GC}\,(-)\,-\text{Pt}\,(+) \\ 80\,°\text{C}}]{2e} \text{PhCH(OH)COOH}$$

91% yield
(57% current efficiency)

Scheme 8.19

Fry found that 1,4- and 1,2-dinitrobenzenes exhibit two one-electron waves in acetonitrile but a single two-electron wave in the ionic liquid [BMIM][BF$_4$] [47]. He explained that latter effect is ascribed to strong ion pairing between the imidazolium cation and the dinitrobenzene dianion. However, it was clarified that this interesting effect is attributable to the hydrogen bonding between the hydrogen at the 2-position of the imidazolium cation and the dinitrobenzene dianion [48].

8.8 CONCLUSION

Organic electrochemistry is an inherently hybrid and interdisciplinary area. Electro-organic synthesis has been recognized as a environmentally friendly process since the 1970s. However, new methodologies must be developed to achieve modern electrosynthsis as a green and sustainable chemistry. As studies on electrolytic reactions and electroorganic synthesis in ionic liquids continue, it is hoped that, this chapter will inspire not only electrochemists but also organic chemists to turn to the new and exciting area of organic electrochemistry in ionic liquids.

REFERENCES

1. T. Welton, *Chem. Rev.*, **99**, 207 (1999).
2. P. Wassersheid, W. Keim, *Angew. Chem. Int. Ed.*, **39**, 3772 (2000).
3. *Green Industrial Applications of Ionic Liquids*, R. D. Rogers, K. R. Seddon, S. Volkov, eds., Kluwer Academic, 2002.
4. H. Ishii, T. Fuchigami, *Electrochemistry*, **70**, 46 (2002).
5. T. Fuchigami, M. Atobe, *Mater. Integr.*, **16**, 20 (2003).
6. M. R. Rifi, F. H. Covitz, *Introduction to Organic Electrochemistry*, Dekker, 1974.
7. J. Simonet, J.-F. Pilard, in *Organic Electrochemistry, 4th ed.*, H. Lund, O. Hammerich, eds., Dekker, 2000, ch. 29.

8. J. Heinze, in *Topics in Current Chemistry: Electrochemistry IV*, E. Steckhan, ed., Springer, 1990, p. 1.

9. T. Fuchigami, *Advances in Electron Transfer Chemistry*, Vol. 6, P. S. Mariano, ed., JAI Press, 1999, p. 41.

10. T. Fuchigami, in *Organic Electrochemistry*, 4th ed., H. Lund, O. Hammerich, ed., Dekker, 2000, ch. 25.

11. J. H. H. Meurs, W. Eilenberg, *Tetrahedron*, **47**, 705 (1991).

12. T. Fuchigami, M. Shimojo, A. Konno, K. Nakagawa, *J. Org. Chem.*, **55**, 6074 (1990).

13. T. Brigaud, E. Laurent, *Tetrahedron Lett.*, **31**, 2287 (1990).

14. T. Fuchigami, A. Konno, M. Shimojo, *J. Org. Chem.*, **59**, 3459 (1994).

15. T. Fuchigami, T. Sunaga, H. Ishii, M. Atobe, *100th ECS Meeting*, Filadelphia, May (2002), Abstr. 1220.

16. K. Momota, M. Morita, Y. Matsuda, *Electrochim. Acta*, **38**, 1123 (1993).

17. K. Momota, T. Yonezawa, Y. Hayakawa, K. Kato, M. Morita, Y. Matsuda, *J. Appl. Electrochem.*, **25**, 651 (1995).

18. K. Momota, H. Horio, K. Kato, M. Morita, Y. Matsuda, *Electrochim. Acta*, **40**, 233 (1995).

19. K. Momota, H. Horio, K. Kato, M. Morita, Y. Matsuda, *Denki Kagaku (presently, Electrochemistry)*, **62**, 1106 (1994).

20. K. Momota, K. Mukai, K. Kato, M. Morita, *Electrochim. Acta*, **43**, 2503 (1998).

21. K. Momota, K. Mukai, K. Kato, M. Morita, *J. Fluorine Chem.*, **87**, 173 (1998).

22. K. Saeki, M. Tmomizitsu, Y. Kawazoe, K. Momota, H. Kimoto, *Chem. Phram. Bull.*, **44**, 2254 (1996).

23. S.-Q. Chen, T. Hatakeyama, T. Fukuhara, S. Hara, N. Yoneda, *Electrochim. Acta*, **42**, 1951 (1997)

24. S. Hara, S. Q. Chen, T. Hoshio, T. Fukuhara, N, Yoneda, *Tetrahedron Lett.*, **37**, 8511 (1996).

25. M. Hasegawa, H. Ishii, T. Fuchigami, *Tetrahedron Lett.*, **43**, 1502 (2002).

26. M. Hasegawa, H. Ishii, T. Fuchigami, *Green Chem.*, **5**, 512 (2003).

27. M. Hasegawa, T. Fuchigami, *Electrochim. Acta*, **49**, 3367 (2004).

28. D. C. Trivedi, *Chem. Commun.*, 544 (1989).

29. L. M. Goldenberg, G. E. Pelekh, V. I. Krinichnyi, O. S. Roshchupkina, A. F. Zueva, R. N. Lyubovskaya, O. N. Efimov, *Synth. Met.*, **36**, 217 (1990).

30. V. M. Kobryanskii, S. A. Arnautov, *Makromol. Chem.*, **193**, 455 (1992).

31. V. M. Kobryanskii, S. A. Arnautov, *Synth. Met.*, **55**, 1371 (1993).

32. P. G. Pickup, R. A. Osteryoung, *J. Am. Chem. Soc.*, **106**, 2294 (1984).

33. P. G. Pickup, R. A. Osteryoung, *J. Electroanal. Chem.*, **195**, 271 (1985).

34. T. A. Zawodzinski, Jr., L. Janiszewska, R. A. Osteryoung, *J. Electroanal. Chem.*, **255**, 111 (1988).

35. L. Janiszewska, R. A. Osteryoung, *J. Electrochem. Soc.*, **134**, 2787 (1987).

36. N. Koura, H. Ejiri, K. Takeishi, *Denki Kagaku* (presently *Electrochemistry*), **59**, 74 (1991).

37. N. Koura, H. Ejiri, K. Takeishi, *J. Electrochem. Soc.*, **140**, 602 (1993).

38. W. Lu, A. G. Fadeev, B. Qi, E. Smela, B. R. Mattes, J. Ding, G. M. Spinks, J. Mazurkiewicz, D. Zhou, G. G. Wallace, D. R. MacFarlane, S. A. Forsyth, M. Forsyth, *Science*, **297**, 983 (2002).

39. K. Sekiguchi, M. Atobe, T. Fuchigami, *Electrochem. Commun.*, **4**, 881 (2002).

40. K. Sekiguchi, M. Atobe, T. Fuchigami, *J. Electroanal. Chem.*, **557**, 1 (2003).

41. N. L. Weinberg, A. K. Hoffmann, T. B. Reddy, *Tetrahedron Lett.*, **12**, 2271 (1971).

42. H. Yang, Y. Gu, Y. Deng, F. Shi, *Chem. Commun.*, 274 (2002).

43. B. K. Sweeny, D. G. Peters, *Electrochem. Commun.*, **3**, 712 (2001).

44. M. Mellah, S. Gmouh, M. Vaultier, V. Jouikov, *Electrochem. Commun.*, **5**, 591 (2003).

45. R. Barhdadi, C. Courtinard, J. Y. Nedelec, M. Troupel, *Chem. Commun.*, 1434 (2003).

46. J.-X. Lu, Q. Sun, M.-Y. He, *Chinese J. Chem.*, **21**, 1229 (2003).

47. A. J. Fry, *J. Electroanal. Chem.*, **546**, 35 (2003).

48. T. Iwasa, M. Atobe, T. Fuchigami, unpublished work.

Chapter 9

Electrodeposition of Metals in Ionic Liquids

Yasushi Katayama

Electrodeposition of metals and alloys in condensed media, including electro- and electrolessplating, has been an essential technology for the advanced electronics industry, besides its conventional industrial uses in electrorefining, decoration, corrosion protection, and other functional surface finishing. Most practical wet processes are based on aqueous media, mainly because of the knowledge and experience accumulated over many years. However, the metals that can be deposited in aqueous media are limited by the electrochemical potential window of water. Thus such nonaqueous media as aprotic organic electrolytes and high temperature molten salts have been investigated as the alternative electrolytes for the electrodeposition of base metals.

Room-temperature ionic liquids are the promising electrolytes for the electrodeposition of various metals because they have the merits of both organic electrolytes and high-temperature molten salts. Ionic liquids can be used in a wide temperature range, so temperatures can be elevated to accelerate such phenomena as nucleation, surface diffusion and crystallization associated with the electrodeposition of metals. In addition the process can be safely constructed because ionic liquids are neither flammable nor volatile if they are kept below the thermal decomposition temperature of the organic cations.

Electrochemical Aspects of Ionic Liquids   Edited by Hiroyuki Ohno
ISBN 0-471-64851-5   Copyright © 2005 John Wiley & Sons, Inc.

A large number of studies on the electrodeposition of metals have been reported for chloroaluminate ionic liquids, but so far only a few studies have explored for nonchloroaluminate ionic liquids. Because chloroaluminate ionic liquids are highly hygroscopic, and upon undergoing hydrolysis produce toxic and corrosive hydrogen chloride, these ionic liquids are not suited for practical use. However, various chemical substances, especially metal chlorides, can be dissolved by controlling the Lewis acidity or basicity of the chloroaluminate ionic liquids, and these systems are worth investigating. Nonchloroaluminate ionic liquids, especially those stable against moisture, have a potential for practical use. In developing ionic liquids for practical applications, it will be necessary to accumulate much fundamental data on the solubility of metal compounds, redox potentials of metals, and mass transport of dissolved species.

There are a number of studies describing the electrodeposition of metals and alloys in the "ionic liquids" diluted with certain aprotic organic solvents. It is important to recognize that although from a practical point of view the addition of the organic solvents improves the mobility of the electroactive species and the morphology of the electrodeposits, these electrolytes should not be classified equally as ionic liquids but be treated as separate entities. The addition of the organic solvents not only weakens the important properties of ionic liquids, such as non-flammability and nonvolatility, but at the same time also significantly alters the environment of the species' participating electrode reactions. Therefore this chapter will not deal with these electrolytes.

There are a small number of studies on the electrochemistry and electrodeposition of metals in the ionic liquids [1, 2]. The following sections review examples pertaining to the electrodeposition of metals in ionic liquids based on organic cations. The organic cations in these sections are abbreviated as follows:

EMI	1-Ethyl-3-methylimmidazolium
BMI	1-Butyl-3-methylimidazolium
DMPI	1,2-Dimethyl-3-propylimidazolim
BP	*N*-Butylpyridinium
MP	Methylpyridinium
BTMA	Benzyltrimethylammonium
TMHA	Trimethylhexylammonium
TEA	Tetraethylammonim

9.1 CHLOROALUMINATE IONIC LIQUIDS

The electrodeposition of metals and alloys has been investigated extensively in the chloroaluminate ionic liquids. Many kinds of metal salts, mostly chlorides, can be dissolved in ionic liquids with their Lewis acidity or basicity controlled by changing the composition of $AlCl_3$. In the case of acidic ionic liquids that contain $AlCl_3$ at more than 50 mol%, a dimeric chloroaluminate anion, $Al_2Cl_7^-$, acts as a Lewis acid and withdraws a chloride ion from a metal chloride, MCl_n to form a cationic species, MCl_{n-m}^{m+}, when the Lewis acidity of MCl_n is weaker than that of $Al_2Cl_7^-$:

$$MCl_n + mAl_2Cl_7^- = MCl_{n-m}^{m+} + 2mAlCl_4^- \tag{9.1}$$

The cationic species, MCl_{n-m}^{m+}, is reduced, possibly to its metallic state under cathodic polarization. In the acidic ionic liquids, the electrodeposition of Al occurs by the reduction of $Al_2Cl_7^-$:

$$4Al_2Cl_7^- + 3e^- = Al + 7AlCl_4^- + 3Cl^- \tag{9.2}$$

There are many studies reporting the electrodeposition of Al in acidic chloroaluminate ionic liquids [3–9]. When the reduction potential of MCl_{n-m}^{m+} is more positive than that of $Al_2Cl_7^-$, it is possible to obtain the electrodeposit of pure metal. Alloys of the metal with Al are obtained if the reduction potential of MCl_{n-m}^{m+} is more negative than that of $Al_2Cl_7^-$. Moreover the underpotential deposition of Al sometimes leads to the formation of Al–M alloys, even if the reduction potential of MCl_{n-m}^{m+} is more positive than that of $Al_2Cl_7^-$.

In the case of basic ionic liquids that contain $AlCl_3$ at less than 50 mol%, the metal salt acts as a Lewis acid by accepting the chloride anions in the ionic liquids to form a chlorocomplex anion:

$$MCl_n + mCl^- = MCl_{n+m}^{m-} \tag{9.3}$$

Polymeric chlorocomplex anions are also known for some metals. Since the electrodeposition of Al does not occur within the electrochemical potential window of known alkylated imidazolium and pyridinium cations under the basic condition of room temperature, an electrodeposit of the pure metal can be obtained.

It is possible to prepare the neutral ionic liquids by mixing the equimolar amounts of the organic chloride and $AlCl_3$. When a metal salt is added to a neutral ionic liquid, the metal salt will act as either a Lewis base or acid to affect the Lewis acidity or basicity of the ionic liquid. Based on this, excess LiCl, NaCl, or HCl is added to maintain the neutrality of the ionic liquid.

Figure 9.1 shows the metals that can be electrodeposited as pure metals or alloys in chloroaluminate ionic liquids. In the discussion below, the potentials are given in relation to the Al/Al(III) electrode. This electrode is composed of Al immersed in an acidic ionic liquid of 66.7 or 60.0 mol% $AlCl_3$. The formal potentials of some redox couples in ionic liquids are illustrated in Figure 9.2 and summarized in Tables 9.1 and 9.2.

9.1.1 Electrodeposition of Main Group Elements

The electrodeposition of alkaline and alkaline earth metals has been investigated in the neutral (buffered) ionic liquids for the purpose of applying the ionic liquids to electrolytes of rechargeable batteries using these metals as anodes. However, the

	1	2	3	4	5	6	7	8	9	10	11	12	13/III	14/IV	15/V	16/VI	17/VII	18/VIII
1	H																	He
2	Li	Be											B	C	N	O	F	Ne
3	Na	Mg _A_											Al _A_	Si	P	S	Cl	Ar
4	K	Ca _A_	Sc	Ti _A_	V	Cr _A_	Mn	Fe _A_	Co _A_	Ni _A_	Cu _AB_	Zn _A_	Ga _A_	Ge	As	Se	Br	Kr
5	Rb	Sr _A_	Y	Zr	Nb _A_	Mo	Tc	Ru	Rh	Pd _B_	Ag _A_	Cd _B_	In _B_	Sn _AB_	Sb _AB_	Te _B_	I	Xe
6	Cs	Ba	Ln	Hf	Ta	W	Re	Os	Ir	Pt _A_	Au _B_	Hg _AB_	Tl _B_	Pb _A_	Bi _A_	Po	At	Rn
7	Fr	Ra	An	Rf	Db	Sg	Bh	Hs	Mt	Uun	Uuu	Uub						

Lanthanide	La _A_	Ce	Pr	Nd	Pm	Sm	Eu	Gd	Tb	Dy	Ho	Er	Tm	Yb	Lu
Actinide	Ac	Th	Pa	U	Np	Pu	Am	Cm	Bk	Cf	Es	Fm	Md	No	Lr

Figure 9.1 Grayed elements can be deposited in the acidic (A) and/or basic (B) chloroaluminate ionic liquids. The hatched elements are deposited only as a constituent of alloys.

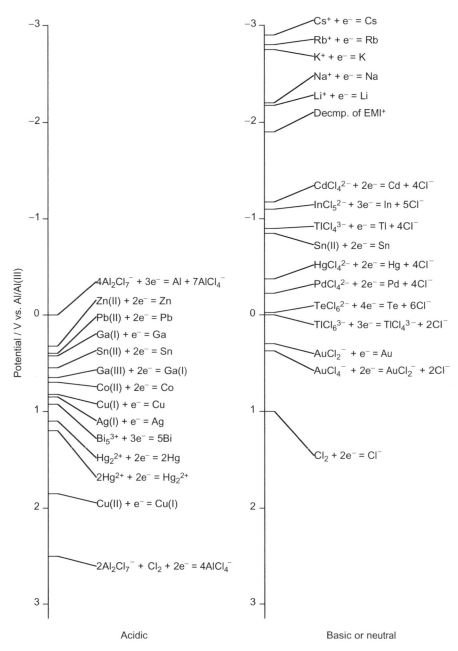

Figure 9.2   *The formal potentials of some redox couples in the chloroaluminate ionic liquids.*

TABLE 9.1 Formal potentials of certain redox couples in acidic chloroaluminate ionic liquids

Redox Couple	$E^{0'}/V^a$	$AlCl_3$ (mol%)	Cation	Temperature (°C)	Reference
$Ga(III) + 2e^- = Ga(I)$	0.655	60.0	EMI	30	23
$Ga(I) + e^- = Ga$	0.437	60.0	EMI	30	23
$Sn(II) + 2e^- = Sn$	0.55	66.7	EMI	40	26
$Pb(II) + 2e^- = Pb$	0.400	66.7	EMI	40	27
$SbCl_2^+ + 3e^- = Sb + 2Cl^-$	0.389**	—	BP	40	28
$Bi_5^{3+} + 3e^- = 5Bi$	0.925	66.7	BP	25	30
$Fe(III) + e^- = Fe(II)$	2.036*	66.7	BP	40	39
$Fe(II) + 2e^- = Fe$	0.773*	66.7	BP	40	39
$Co(II) + 2e^- = Co$	0.71	60.0	EMI	—	41
	0.894*	66.7	BP	36	40
$Ni(II) + 2e^- = Ni$	0.800*	60.0	BP	40	46
	1.017*	66.7	BP	40	46
$Cu(I) + e^- = Cu$	0.784	66.7	BP	40	50
	0.837	60.0	EMI	40	52
	0.843	66.7	EMI	40	52
	0.777	66.7	MP	30	49
$Cu(II) + e^- = Cu(I)$	1.825	66.7	BP	40	50
	1.851	66.7	MP	30	49
$Zn(II) + 2e^- = Zn$	0.322	60.0	EMI	40	57
$Ag(I) + e^- = Ag$	0.844	66.7	EMI	25	66
$2Hg^{2+} + 2e^- = Hg_2^{2+}$	1.21	66.7	EMI	40	73
$Hg_2^{2+} + 2e^- = 2Hg$	1.093	66.7	EMI	40	73

"The potential is represented against Al/Al(III) on molar scale except those couples noted with * and ** (see text).

TABLE 9.2 Formal potentials of certain redox couples in basic chloroaluminate ionic liquids

Redox Couple	$E^{0'}/V^a$	$AlCl_3$ (mol%)	Cation	Temperature (°C)	Reference
$InCl_5^{2-} + 3e^- = In + 5Cl^-$	-1.009	49.0	DMPI	27	24
	-1.096	44.0	DMPI	27	24
$TiCl_6^{3-} + 3e^- = TiCl_4^{3-} + 2Cl^-$	-0.025	40.0	EMI	30	25
	-0.005	44.4	EMI	30	25
$TiCl_4^{3-} + e^- = Tl + 4Cl^-$	-0.965	40.0	EMI	30	25
	-0.900	44.4	EMI	30	25
$Sn(II) + 2e^- = Sn$	-0.85	44.4	EMI	40	26
$SbCl_4^- + 3e^- = Sb + 4Cl^-$	-0.523*	—	BP	40	28
$TeCl_6^{2-} + 4e^- = Te + 6Cl^-$	-0.013	44.4	EMI	30	31
	0.077	49.0	EMI	30	31
$Te + 2e^- = Te(-II)$	-1.030	44.4	EMI	30	31
	-1.036	49.0	EMI	30	31
$Cu(I) + e^- = Cu$	-0.647	42.9	BP	40	50
$Cu(II) + e^- = Cu(I)$	0.046	42.9	BP	40	50
$PdCl_4^{2-} + 2e^- = Pd + 4Cl^-$	-0.230	44.4	EMI	40	61
	-0.110	49.0	EMI	40	61
$AuCl_2^- + e^- = Au$	0.310	44.4	EMI	40	72
$AuCl_4^- + 2e^- = AuCl_2^- + 2Cl^-$	0.374	44.4	EMI	40	72
$HgCl_4^{2-} + 2e^- = Hg + 4Cl^-$	-0.370	44.4	EMI	40	73

"The potentials are given against Al/Al(III) on the molar scale. The formal potentials are given at the specified $AlCl_3$ compositions except for Sb(III)/Sb (marked with *).

reduction potentials of alkaline metals (Li, Na, K, Rb, and Cs) were found to be more negative than those of alkylated imidazolium cations [10]. Reversible deposition and dissolution of some alkaline metals have been reported in the presence of such additives as HCl [11–17] and $SOCl_2$ [18, 19], as these are considered to produce a passivation film on the deposited alkaline metals.

The anodic dissolution of magnesium, Mg, has been reported as possible in the basic or buffered ionic liquids that consist of EMI^+ and $DMPI^+$ [20]. However, the deposition of Mg is impossible due to the instability of metallic Mg against these organic cations, whereas the formation of Al–Mg alloys containing Mg up to 2.2 at% has been observed in the acidic $EMICl–AlCl_3$ ionic liquid [21]. Calcium and strontium dichlorides, $CaCl_2$ and $SrCl_2$, are soluble in an acidic $EMICl–AlCl_3$ ionic liquid and the co-deposition of calcium and strontium with bismuth and copper has been examined in the acidic ionic liquid [22].

The electrodeposition of gallium, Ga, has been investigated in an acidic $EMICl–AlCl_3$ ionic liquid [23]. Gallium species can be introduced by the anodic dissolution of metallic Ga or dissolution of $GaCl_3$. Metallic Ga can be obtained by reducing monovalent gallium species, Ga(I), which can be produced by reduction of trivalent gallium species, Ga(III):

$$Ga(III) + 2e^- = Ga(I) \qquad (9.4)$$

and

$$Ga(I) + 2e^- = Ga \qquad (9.5)$$

The formal potential of Ga(III)/Ga(I) and Ga(I)/Ga is reported as 0.655 and 0.437 V, respectively.

The electrodeposition of indium, In, has been reported in a basic $DMPICl–AlCl_3$ ionic liquid [24]. Indium trichloride, $InCl_3$, dissolves in the basic ionic liquid and forms a trivalent indium chlorocomplex anion, $InCl_5^{2-}$, which can be reduced to metallic In by a three-electron transfer reaction:

$$InCl_5^{2-} + 3e^- = In + 5Cl^- \qquad (9.6)$$

The formal potentials of In(III)/In in the basic ionic liquids of 49.0 and 44.4 mol% $AlCl_3$ are reported as -1.009 and -1.096 V, respectively. A monovalent indium species, In(I), is not stable in this melt forming metallic In and $InCl_5^{2-}$ as shown in the following disproportionation reaction:

$$3InCl + 2Cl^- = 2In + InCl_5^{2-} \qquad (9.7)$$

Metallic thallium, Tl, can be electrodeposited in a basic EMICl-AlCl$_3$ ionic liquid [25]. It is possible to dissolve thallium trichloride, $TlCl_3$, to form a trivalent thallium chlorocomplex anion, $TlCl_6^{3-}$, which is reducible to a monovalent thallium chlorocomplex anion, $TlCl_4^{3-}$:

$$TlCl_6^{3-} + 3e^- = TlCl_4^{3-} + 2Cl^- \qquad (9.8)$$

The formal potentials of Tl(III)/Tl(I) in the basic ionic liquids of 40.0 and 44.4 mol% $AlCl_3$ are reported as -0.025 and -0.005 V, respectively. Thallium metal is obtained by further reduction of $TlCl_4^{3-}$:

$$TlCl_4^{3-} + e^- = Tl + 4Cl^- \tag{9.9}$$

The formal potentials of Tl(I)/Tl in the basic ionic liquids of 40.0 and 44.4 mol% $AlCl_3$ are reported as -0.965 and -0.900 V, respectively.

The electrodeposition of tin, Sn, has been reported in both basic and acidic $EMICl$–$AlCl_3$ ionic liquid [26]. A divalent tin species, Sn(II), can be introduced by the anodic dissolution of metallic tin. The introduction of a tetravalent tin species, Sn(IV), is also possible by dissolving tin tetrachloride, $SnCl_4$. However, the evaporation of $SnCl_4$ occurs in the case of an acidic ionic liquid. The irreversible reduction of Sn(IV) to Sn(II) occurs at around 0.91 and -0.9 V in the acidic and basic ionic liquids, respectively. The electrodeposition of metallic Sn is possible by the reduction of Sn(II):

$$Sn(II) + 2e^- = Sn \tag{9.10}$$

The formal potentials of Sn(II)/Sn in the acidic (66.7 mol% $AlCl_3$) and basic (44.4 mol% $AlCl_3$) ionic liquids are reported as 0.55 and -0.85 V, respectively.

The electrodeposition of lead, Pb, has been investigated in an acidic $EMICl$–$AlCl_3$ ionic liquid [27]. A divalent lead species, Pb(II), can be introduced by dissolving lead dichloride, $PbCl_2$, or the anodic dissolution of metallic Pb. Metallic Pb is obtained by the reduction of Pb(II):

$$Pb(II) + 2e^- = Pb \tag{9.11}$$

The formal potential of Pb(II)/Pb is reported as 0.400 V.

The electrodeposition of antimony, Sb, has been reported in both acidic and basic $BPCl$–$AlCl_3$ ionic liquid [28, 29]. Antimony trichloride, $SbCl_3$, is soluble in both acidic and basic ionic liquids. In the case of an acidic ionic liquid, a cationic trivalent antimony species, $SbCl_2^+$, is formed and can be reduced to the metallic state:

$$SbCl_2^+ + 3e^- = Sb + 2Cl^- \tag{9.12}$$

The formal potential is dependent on the concentration of Cl^- and given as follows:

$$E^{0'} = E_{eq} - \frac{RT}{3F}\ln[SbCl_2^+] + \frac{2RT}{3F}\ln[Cl^-] \tag{9.13}$$

The experimental $E^{0'}$ value is reported as 0.389 V. On the other hand, an trivalent antimony chlorocomplex anions, $SbCl_4^-$, is dominant in the basic ionic liquid and is also reducible to metallic Sb at more negative potential:

$$SbCl_4^- + 3e^- = Sb + 4Cl^- \tag{9.14}$$

The formal potential for this equilibrium is given by

$$E^{0'} = E_{eq} - \frac{RT}{3F}\ln[\text{SbCl}_4^-] + \frac{4RT}{3F}\ln[\text{Cl}^-]$$ (9.15)

and is reported as -0.523 V.

The electrodeposition of bismuth, Bi, has been investigated in an acidic BPCl–AlCl$_3$ ionic liquid [30]. Bismuth trichloride, BiCl$_3$, is soluble in the acidic ionic liquid forming a trivalent bismuth species, Bi(III), which can be reduced to a cluster cation, Bi$_5^{3+}$, at 0.95 V. Metallic Bi can be obtained by reducing Bi$_5^{3+}$ via an intermediate species, Bi$_5^+$:

$$\text{Bi}_5^{3+} + 2e^- = \text{Bi}_5^+$$ (9.16)

and

$$\text{Bi}_5^+ + e^- = 5\text{Bi}$$ (9.17)

The formal potential of Bi$_5^{3+}$/Bi is reported as 0.925 V.

The electrodeposition of tellurium, Te, has been reported in a basic EMICl–AlCl$_3$ ionic liquid [31]. Tellurium tetrachloride, TeCl$_4$, is soluble in the basic ionic liquid to form a tetravalent tellurium chlorocomplex anion, TeCl$_6^{2-}$, which can be reduced to metallic Te by four electron transfer reaction:

$$\text{TeCl}_6^{2-} + 4e^- = \text{Te} + 6\text{Cl}^-$$ (9.18)

The formal potentials of Te(IV)/Te in the basic ionic liquids of 44.4 and 49.0 mol% AlCl$_3$ are reported as -0.013 and 0.077 V, respectively. Metallic Te is further reduced to Te($-$II), which is soluble in the ionic liquid:

$$\text{Te} + 2e^- = \text{Te}(-\text{II})$$ (9.19)

The formal potentials of Te/Te($-$II) in the basic ionic liquids of 44.4 and 49.0 mol% AlCl$_3$ are reported as -1.030 and -1.036 V, respectively. It is also possible to obtain metallic Te by oxidizing Te($-$II).

9.1.2 Electrodeposition of Transition Metals

The electrochemical behavior of titanium species has been investigated in an acidic EMICl–AlCl$_3$ ionic liquid [32]. Titanium dichloride, TiCl$_2$, dissolves up to 60 and 170 mmol dm^{-3} at 80°C in the acidic ionic liquids of 60.0 and 66.7 mol% AlCl$_3$, respectively. The divalent titanium species, Ti(II), is less stable in the weakly acidic ionic liquid, leading to the disproportionation into insoluble titanium trichloride and metallic Ti. Metallic Ti is impossible to obtain probably because of the large overpotential of the electrodeposition of metallic Ti whose reduction potential is expected to be more positive than that of Al. However, Al–Ti alloys containing Ti up to about 19 at% can be obtained by the co-deposition of Ti and Al.

The electrodeposition of chromium, Cr, has been investigated in an acidic BPCl–AlCl$_3$ ionic liquid [33]. Chromium dichloride, CrCl$_2$, dissolves in the ionic liquid up to 0.31 mol dm^{-3} at 80°C. It is reported that the reduction of a divalent chromium species, Cr(II), occurs on a platinum electrode at the potential close to that of the bulk electrodeposition of Al, to give the electrodeposit containing Cr at 94 at%. The Al content in the deposit increases with the decrease in the deposition potential. Al–Cr alloys containing Cr at 0 to 94 at% have been prepared by controlling the deposition potential. In case of a basic EMICl– or DMPICl–AlCl$_3$ ionic liquid, CrCl$_2$ dissolves in the basic ionic liquid and forms a divalent chromium chlorocomplex anion, CrCl$_4^{2-}$, which is not, in this case, reduced to metallic state within the electrochemical potential window of the ionic liquid [34].

The electrodeposition of iron, Fe, has been investigated in BPCl– and EMICl–AlCl$_3$ ionic liquids [35–38]. In case of the basic ionic liquids, iron dichloride, FeCl$_2$, dissolves and forms a divalent iron chlorocomplex anion, FeCl$_4$ [2–39], which is not reducible to metallic Fe within the electrochemical potential window of the EMICl–AlCl$_3$ ionic liquid [37]. In the neutral and neutral-like ionic liquids, metallic Fe can be obtained by the reduction of a divalent iron species, but the dissolved iron species in these ionic liquids are not identified [36, 37]. In the acidic ionic liquids, FeCl$_2$ dissolves and forms a divalent iron species, which is reducible to metallic Fe at 0.3 to 0.4 V prior to the electrodeposition of Al [35, 37]. The formal potential of Fe(II)/Fe in the acidic BPCl–AlCl$_3$ ionic liquid is reported as 0.773 V on the mole fraction scale [39]. The formation of Al–Fe alloys containing Fe at 0 to 61 at% has been also reported in an acidic EMICl–AlCl$_3$ ionic liquid [38].

The electrodeposition of cobalt, Co, has been studied extensively in acidic ionic liquids [38, 40–45]. A divalent cobalt species, Co(II), can be introduced into the ionic liquids by dissolving cobalt dichloride, CoCl$_2$ or the anodic dissolution of Co metal. The formal potentials of Co(II)/Co in the acidic EMICl–AlCl$_3$ and BPCl–AlCl$_3$ ionic liquids are reported as 0.71 [41] and 0.894 [40] (on the mole fraction scale) V, respectively. The electrodeposition of Al–Co alloys containing Co at 33 to 50 at% occurs at the potential more negative than the formal potential of Co(II)/Co. The nucleation and growth processes of Co and Al–Co alloys are studied on a glassy carbon electrode in EMICl–AlCl$_3$ ionic liquid [41] and a single crystal Au electrode in the BMICl–AlCl$_3$ ionic liquid [44, 45]. In case of the basic BPCl–AlCl$_3$ ionic liquid, CoCl$_2$ dissolves and forms a divalent cobalt chlorocomplex anion, CoCl$_4^{2-}$, which is not reducible to metallic Co [40].

The electrodeposition of nickel, Ni, has been studied in acidic ionic liquids [37, 38, 45–48]. A divalent nickel species, Ni(II), can be introduced into the ionic liquids by dissolving nickel dichloride, NiCl$_2$, or the anodic dissolution of Ni metal. The reduction of Ni(II) occurs at around 0.38 V, producing metallic nickel in the acidic EMICl–AlCl$_3$ ionic liquid [47]. The formal potential of Ni(II)/Ni in an acidic BPCl–AlCl$_3$ ionic liquid is reported as 0.800 V on the mole fraction scale [46]. The electrodeposition of Al–Ni alloys occurs at the potential more negative than the deposition potential of metallic Ni. The nucleation and growth processes of Ni and Al–Ni alloys are studied on a glassy carbon electrode in the EMICl–AlCl$_3$ ionic liquid [47] and a single crystal Au electrode in the BMICl–AlCl$_3$ ionic liquid [45]. In a neutral buffered EMICl–AlCl$_3$ ionic liquid, the anodic oxidation of Ni metal results in forming NiCl$_2$, which is insoluble in this ionic liquid [37]. In the case of a basic BPCl–AlCl$_3$ ionic liquid, NiCl$_2$ dissolves and forms a divalent nickel chlorocomplex anion, NiCl$_4^{2-}$, which cannot be reduced to metallic Ni [46].

The electrodeposition of copper, Cu, has been studied in MPCl– [49], BPCl– [50], EMICl– [22, 37, 38, 51–54], and BMICl–AlCl$_3$ [55] ionic liquids. In the case of acidic ionic liquids, it is possible to introduce monovalent and divalent copper species, Cu(I) and Cu(II), by dissolving copper chloride and dichloride, respectively, and by the anodic dissolution of metallic copper. Metallic Cu can be obtained by the reduction of Cu(I). The formal potentials of Cu(I)/Cu in an acidic MPCl–, BPCl–, EMICl–AlCl$_3$ are reported as 0.777 [49], 0.784 [50], and 0.837 [52] V, respectively. Cu(I) can be also obtained by the reduction of Cu(II), of which the formal potentials in the acidic MPCl– and BPCl–AlCl$_3$ are reported as 1.851 [49] and 1.825 [50] V, respectively. The nucleation and growth process of Cu are investigated on a single crystal Au electrode [55]. The formation of Al–Cu alloys containing Al up to 43 at% occurs at the potential more negative than the formal potential of Cu(I)/Cu. However, it is reported that the displacement of Al with Cu occurs when the Al–Cu alloy is immersed in the ionic liquid containing Cu(I) [54]. In the case of basic ionic liquids, it is possible to introduce the monovalent copper chlorocomplex anions, CuCl$_2$ and CuCl$_4^{3-}$, by dissolving CuCl and by the anodic dissolution of Cu metal. CuCl$_2$ also dissolves in the basic ionic liquids and forms a divalent copper chlorocomplex anion, CuCl$_4^{2-}$ [50]. The formal potentials of Cu(I)/Cu and Cu(II)/Cu(I) in the BPCl–AlCl$_3$ ionic liquid are reported as -0.647 and 0.046 V, respectively [50]. The electrodeposition of copper is also possible in a basic EMICl–AlCl ionic liquid [53].

The electrochemical study on zinc, Zn, has been reported in EMICl– and BMICl–AlCl$_3$ ionic liquids [56–59]. A divalent zinc species, Zn(II), can be introduced by the anodic dissolution of Zn metal in the acidic ionic liquid. The electrodeposition of metallic Zn is possible in the acidic ionic liquid. The formal potential of Zn(II)/Zn is reported as 0.322 V in the EMICl–AlCl$_3$ ionic liquid [57]. The nucleation and growth processes have been studied on some polycrystalline metal electrodes in the EMICl–AlCl$_3$ ionic liquid [57] and on a single crystal gold electrode in the BMICl–AlCl$_3$ ionic liquid [59]. The formation of Al–Zn alloys was not reported. The electrodeposition of Zn has been investigated in the neutral EMICl–AlCl$_3$ ionic liquid mixed with the 1:1 mixture of EMICl and ZnCl$_2$ [58]. The divalent zinc species in this ionic liquid is considered to be ZnCl$_3^-$, which can be reduced to metallic Zn at around -0.8 V on a glassy carbon electrode. In a basic EMICl–AlCl$_3$ ionic liquid the anodic dissolution of Zn metal results in the formation of a divalent zinc chlorocomplex anion, ZnCl$_4^-$, which cannot be reduced within the electrochemical potential window of the ionic liquid [56].

The electrodeposition of Al–Nb alloys has been reported in a BPCl–AlCl$_3$ ionic liquid containing NbCl$_5$ or Nb$_3$Cl$_8$ [60]. A pentavalent niobium species is reduced chemically by Al powder prior to the electrodeposition of Al–Nb alloys. The Al–Nb alloys containing Nb up to 29.3 wt% have been obtained at 90° to 140°C.

The electrodeposition of palladium, Pd, has been studied in a EMICl–AlCl$_3$ ionic liquid [61, 62]. Palladium dichloride, PdCl$_2$, is soluble in the basic ionic liquid but less soluble in the neutral and acidic ionic liquids. In the basic ionic liquid, the divalent palladium species is considered to be PdCl$_4^{2-}$, which is reducible to metallic Pd:

$$PdCl_4^{2-} + 2e^- = Pd + 4Cl^- \tag{9.20}$$

The formal potential of Pd(II)/Pd in the basic ionic liquids of 44.4 and 49.0 mol% AlCl$_3$ are reported as -0.230 and -0.110 V, respectively. The electrodeposition of Pd in the acidic ionic liquid is also reported as occurring at 1.3 to 1.6 V.

The electrodeposition of silver, Ag, has been investigated in acidic EMICl– and BMICl–AlCl$_3$ ionic liquids [38, 63–66]. Silver chloride, AgCl, is soluble in the acidic ionic liquids. The formal potential of Ag(I)/Ag in the EMICl–AlCl$_3$ ionic liquid is reported as 0.844 V [66]. The nucleation and growth processes of metallic silver have been studied on some polycrystalline metal electrodes in the EMICl–AlCl$_3$ ionic liquid [63] and the HOPG (highly oriented pyrolytic graphite) electrode in the BMICl–AlCl$_3$ ionic liquid [64, 65]. The electrodeposition of Al–Ag alloys is reported in the acidic EMICl–AlCl$_3$ ionic liquid [38, 66]. AgCl is soluble also in a basic EMICl–AlCl$_3$ ionic liquid to form AgCl$_4^{3-}$ as a predominant species [67]. However, there is no description of the deposition of metallic Ag in the basic ionic liquid.

The electrodeposition of cadmium, Cd, is possible in the basic and neutral EMICl–AlCl$_3$ ionic liquids [56, 58]. In the basic ionic liquid, cadmium dichloride, CdCl$_2$, dissolves and forms a divalent cadmium chlorocomplex anion, CdCl$_4^{2-}$ [69], which is reducible to metallic Cd at around -1.2 V. In the case of a neutral buffered ionic liquid, the electrodeposition of Cd is observed at around -1.0 V, which is slightly positive than in the basic ionic liquid probably caused by the formation of more reducible species, CdCl$_3^-$.

The electrodeposition of lanthanum, La, has been attempted in neutral and acidic EMICl–AlCl$_3$ ionic liquids [70, 71]. Lanthanum trichloride, LaCl$_3$, is soluble in the acidic ionic liquid. The solubility of LaCl$_3$ in the acidic ionic liquid containing AlCl$_3$ at 66.7 mol% is reported as 45 mmol kg^{-1} at 25°C. The deposition of metallic La is not confirmed, while the reduction of a trivalent lanthanum species is suggested in the presence of SOCl$_2$. The electrodeposition of Al–La alloys has been also examined in the acidic ionic liquid where the content of La in the Al–La alloys is less than 0.1 at%.

The electrodeposition of platinum, Pt, has been reported in an acidic BTMACl–AlCl$_3$ ionic liquid [9]. A divalent platinum species, PtCl$_6^{2-}$, is introduced by dissolving either (BTMA)$_2$PtCl$_6$ or (TEA)$_2$PtCl$_6$. The deposition of metallic Pt occurs at the potential more positive ($\sim$100 mV) than that of the bulk deposition of Al. The formation of Al–Pt alloys containing Pt up to 13 at% is also reported.

The electrodeposition of gold, Au, has been studied in a basic EMICl–AlCl$_3$ ionic liquid [72]. Gold trichlolride, AuCl$_3$, is soluble in the ionic liquid and forms a trivalent gold chlorocomplex anion, AuCl$_4^-$, which is reducible to a monovalent gold chlorocomplex anion, AuCl$_2^-$, electrochemically:

$$AuCl_4^- + 2e^- = AuCl_2^- + 2Cl^- \qquad (9.21)$$

It is also possible to introduce AuCl$_2^-$ by the anodic dissolution of Au metal. Metallic Au can be obtained by the reduction of AuCl$_2^-$:

$$AuCl_2^- + e^- = Au + 2Cl^- \qquad (9.22)$$

The formal potentials of Au(III)/Au(I) and Au(I)/Au are reported as 0.374 and 0.310 V, respectively.

The electrodeposition of mercury, Hg, has been studied in both basic and acidic EMICl–AlCl$_3$ ionic liquids [73]. Mercury dichloride, HgCl$_2$, is soluble in the acidic ionic liquid to form a divalent mercury species, Hg^{2+}, which can be reduced irreversibly to a cluster cation, Hg$_2^{2+}$:

$$2Hg^{2+} + 2e^- = Hg_2^{2+} \qquad (9.23)$$

It is also possible to introduce Hg$_2^{2+}$ by dissolving Hg$_2$Cl$_2$. Metallic Hg can be obtained by the reduction of Hg$_2^{2+}$:

$$Hg_2^{2+} + 2e^- = 2Hg \qquad (9.24)$$

The formal potentials of Hg^{2+}/Hg$_2^{2+}$ and Hg$_2^{2+}$/Hg are reported as 1.21 and 1.093 V, respectively. In case of the basic ionic liquid, HgCl$_2$ dissolves and forms a divalent mercury chlorocomplex anion, HgCl$_4^{2-}$. The reduction of HgCl$_4^{2-}$ results in the deposition of metallic Hg:

$$HgCl_4^{2-} + 2e^- = Hg + 4Cl^- \qquad (9.25)$$

The formal potential of Hg(II)/Hg in the basic ionic liquid is reported as -0.370 V.

9.2 NONCHLOROALUMINATE IONIC LIQUIDS

9.2.1 Tetrafluoroborate (BF$_4^-$) and Hexafluorophosphate (PF$_6^-$) Ionic Liquids

The electrodeposition of several metals and alloys has been investigated in tetrafluoroborate ionic liquids. In contrast to the chloroaluminate ionic liquids, the tetrafluoroborate ionic liquids are considerably more stable against moisture and are expected to be applicable to practical use. Moreover the co-deposition of

the metals derived from the ionic liquids does not occur because the tetrafluoroborate anion is not reducible within the cathodic potential limits of known organic
cations. On the other hand, the introduction of metal ions is not straightforward
because the ionic liquids are neutral in the concept of Lewis acid and base.
Some metal ions can be introduced simply by dissolving their tetrafluoroborate
salts. However, this method is limited to a small number of metals, for which the
anhydrous tetrafluoroborate salts are available. It is difficult to adjust the Lewis
acidity or basicity by the addition of BF_3 or other binary fluorides. Thus organic
chlorides are usually added as a Lewis base in order to dissolve metal chlorides,
which act as Lewis acids to form their chlorocomplex anions, the same as observed
in the case of basic chloroaluminate ionic liquids. The hexafluorophosphate ionic
liquids are similar to the tetrafluoroborate ones ionic liquids, but the former is not
completely miscible with water. Thus the PF_6^- ionic liquids are sometimes regarded
as hydrophobic. However, it should be noted that the PF_6^- anion can undergo hydrolysis when it contacts with an acidic aqueous solution.

There is no common reference electrode in these ionic liquids, but the Al/Al(III)
electrode is used in $EMICl$–$EMIBF_4$ ionic liquids.

The Electrodeposition of Main Group Metals. The electrodeposition of germanium, Ge, has been studied in a $BMIPF_6$ ionic liquid [45, 74–76]. Germanium
halides, $GeCl_4$, $GeBr_4$, and GeI_4, dissolve very slowly in the ionic liquid. It is pre-
sumed that the dissolved tetravalent germanium species is reduced to metallic Ge
via an intermediate divalent germanium species. The electrodeposition process of
nano-scaled Ge has been investigated by in situ STM (scanning tunneling microscopy).

The electrodeposition of antimony [77] and indium-antimony [78] alloys has
been reported in a basic $EMICl$–$EMIBF_4$ ionic liquid. Antimony trichloride,
$SbCl_3$, dissolves in the ionic liquid and forms $SbCl_4^-$, the same as in the basic chloroaluminate ionic liquid. Metallic Sb can be obtained by the cathodic reduction of
$SbCl_4^-$, as shown in Eq. (9.14). The formal potential of Sb(III)/Sb is reported as
-0.27 V vs. Al/Al(III) in the ionic liquid containing Cl^- at 0.11 M. In addition
the oxidation of $SbCl_4^-$ leads to the formation of a pentavalent antimony species,
$SbCl_6^-$:

$$SbCl_6^- + 2e^- = SbCl_4^- + 2Cl^- \qquad (9.26)$$

The formal potential of Sb(V)/Sb(III) is reported as 0.60 V vs. Al/Al(III) in the
ionic liquid containing Cl^- at 0.11 M. On the other hand, a trivalent indium species,
$InCl_5^{2-}$, can be introduced by dissolving indium trichloride, $InCl_3$. The reduction of
$InCl_5^{2-}$ occurs at around -0.9 V vs. Al/Al(III) in the ionic liquid containing Cl^- at
0.35 M. The electrodeposition of InSb alloys has been reported as produced from
the ionic liquid containing both Sb(III) and In(III).

Electrodeposition of Transition Metals. The electrodeposition of copper,
Cu, has been reported in a basic $EMICl$–$EMIBF_4$ ionic liquid [79]. A monovalent

copper chlorocomplex anion, $CuCl_2^-$, is introduced by the anodic dissolution of Cu metal. The electrodeposition of metallic Cu is possible by reducing $CuCl_2^-$:

$$CuCl_2^- + e^- = Cu + 2Cl^- \tag{9.27}$$

The formal potential of Cu(I)/Cu is reported as -0.633 V vs. Al/Al(III). The oxidation of $CuCl_2^-$ leads to the formation of a divalent copper chlorocomplex anion, $CuCl_4^{2-}$:

$$CuCl_4^{2-} + e^- = CuCl_2^- + 2Cl^- \tag{9.28}$$

The formal potential of Cu(II)/Cu(I) is reported as 0.192–0.202 V vs. Al/Al(III).

The electrodeposition of silver, Ag, has been reported in a neutral $EMIBF_4$ ionic liquid [80]. Silver tetrafluoroborate, $AgBF_4$, dissolves in the ionic liquid up to about 0.2 M at room temperature and forms a monovalent silver species. The reduction of the monovalent silver species occurs at around 1 V vs. Al/Al(III).

The electrodeposition of cadmium, Cd, has been studied in a basic EMICl– $EMIBF_4$ ionic liquid [81]. Cadmium dichloride, $CdCl_2$, dissolves in the ionic liquid and forms a divalent cadmium chlorocomplex anion, $CdCl_4^{2-}$, which is reducible to metallic Cd at around -1.1 V vs. Al/Al(III):

$$CdCl_4^{2-} + 2e^- = Cd + 4Cl^- \tag{9.29}$$

9.2.2 Chlorozincate Ionic Liquids

It has been reported that the mixture of EMICl and zinc dichloride, $ZnCl_2$, forms an ionic liquid, which melts at around room temperature and is less reactive against water than the chloroaluminate ionic liquids. Since $ZnCl_2$ acts as a Lewis acid, it is possible to adjust the Lewis acidity or basicity by changing the ratio of $ZnCl_2$ to EMICl as well as the chloroaluminate ionic liquids. The electrochemical potential window of this ionic liquid is dependent on the molar fraction of $ZnCl_2$ [82]. In the basic ionic liquid, which contains $ZnCl_2$ at less than 33.3 mol%, the predominant ionic species are EMI^+, $ZnCl_4^{2-}$, and Cl^-. Thus the cathodic and anodic limit are determined by the reduction of EMI^+ and the oxidation of Cl^-, respectively. The deposition of metallic zinc does not occur in the basic condition because $ZnCl_4^{2-}$ is not reducible at a potential more positive than the cathodic decomposition of EMI^+. In the case of an acidic ionic liquid that contains $ZnCl_2$ at more than 33.3 mol%, the predominant ionic species are EMI^+, $ZnCl_4^{2-}$, $ZnCl_3^-$, $Zn_2Cl_5^-$, and $Zn_3Cl_7^-$. The cathodic limit is determined by the electrodeposition of metallic zinc as a result of the reduction of the chlorozincate anions, $ZnCl_4^{2-}$, $ZnCl_3^-$, $Zn_2Cl_5^-$, and $Zn_3Cl_7^-$. The electrodeposition of zinc has been studied in an acidic EMICl–$ZnCl_2$ ionic liquid that contains $ZnCl_2$ at 50mol% [83]. The anodic limit is considered to be the chlorine evolution derived from the oxidation of the chlorozincate anions. The dissolution of metal chlorides is expected to be the same as in the chloroaluminate ionic liquids. The electrode potentials are usually measured against a Zn/Zn(II) reference electrode, which consists of a zinc electrode immersed in the EMICl–ZnCl$_2$ ionic liquid containing ZnCl$_2$ at 50 mol%.

Electrodeposition of Main Group Elements and Transition Metals. The electrodeposition of tellurium, Te, has been investigated in an acidic EMICl–ZnCl$_2$ ionic liquid [84]. Tellurium tetrachloride, TeCl$_4$, is soluble in the ionic liquid. The electrodeposition of metallic tellurium occurs at around 1 V vs. Zn/Zn(II). The formation of the Zn–Te alloys also occurs at more negative potentials. The further reduction of metallic tellurium to Te($-$II) is reported as well just as in the chloroaluminate ionic liquid.

The electrodeposition of iron, Fe, has been reported in an acidic EMICl–ZnCl$_2$ ionic liquid [85]. Iron dichloride, FeCl$_2$, is soluble in the ionic liquid which contains ZnCl$_2$ at 40 mol%. The dissolved iron species is reducible at around 0.1 V vs. Zn/Zn(II), resulting in the electrodeposition of metallic iron. Zn–Fe alloys are obtained in this ionic liquid.

The electrodeposition of cobalt, Co, has been investigated in an acidic EMICl–ZnCl$_2$ ionic liquid [86, 87]. Cobalt dichloride, CoCl$_2$, dissolves in the ionic liquid at 80°C. The electrodeposition of metallic cobalt is observed at around 0.2 V vs. Zn/Zn(II). The formation of Zn–Co alloys is also reported at more negative potentials. The electrodeposition of Zn–Co alloys has been reported in BPCl– and EMICl–CoCl$_2$–ZnCl$_2$ ionic liquids [88, 89]. It was reported that Zn–Co–Dy alloys can be obtained from an EMICl–ZnCl$_2$ ionic liquid containing both CoCl$_2$ and DyCl$_3$ [87].

The electrodeposition of Zn–Ni alloys has been reported in a EMICl–ZnCl$_2$–NiCl$_2$ ionic liquid [90]. The Zn–Ni alloys containing Ni at 12.3 to 98.6 mol% were obtained by changing the deposition potential and the composition of the ionic liquid.

The electrodeposition copper, Cu, has been investigated in an acidic EMICl–ZnCl$_2$ ionic liquid [91]. A monovalent copper species, Cu(I), is introduced by dissolving copper chloride, CuCl, or the anodic dissolution of Cu metal. The electrodeposition of metallic cupper is possible by the reduction of Cu(I) at 0.5 $\sim$ 0.6 V vs. Zn/Zn(II). The formation of Zn–Cu alloys was also reported.

The electrodeposition of cadmium, Cd, has been studied in an acidic EMICl–ZnCl$_2$ ionic liquid [92]. Cadmium dichloride, CdCl$_2$, dissolves in the ionc liquid and forms a divalent cadmium species, which is reducible to metallic Cd prior to the bulk deposition of metallic Zn. The electrodeposition of Zn-Cd alloys was also reported.

9.3 OTHER IONIC LIQUIDS

Most of the ionic liquids melting at low temperature are composed of alkylated imidazolium cations, and there is no good explanation of why the melting point and viscosity become lower in the presence of these imidazolium cations. However, the cathodic stability of the imidazolium cations is not sufficient for the electrodeposition of the base metals, which have not been obtained in room-temperature ionic liquids. It has been known that many kinds of room-temperature ionic liquids can be prepared by combining various quaternary ammonium cations with bis(tri-fluoromethylsulfonyl)imide, $N(CF_3SO_2)_2^-$ [93–97]. Furthermore these $N(CF_3-SO_2)_2^-$ ionic liquids are not only stable against moisture but also immiscible with water, whereas a certain amount of water dissolves in the $N(CF_3SO_2)_2^-$ ionic liquids. Therefore these $N(CF_3SO_2)_2^-$ ionic liquids are the promising candidates as the supporting electrolytes for practical electrodeposition processes. However, there are only a few reports on the electrodeposition of metals.

The electrodeposition of lithium has been studied in some $N(CF_3SO_2)_2^-$ ionic liquids consisting of the aliphatic and alicyclic ammonium cations [98–100]. Since these ammonium cations are more stable against lithium metal than alkylated imi-dazolium cations, the reversible deposition and dissolution of lithium metal are reported to be possible in these ionic liquids.

The electrodeposition of silver, Ag, has been investigated in a $EMIN(CF_3SO_2)_2$ ionic liquid [101–103]. A monovalent silver species is introduced by dissolving $AgN(CF_3SO_2)_2$, which is prepared by the reaction of Ag_2O with $HN(CF_3SO_2)_2$. The electrodeposition of metallic Ag is possible by the reduction of Ag(I). The formal potential of Ag(I)/Ag is estimated as around 0.5 V vs. Fc/Fc^+ [103].

The electrodeposition of copper, Cu, has been reported in a $TMHAN(CF_3SO_2)_2$ ionic liquid [104]. A divalent copper species, Cu(II), is introduced by dissolving $Cu[N(CF_3SO_2)_2]_2$, which is prepared by the reaction of CuO with $HN(CF_3SO_2)_2$. The reduction of Cu(II) to a monovalent copper species, Cu(I), occurs at around 0.6 V vs. I^-/I_3^-. The formation of Cu(I) is also observed by the proportionation reaction of Cu(II) and Cu metal. Metallic Cu is obtained by the reduction of Cu(I) at around -0.2 V vs. I^-/I_3^-.

There are some reports on the electrodeposition of cobalt [105], nickel, zinc, and magnesium [106] in $N(CF_3SO_2)_2^-$ ionic liquids. The electrochemical behavior of lanthanide elements has been reported in $EMIN(CF_3SO_2)_2$ and $BMPN(CF_3SO_2)_2$ ionic liquids, though their metals were not obtained within the electrochemical potential windows of these ionic liquids [107].

Some ionic liquids composed of organic halides and metal halides have been studied for the electrodeposition of the metals and alloys. Most of these ionic liquids are hygroscopic and unstable against water, the same as the chloroaluminate ionic liquids. Moreover some of these ionic liquids are viscous to the extent that they need elevating temperature and/or addition of co-solvents. The electrodeposition of Ga [108], Ga–As [109], In–Sb [110], Sn [111], and Nb–Sn [112–114] has been reported in these ionic liquids.

REFERENCES

1. F. Endres, *Chem. Phys. Phys. Chem.*, **3**, 144 (2002).

2. G. Mamantov, A. I. Popov, eds., *Chemistry of Nonaqueous Solvents Recent Advances*, VCH, 1994.

3. R. T. Carlin, W. Crawford, M. Bersch, *J. Electrochem. Soc.*, **139**, 2720 (1992), and references therein.

4. Q. Liao, W. R. Pinter, G. Stewart, C. L. Hussey, G. R. Stafford, *J. Electrchem. Soc.*, **144**, 936 (1997).

5. Y. Zhao, T. J. VanderNoot, *Electrochim. Acta*, **42**, 1639 (1997).

6. C. A. Zell, F. Endres, W. Freyland, *Phys. Chem. Chem. Phys.*, **1**, 697 (1999).

7. J.-J. Lee, I.-T. Bae, D. A. Scherson, B. Miller, K. A. Wheeler, *J. Electrochem. Soc.*, **147**, 562 (2000).

8. J.-J. Lee, B. Miller, X. Shi, R. Kalish, K. A. Wheeler, *J. Electrochem. Soc.*, **147**, 3370 (2000).

9. A. P. Abbot, C. A. Eardley, N. R. S. Farley, G. A. Griffith, A. Pratt, *J. Appl. Electrochem.*, **31**, 1345 (2001).

10. C. Scordilis-Kelley, J. Fuller, R. T. Carlin, *J. Electrochem. Soc.*, **139**, 694 (1992).

11. T. L. Riechel, J. S. Wilkes, *J. Electrochem. Soc.*, **139**, 977 (1992).

12. C. Scordilis-Kelley, R. T. Carlin, *J. Electrochem. Soc.*, **140**, 1606 (1993).

13. C. Scordilis-Kelley, R. T. Carlin, *J. Electrochem. Soc.*, **141**, 873 (1994).

14. G. E. Gray, P. A. Kohl, J. Winnick, *J. Electrochem., Soc.*, **142**, 3636 (1995).

15. D. M. Ryan, E. R. Schumacher, T. L. Riechel, *J. Electrochem. Soc.*, **143**, 908 (1996).

16. G. E. Grey, J. Winnick, P. A. Kohl, *J. Electrochem. Soc.*, **143**, 2262 (1996).

17. G. E. Grey, J. Winnick, P. A. Kohl, *J. Electrochem. Soc.*, **143**, 3820 (1996).

18. J. Fuller, R. A. Osteryoung, R. T. Carlin, *J. Electrochem. Soc.*, **142**, 3632 (1995).

19. J. Fuller, R. T. Carlin, R. A. Osteryoung, *J. Electrochem. Soc.*, **143**, L145 (1996).

20. J. Fuller, R. T. Carlin, R. A. Osteryoung, P. Koranaios, R. Mantz, *J. Electrochem. Soc.*, **145**, 24 (1998).

21. M. Morimitsu, N. Tanaka, M. Matsunaga, *Chem. Lett.*, 1028 (2000).

22. Y. S. Fung, W. B. Zhang, *J. Appl. Electrochem.*, **27**, 857 (1997).

23. P.-Y. Chen, Y.-F. Lin, I.-W. Sun, *J. Electrochem. Soc.*, **146**, 3290 (1999).

24. J. S.-Y. Liu, I.-W. Sun, *J. Electrochem. Soc.*, **144**, 140 (1997).

25. E. G.-S. Jeng, I.-W. Sun, *J. Electrochem. Soc.*, **145**, 1196 (1998).

26. X.-H. Xu, C. L. Hussey, *J. Electrochem. Soc.*, **140**, 618 (1993).

27. C. L. Hussey, X. Xu, *J. Electrochem. Soc.*, **138**, 1886 (1991).

28. D. A. Habboush, R. A. Osteryoung, *Inorg. Chem.*, **23**, 1726 (1984).

29. M. Lipsztajn, R. A. Osteryoung, *Inorg. Chem.*, **24**, 3492 (1985).

30. L. Heerman, W. D'Olieslager, *J. Electrochem. Soc.*, **138**, 1372 (1991).

31. E. G.-S. Jeng, I.-W. Sun, *J. Electrochem. Soc.*, **144**, 2369 (1997).

32. T. Tsuda, C. L. Hussey, G. R. Stafford, J. E. Bonevich, *J. Electrochem. Soc.*, **150**, C234 (2003).

33. M. R. Ali, A. Nishikata, T. Tsuru, *Electrochim. Acta*, **42**, 2347 (1997).

34. J. S.-Y. Liu, P.-Y. Chen, I.-W. Sun, *J. Electrochem. Soc.*, **144**, 2388 (1997).

35. C. Nanjundiah, K. Shimizu, R. A. Osteryoung, *J. Electrochem. Soc.*, **129**, 2474 (1982).

36. M. Lipsztajn, R. A. Osteryoung, *Inorg. Chem.*, **24**, 716 (1985).

37. S. Pye, J. Winnick, P. A. Kohl, *J. Electrochem. Soc.*, **144**, 1933 (1997).

38. R. T. Carlin, H. C. De Long, J. Fuller, P. C. Trulove, *J. Electrochem. Soc.*, **145**, 1598 (1998).

39. T. M. Laher, C. L. Hussey, *Inorg. Chem.*, **21**, 4079 (1982).

40. C. L. Hussey, T. M. Laher, *Inorg. Chem.*, **20**, 4201 (1981).

41. R. T. Carlin, P. C. Trulove, H. C. De Long, *J. Electrochem. Soc.*, **143**, 2747 (1996).

42. J. A. Mitchell, W. R. Pitner, C. L. Hussey, G. R. Stafford, *J. Electrochem. Soc.*, **143**, 3448 (1996).

43. M. R. Ali, A. Nishikata, T. Tsuru, *Electrochim. Acta*, **42**, 1819 (1997).

44. C. A. Zell, W. Freyland, *Langmuir*, **19**, 7445 (2003).

45. W. Freyland, C. A. Zell, S.-Z. El Abedin, F. Endres, *Electrochim. Acta*, **48**, 3053 (2003).

46. R. J. Gale, B. Gilbert, R. A. Osteryoung, *Inorg. Chem.*, **18**, 2723 (1979).

47. W. R. Pitner, C. L. Hussey, G. R. Stafford, *J. Electrochem. Soc.*, **143**, 130 (1996).

48. M. R. Ali, A. Nishikata, T. Tsuru, *J. Electroanal. Chem.*, **513**, 111 (2001).

49. C. L. Hussey, L. A. King, R. A. Carpio, *J. Electrochem. Soc.*, **126**, 1029 (1979).

50. C. Nanjunsiah, R. A. Osteryoung, *J. Electrochem. Soc.*, **130**, 1312 (1983).

51. T. M. Laher, C. L. Hussey, *Inorg. Chem.*, **22**, 3247 (1983).

52. B. J. Tierney, W. R. Pitner, J. A. Mitchell, C. L. Hussey, G. R. Stafford, *J. Electrochem. Soc.*, **145**, 3110 (1998).

53. J.-J. Lee, B. Miller, X. Shi, R. Kalish, K. A. Wheeler, *J. Electrochem. Soc.*, **148**, C183 (2001).

54. Q. Zhu, C. L. Hussey, *J. Electrochem. Soc.*, **148**, C395 (2001).

55. F. Endres, A. Schweizer, *Phys. Chem. Chem. Phys.*, **2**, 5455 (2000).

56. C. J. Dymek Jr., G. F. Reynolds, J. S. Wilkes, *J. Electrochem. Soc.*, **134**, 1658 (1987).

57. W. R. Pitner, C. L. Hussey, *J. Electrochem. Soc.*, **144**, 3095 (1997).

58. Y.-F. Lin, I.-W. Sun, *J. Electrochem. Soc.*, **146**, 1054 (1999).

59. J. Dogel, W. Freyland, *Phys. Chem. Chem. Phys.*, **5**, 2484 (2003).

60. N. Koura, T. Kato, E. Yumoto, *Hyomen Gijutsu*, **45**, 805 (1994).

61. I.-W. Sun, C. L. Hussey, *J. Electroanal. Chem.*, **274**, 325 (1989).

62. H. C. De Long, J. S. Wilkes, R. T. Carlin, *J. Electrochem. Soc.*, **141**, 1000 (1994).

63. X.-H. Xu, C. L. Hussey, *J. Electrochem. Soc.*, **139**, 1295 (1992).

64. F. Endres, W. Freyland, *Ber. Bunsenges. Phys. Chem.*, **101**, 1075 (1997).

65. F. Endres, W. Freyland, *J. Phys. Chem. B*, **102**, 10229 (1998).

66. Q. Zhu, C. L. Hussey, G. R. Stafford, *J. Electrochem. Soc.*, **148**, C88 (2001).

67. T. M. Laher, C. L. Hussey, *Inorg. Chem.*, **22**, 1279 (1983).

68. M. A. M. Noel, R. A. Osteryoung, *J. Electroanal. Chem.*, **293**, 139 (1990).

69. M. A. M. Noel, R. A. Osteryoung, *J. Electroanal. Chem.*, **284**, 413 (1990).

70. T. Tsuda, T. Nohira, Y. Ito, *Electrochim. Acta*, **46**, 1891 (2001).

71. T. Tsuda, T. Nohira, Y. Ito, *Electrochim. Acta*, **47**, 2817 (2002).

72. X.-H. Xu, C. L. Hussey, *J. Electrochem. Soc.*, **139**, 3103 (1992).

73. X.-H. Xu, C. L. Hussey, *J. Electrochem. Soc.*, **140**, 1226 (1993).

74. F. Endres, C. Schrodt, *Phys. Chem. Chem. Phys.*, **2**, 5517 (2000).

75. F. Endres, *Phys. Chem. Chem. Phys.*, **3**, 3165 (2001).

76. F. Endres, S. Z. El Abedin, Phys. Chem. Chem. Phys., **4**, 1640 (2002).

77. M.-H. Yang, I.-W. Sun, *J. Appl. Electrochem.*, **33**, 1077 (2003).

78. M.-H. Yang, M.-C. Yang, I.-W. Sun, *J. Electrochem. Soc.*, **150**, C544 (2003).

79. P.-Y. Chen, I.-W. Sun, *Electrochim. Acta*, **45**, 441 (1999).

80. Y. Katayama, S. Dan, T. Miura, T. Kishi, *J. Electrochem. Soc.*, **148**, C102 (2001).

81. P.-Y. Chen, I.-W. Sun, *Electrochim. Acta*, **45**, 3163 (2000).

82. S.-I. Hsiu, J.-F. Huang, I.-W. Sun, C.-H. Yuan, J. Shiea, *Electrochim. Acta*, **47**, 4367 (2002).

83. Y.-F. Lin, I.-W. Sun, *Electrochim. Acta*, **44**, 2771 (1999).

84. M.-C. Lin, P.-Y. Chen, I.-W. Sun, *J. Electrochem. Soc.*, **148**, C653 (2001).

85. J.-F. Huang, I.-W. Sun, *J. Electrochem. Soc.*, **151**, C8 (2004).

86. P.-Y. Chen, I.-W. Sun, *Electrochim. Acta*, **46**, 1169 (2001).

87. H.-Y. Hsu, C.-C. Yang, *Z. Naturforsch.*, **58b**, 139 (2003).

88. N. Koura, T. Endo, Y. Idemoto, J. Non-cryst. Solids, **205–207**, 650 (1996).

89. N. Koura, S. Matsumoto, Y. Idemoto, *Hyomen Gijutsu*, **49**, 1215 (1998).

90. N. Koura, Y. Suzuki, Y. Idemoto, T. Kato, F. Matsumoto, *Surf. Coat. Technol.*, **169**, 120 (2003).

91. P.-Y. Chen, M.-C. Lin, I.-W. Sun, *J. Electrochem. Soc.*, **147**, 3350 (2000).

92. J.-F. Huang, I.-W. Sun, *J. Electrochem. Soc.*, **149**, E348 (2002).

93. V. R. Koch, C. Nanjundiah, G. Battista Appetecchi, B. Scrosati, *J. Electrochem. Soc.*, **142**, L116 (1995).

94. P. Bonhôte, A.-P. Dias, N. Papageorgiou, K. Kalyanasundaram, M. Grätzel, *Inorg. Chem.*, **35**, 1168 (1996).

95. D. R. MacFarlane, J. Sun, J. Golding, P. Meakin, M. Forsyth, *Electrochim. Acta*, **45**, 1271 (2000).

96. H. Matsumoto, M. Yanagida, K. Tanimoto, M. Nomura, Y. Kitagawa, Y. Miyazaki, *Chem. Lett.*, **2000**, 922.

97. J. Sun, D. R. MacFalane, M. Forsyth, *Electrochim. Acta*, **48**, 1707 (2003).

98. H. Matsumoto, M. Yanagida, K. Tanimoto, T. Kojima, Y. Tamiya, Y. Miyazaki, *in Molten Salts XII*, P. C. Trulove, H. C. De Long, G. R. Stafford, S. Deki, eds., Proc. Vol. 99–41, Electrochemical Society, 1999, p. 186.

99. H. Sakaebe, H. Matsumoto, *Electrochem. Commun.*, **5**, 594 (2003).

100. Y. Katayama, T. Morita, M. Yamagata, T. Miura, *Electrochemistry*, **71**, 186 (2003).

101. Y. Katayama, T. Miura, T. Kishi, *Hyomen Gijutsu*, **52**, 64 (2001).

102. Y. Katayama, T. Miura, *Yoyuen*, **45**, 61 (2002).

103. Y. Katayama, M. Yukumoto, M. Yamagata, T. Miura, T. Kishi, in *Proc. 6th Int. Symp. Molten Salt Chemistry, Technology*, C. Nianyi, Q. Zhiyu eds., Shanghai University Press, 2001, p. 190.

104. K. Murase, K. Nitta, T. Hirato, Y. Awakura, *J. Appl. Electrochem.*, **31**, 1089 (2001).

105. Y. Katayama, Y. Moteki, T. Miura, unpublished result.

106. N. Shinohara, K. Sato, K. Murase, T. Hirato, Y. Awakura, unpublished result.

107. M. Yamagata, Y. Katayama, T. Miura, in *Molten Salt XIII*, H. C. De Long, R. W. Bradshaw, M. Matsunaga, G. R. Stafford, P. C. Trulove, eds., Proc. Vol. 2002–19, Electrochemical Society 2002, p. 640.

108. S. P. Wicelinski, R. J. Gale, J. S. Wilkes, *J. Electrochem. Soc.*, **134**, 262 (1987).

109. M. K. Carpenter, M. W. Verbrugge, *J. Electrochem. Soc.*, **137**, 123 (1990).

110. M. K. Carpenter, M. W. Verbrugge, J. Mater. Res., **9**, 2584 (1994).

111. G. Lin, N. Koura, *Denki Kagaku (presently Electrochemistry)*, **65**, 149 (1997).

112. N. Koura, G. Ling, H. Ito, *Hyomen Gijutsu*, **46**, 1162 (1995).

113. G. Ling, N. Koura, *Hyomen Gijutsu*, **48**, 454 (1997).

114. N. Koura, T. Umebayashi, Y. Idemoto, G. Ling, *Electrochemistry*, **67**, 684 (1999).

Part II

Bioelectrochemistry

Chapter *10*

Enzymatic Reactions

Tomoya Kitazume

Studies of reaction media, and of conditions in organic reactions that improve selectivity and minimize energy to decrease the damage to the endocrine system, are having a big impact. A recent finding of enzyme and/or antibody activity in ionic liquids is expected to open a road to the rich use of environmentally benign solvents [1]. Especially interesting is the reuse of the enzyme or antibody in an ionic liquid solvent system that enables the asymmetric resolution and/or transesterification [2].

10.1 LIPASE (*CANDIDA ANTARCTICA*) CATALYZED REACTIONS

The activity of microorganisms in organic solvents has been observed to have useful applications in the synthesis of organic materials. Using an ionic liquid solvent system of ionic liquid and water, Cull and coworkers have reported the a two-phase biotransformation with cells of *Rhodococcus* R312 [3]. Extensive research of *Candida antarctica*-catalyzed reactions has also been carried out for catalyzing alcoholysis, ammoniolysis, and perhydrolysis reactions in ionic liquids ([bmim][BF$_4$] or [bmim][PF$_6$]), Scheme 10.1.

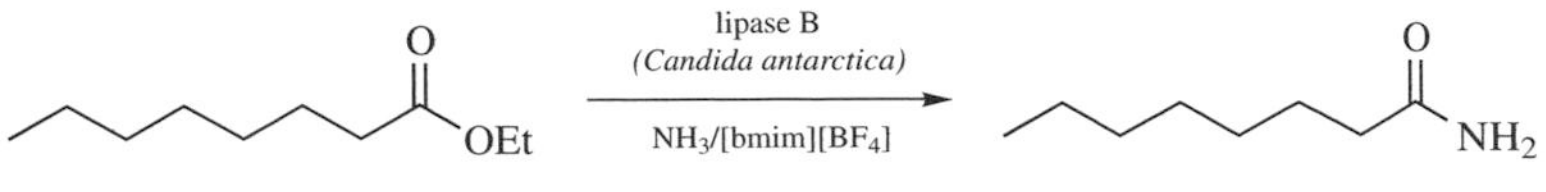

Scheme 10.1

Electrochemical Aspects of Ionic Liquids   Edited by Hiroyuki Ohno
ISBN 0-471-64851-5   Copyright © 2005 John Wiley & Sons, Inc.

Lau and coworkers the epoxidation of cyclohexene by peracid, Scheme 10.2, generated in situ by *Candida Antarctica* lipase-catalyzed reaction of acid with 60% aq H_2O_2 in [bmim][BF$_4$], to yield 83% [4].

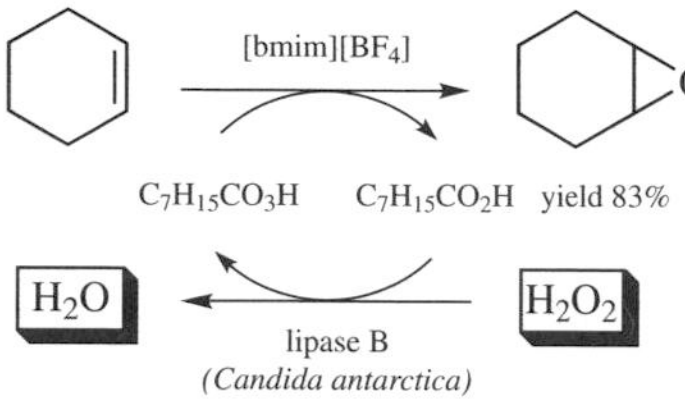

Scheme 10.2

10.2 ENZYME-CATALYZED SYNTHESIS OF *Z*-ASPARTAME IN IONIC LIQUID

Erbeldinger, Mesiano, and Russell carried out a thermolysin-catalyzed reaction of L-aspartate, protected with a carboxybenzoxy group with L-phenylalanine methyl ester, in order to synthesize Z-aspartama [5]. The thermolysin-catalyzed condensation reaction is reported in Scheme 10.3.

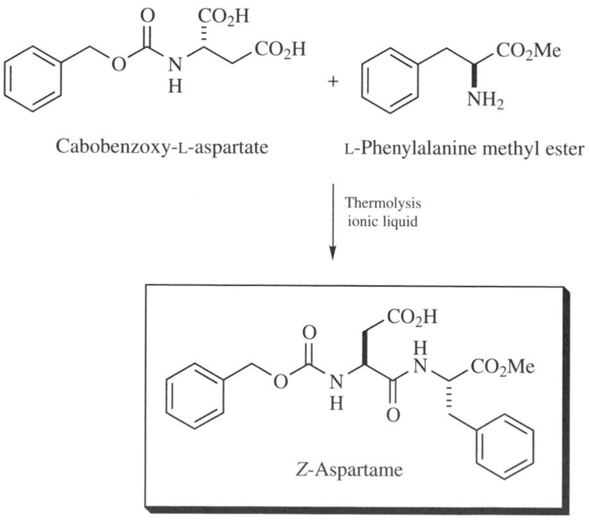

Cabobenzoxy-L-aspartate   L-Phenylalanine methyl ester

Z-Aspartame

Scheme 10.3

10.3 LIPASE-CATALYZED ENANTIOSELECTIVE ACYLATION

The fact that microorganisms discriminate among enantiomers has important consequences for resolution and asymmetric synthesis. However, microorgamisms have been reported to exert prochirality stereospecific control in their catalyses, and thus the problem has been overcome with direct asymmetric synthesis now permitted of chiral products from symmetric starting materials [6]. The enzymatic resolution of carbinols has proved to be a valuable synthetic way to obtain the chiral material. The enzymatic resolution of carbinols in an ionic liquid is shown in Scheme 10.4. Reaction a yields lipase PS (*Pseudomonas cepacia*, Amano Pharmacetical Co., Ltd.) and vinyl acetate, which breaks down into [emim][NTf$_2$] a 0.5 *N* NaOH an aqueous solution in b. The authors have examined the esterification of α-methyl benzyl alcohol with vinyl acetate in several kinds of ionic liquids to construct the reusable ionic liquid-enzyme reaction media [5]. As the results in Table 10.1 show, ionic liquids do not accelerate the esterification of alcohol more than organic solvents (toluene and *n*-hexane) under the same reactions conditions (the mixture was stirred at 30°C for 5 h); however, an ionic liquid ([emim][OTf]) can be used as the solvent in an enzymatic esterification. In the asymmetric esterification of α-methyl benzyl alcohol with the system of lipase PS (*Pseudomonas cepacia*, Amano Pharmaceutical Co., Ltd.)-ionic liquid ([emim][NTf$_2$]), at 27% conversion from *rac*-alcohol, >99% ee was obtained of the corresponding acetate after extracting with diethyl ether. The optical purities were determined by gas chromatography (column). Scheme 10.5 shows the lipase-catalyzed transesterification in an ionic liquid. Further, as the ionic liquid-lipase PS system was recovered easily after the extraction of the starting substrates and products with diethyl ether, successive reuse of the recovered system and in the same reaction yielded amounts and stereoselectivity of product as high as in the first cycle shown in Table 10.2 [7].

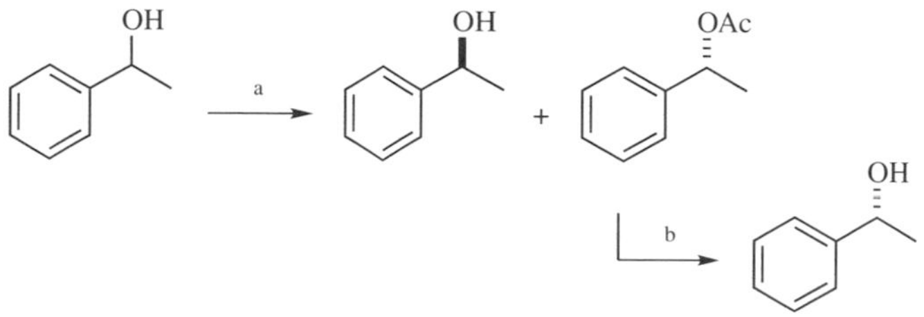

Scheme 10.4

TABLE 10.1 Conversion of Alcohol at 30°C for 5 hours Ionic Liquid

Ionic Liquid	Conversion of Alcohol (%)
[bmim][BF$_4$]	7.4
[bmim][PF$_6$]	8.7
[emim][NTf$_2$]	14.6
THF	4.6
Et$_2$O	5.5
Toluene	22.0
n-Hexane	32.1

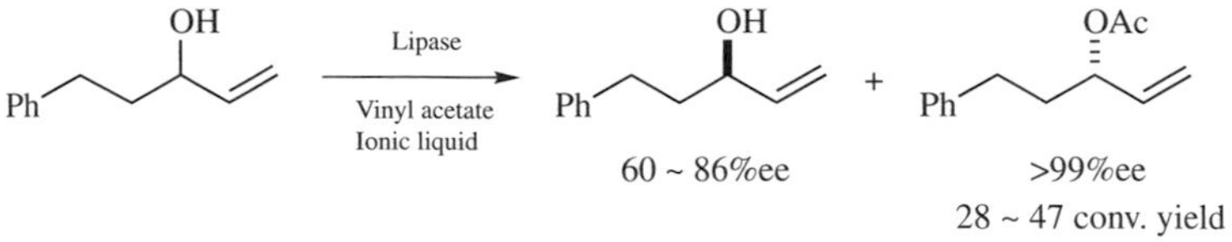

60 ~ 86%ee   >99%ee
                    28 ~ 47 conv. yield

Third cycle : 3 ~ 6 h; fourth: 22 h; fifth cycle: 91 h

Scheme 10.5

TABLE 10.2   Enzymatic Resolution

	Alcohol Yield Optical (%)	Purity (% ee)	Acetate Yield Optical (%)	Purity (% ee)
Cycle 1	34	70	27	>99
Cycle 2	34	72	33	>99
Cycle 3	53	44	29	>99

TABLE 10.3   Biotransformation in an Ionic Liquid

Lipase[a]	Solvent	Time (h)	Acetate Yield (%)	Optical Purity (% ee)
CAL	[bmim][PF_6]	5	45	>99
	[bmim][OTf]	48	19	91
	[bmim][BF_4]	3.5	44	>99
	[bmim][OTf]	24	34	>99
	[bmim][SbF_6]	48	31	>99
QL	[bmim][PF_6]	25	49	94
PS	[bmim][PF_6]	168	17	>99
CRL	[bmim][PF_6]	168	0	0
PPL	[bmim][PF_6]	168	0	0
CAL	i-Pr$_2$O	3	47	>99

[a]CAL (Novozym 435) (*Candida antarctica*): QL (*Alcalgenes* sp.) PS (*Psedomonas cepacia* Amano lipase PS): CRL (*Candida rugosa* Meito lipase OF): PPL Porcine liver lipase (Sigma Type II).

In addition, Itoh and coworkers have reported that acylation of the alcohol was accomplished by three types of enzymes: *Candida Antarctica* lipase (CAL, Novozym 435), lipase QL (*Alcalgenes* sp.), and lipase PS (*Pseudomonas cepacia*), Scheme 10.5. The desired acetate showed extremely high enantioselectivity, but no reaction took place when lipase (CRL, *Candida rugosa*) or Procine liver lipase (PPL) was used as the catalyst in the ionic liquid (Table 10.3).

10.4   ALDOLASE ANTIBODY 38C2 PROMOTED ALDOL REACTION

Although the use of enzymes in ionic liquids has been explored, the enzymatic carbon–carbon bond forming the reactions in ionic liquids has not been studied in detail. Since aldol reactions catalyzed by the aldolase antibody 38C2 in buffer solution are known to proceed at room temperature [9], we decided to carry out an experiment whereby the aldol reaction would be catalyzed by the aldolase antibody 38C2 (Aldrich no 48157-2) in an inonic liquid. We focused on the aldol reaction of hydroxyacetone with 4- or 3-(trifluoromethyl)-benzaldehyde in the antibody aldolase 38C2 (Aldrich no 48157-2, 10 mg)–[bmim][PF_6] system. After standing for two weeks at room temperature, the starting materials and products were extracted with diethyl ether, and the ionic liquid and antibody aldolase 32C2 were recovered. The aldol reaction was promoted by the antibody 38C2, Scheme 10.6. As shown in Table 10.4, the reaction of hydroxyacetone with 4- or 3-(trifluoromethyl)benzalde-hyde in this system produced a 3,4-dihydroxy-4-{4- or 3-(trifluoromethyl)phenyl}-butan-2-one solvent. Acetone, methyl ethyl ketone, methoxyacetone, fluoroacetone, and chloroacetone did not react in this system. Moreover, in the aliphatic and/or α, β-unsaturated aldehydes, the aldol reaction did not proceed. Successive reuse of the recovered Ab32C2-ionic liquid system in the same reaction yielded higher amounts

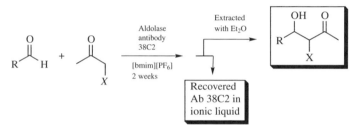

Scheme 10.6

TABLE 10.4   Antibody Aldolase Promoted Aldol Reaction

RCHO	X	Ionic Liquid	Conversion (%)	Diastereomeric Ratio[e]
C_6H_5CHO	OH	[bmim][PF_6]	No reaction	
$4\text{-}CF_3C_6H_4CHO$	OH	[bmim][PF_6]	21 $(59)^a$	37:63
			89^b	42:58
			66^c (75)	36:64
			46^d (58)	35:65
	OH	[emim][OTf]	56	69:31
	OMe	[bmim][PF_6]	No reaction	
	H	[bmim][PF_6]	Trace	
	Cl	[bmim][PF_6]	No reaction	
	F	[bmim][PF_6]	No reaction	
$3\text{-}CF_3C_6H_4CHO$	OH	[bmim][PF_6]	17 (33)	42:58
			45^b (95)	29:71
			22^c (40)	29:71
$2\text{-}CF_3C_6H_4CHO$	CH_3	[bmim][PF_6]	No reaction	
C_6F_5CHO	OH	[bmim][PF_6]	No reaction	

^{a}Conversion yield in parenthesis; ^{b}second cycle; ^{c}third cycle; ^{d}fourth cycle; ^{e}diastereomeric ratio was determined by NMR.

of product than the first cycle in both the cases of 3- and 4-(trifluoromethyl)-bezal-dehyde. In the third and fourth cycles, reuse of this system recovered from the second cycle produced the same alcohol in the same reaction.

10.5   ANTIBODY ALDOLASE PROMOTED MICHAEL REACTION

In the next step we examined the Michael addition reaction of hydroxyacetone to 2-(phenyl)-ethyl-2-(trifluoromethyl)acrylate in the ionic liquid [emim][OTf]-aldolase antibody 38C2 system. After standing for two weeks at room temperature, the products were extracted with diethyl ether, and the ionic liquid and antibody aldolase 38C2 were recovered. On removal of the solvent, the yield of products (diastereomer) was determined by ^{19}F NMR using $C_6H_5CF_3$ as an internal standard.

A proposed mechanism for the Michael addition reaction is shown in Scheme 10.7. Note that enamine, generated from the reaction of hydroxyacetone and aldolase antibody 38C2, reacts with the activated methylene group in 2-(phenyl)ethyl-2-(trifluoromethyl)acrylate.

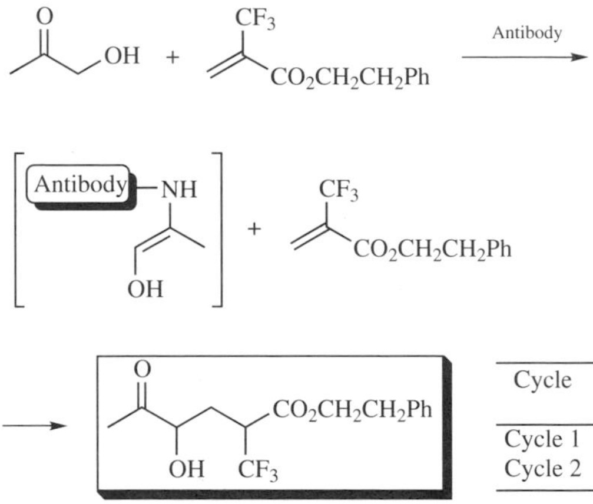

Cycle	Yield (%)
Cycle 1	30
Cycle 2	45

Scheme 10.7

10.6   REACTION OF FLUOROMETHYLATED IMINES USING ANTIBODY ALDOLASE 38C2-IONIC LIQUID SYSTEM

We carried out a reaction of fluoromethylated imines with hydroxyacetone in the antibody aldolase 38C2-ionic liquid system. In this system the reaction did not proceed in the absence of antibody aldolase 38C2. In the presence of antibody aldolase 38C2, the reaction smoothly proceeded, giving a fluoromethylated carbinol.

Proposed reaction mechanism is shown in Scheme 10.8. Note the reaction paths of the fluoromethylated imines with hydroxyacetone in antibody aldolase 38C2-ionic liquid in the scheme. At first an intermediate reaction and H_2O are produced

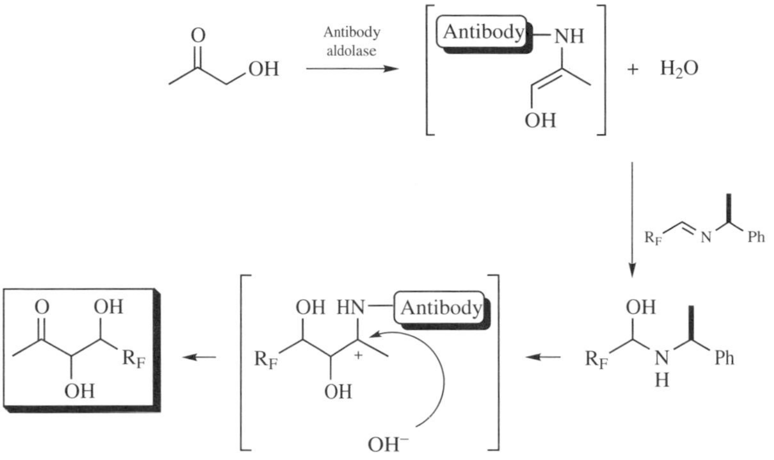

Scheme 10.8

from the reaction of the hydroxyacetone and antibody 38C2. Then the produced H_2O attacks an imine to produce N,O-acetal. The N,O-acetal reacts with the intermediate reaction products to give an intermediate reaction. The intermediate reaction is converted to the product as there is added a small amount of water containing ionic liquids [10]. To construct the fluorinated materials based on the utility of aldolase antibody, we designed a simple synthetic reaction of N,O-acetal with CF_3 group. Initially, we attempted a reaction of trifluoroacetaldehyde N,O-acetal with hydroxyacetone in [emim][OTf]. After standing for two weeks at room temperature, the reaction mixture similarly gave a -2-ethoxy-1,1,1-trifluoro-3-hydroxypentan-4-one in 18% (conversion yield: 57%) yield.

REFERENCES

1. T. Kitazume, T. Yamazaki, in *Topics in Current Chemistry*, Vol. 193, Springer, 1997, p. 91, and references cited therein.

2. M. Freemantle, *C & EN*, **1**, 21–25 (2001).

3. S. G. Cull, J. D. Holbrey, V. Vargas-Mora, K. R. Seddon, G. J. Lye, *Biotechnol. Bioeng.*, **69**, 227 (2000).

4. R. M. Lau, F. V. Rantwijk, K. R. Seddon, R. A. Sheldon, *Org. Lett.*, **2**, 4189 (2000).

5. M. Erbeldinger, A. J. Mesiano, A. J. Russell, *Biotechnol. Prog.*, **16**, 1129 (2000).

6. $\tilde{T}$. Kitazume, T. Ishizuka, M. Takeda, K. Itoh, *Green Chem.*, 1, 221 (1999). (b) A. Zaks, A. M. Klibanov, *Science*, **224**, 1249 (1984). (c) L. Dai, A. M. Klibanov, *Proc. Natl. Acad. Sci. USA*, **96**, 9475 (1999). (d) G. A. Sellek, J. B. Chaudhuri, *Enzyme Microb. Technol.*, **25**, 471 (1999).

7. T. Kitazume, in 16th *Int. Symp. on Fluorine Chemistry*, Durham, UK, 2000, B-36.

8. (a) T. Itoh, E. Arasaki, K. Kudo, S. Shirakami, *Chem. Lett.*, 262 (2001). (b) S. H. Schöfer, N. kaftzik, P. Wasserscheid, U. Kragl, *Chem. Commun.*, 425 (2001). (c) M. T. Reetz, W. Wiesenhöfer, G. Francio, W. Leitner, *Chem. Commun.*, 992 (2002).

9. (a) J. Wagner, R. A. Lerner, C. F. Barbas III, *Science* **270**, 1797 (1995). (b) C. F. Barbas III, H. Heine, G. Zhong, T. Hoffmann, S. Gramatikova, R. Bjömestedt, B. List, J. Anderson, E. A. Stura, I. A Wilson, R. A. Lerner, *Science*, **278**, 2085 (1997). (c) B. List, R. A. Lerner, C. F. Barbas III, *Org. Lett.*, **1**, 59 (1999).

10. T. Kitazume, K. Tamura, Z. Jiang, N. Miyake, I. Kawasaki, *J. Fluorine Chem.*, **115**, 49 (2002).

Chapter 11

Molecular Self-assembly in Ionic Liquids

Nobuo Kimizuka and Takuya Nakashima

Because room-temperature ionic liquids are environmentally benign solvents, they are being rapidly considered for many different applications, including organic chemical reactions [1], separations [2], and electrochemical applications [3, 4], that take advantage of their unique properties: negligible vapor pressure, high ionic conductivity, and limited miscibility with water and common organic solvents [1, 5]. Although the chemistry of ionic liquids is an active and expanding field, two major issues remained unexplored until 2001 [6]. The first was the development of molecularly organized systems or supramolecular assemblies that can be stably dispersed in ionic liquids. The second was the development of ionic liquids that molecularly dissolve biomacromolecules such as carbohydrates and proteins. Although ionic materials composed of DNA and polyether-containing transition metal complexes were reported at that time [7], the studies of enzymatic reactions in ionic liquids employed nonhomogeneous suspensions or aqueous mixtures [8]. Dissolution of carbohydrates and sugar derivatives in ionic liquids is further hindered by their insolubility. In 2001 we reported that ether-containing ionic liquids that carry bromide counter anions can dissolve carbohydrates, glycosylated proteins, and glycolipid bilayer membranes [6]. Our report touched off the succeeding studies on self-assembly in ionic media, including gelation of ionic liquids by molecular assemblies. In this chapter we first give an overview of molecular assemblies in water and in organic media, and then review important recent developments of

Electrochemical Aspects of Ionic Liquids   Edited by Hiroyuki Ohno
ISBN 0-471-64851-5   Copyright © 2005 John Wiley & Sons, Inc.

molecular self-assemblies in ionic liquids. Our discussion complements some recent published work [9].

11.1   MOLECULAR SELF-ASSEMBLIES IN AQUEOUS AND IN ORGANIC MEDIA

Aqueous molecular assemblies such as micelles and bilayer membranes are formed by the self-assembly of amphiphilic compounds (Figure 11.1*a, b*) [10]. Aqueous micelles have been utilized for a variety of applications in surfactant industry, including emulsification, washing, and extraction processes [11]. Bilayer membranes are basic structural components of biomembranes, and their structures are maintained even in dilute aqueous media. This is in contrast to micelles that show dynamic equilibrium between aggregates and monomeric species. Thus bilayers are more stable and sophisticated self-assemblies, and they require suitable molecular design of the constituent amphiphiles. Bilayer membranes and vesicles have wide-ranging applications, as exemplified by drug delivery [12], sensors [13], and bilayer-templated material synthesis [14].

The basic difference between the micelle-forming surfactants and bilayer-forming amphiphiles lies on the chemical structure. Typical micelle-forming surfactants have a single alkyl chain (hydrophobic group) and a hydrophilic group, as exemplified by cetyltrimethylammonium bromide (CTAB; Figure 11.1*c*). In contrast, double-chained amphiphiles $2C_nN^+$ (Figure 11.1*d*; typically, $n = 12$–18) spontaneously give bilayer membranes when dispersed in water [10].

Although the formation of amphiphilic molecular assemblies in water is facilitated by hydrophobic interactions, micelles and bilayers are available even in

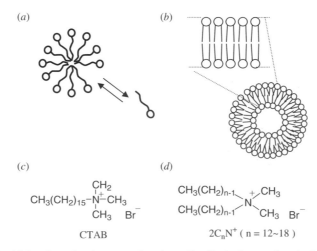

Figure 11.1 *Molecular structures and schematic illustrations of micelle and bilayer membranes: (a) micelle, (b) bilayer membrane and vesicle, (c) CTAB, and (d) $2C_nN^+$.*

organic solvents. For example, single-chain alkylammonium salts form micelles in polar solvents such as hydrazine [15], formamide [16], and glycerol [17]. The ability to form hydrogen bonding is not a mandatory characteristic of the solvents, since micelles and bilayer membranes are also formed in nonpolar, hydrocarbon solvents [10, 18, 19]. These findings clearly indicate that self-assembly in solution is a general phenomenon observed for amphiphilic molecules consisting of solvophobic and solvophilic moieties. Bilayer membranes were also formed in water-alcohol mixtures [20] and in ethanol [21]. In these polar media, enhancement of the intermolecular interactions is required to maintain the bilayer structure. The formation of mesoscopic supramolecular assemblies in chloroform by amphiphilic complementary hydrogen bond networks clearly shows that solvophobic interactions are rationally designable by combining hydrogen bonds and aromatic stacking [22]. Clearly, the formation of molecular assemblies is not limited to aqueous media. The interesting question is how to design solvophilic and solvophobic modules in ionic liquids.

11.2 FORMATION OF MICELLES AND LIQUID CRYSTALS IN IONIC LIQUIDS

The formation of micelles in ionic liquid was observed more than 30 years ago. At that time the ionic liquid was referred to as a low-melting fused salt [23–25]. Bloom and Reinsborough determined the critical micelle concentration (CMC) of CTAB in pyridinium chloride (melting point, 146°C) by means of surface tension studies (CMC, 66 mM at 155°C) [23]. Evans et al. reported the formation of micelles from CTAB in N-ethylammonium nitrate (EAN, melting point, 12°C). N-ethylammonium nitrate was used because it has three hydrogen bonding sites for each solvent molecule, and it has the ability, like water, to form such three-dimensional structures in the liquid state. The CMC of CTAB micelles in EAN was determined to be 2.2 mM at 55°C, which is 7–10 times larger than what is observed in aqueous solution [24]. This is attributable to the considerably lower interfacial tension for oil-fused salt (ca. 20 mN m^{-1}) compared to that observed for oil-water (ca. 50 mN m^{-1}). Thus the ethylammonium cation is a better solvent than water for hydrocarbons. Evidence for micelle formation was also given by light-scattering experiments. The micellar aggregation number and a hydrodynamic radius of hexadecylpyridinium bromide were determined as approximately 26 with and 22 ± 3 Å, respectively [25]. For spherical micelles, these numbers correspond to surface areas of 100 Å^2, which is considerably larger than the corresponding values of 50 to 60 Å^2 observed in an aqueous solution. These values are consistent with either a small classical spherical micelle containing only surfactant or a spherical mixed micelle containing surfactant and ethylammonium ions as a co-surfactant. Recently micelle formation of sodium dodecylsulfate (SDS) and polyoxyethylene-23-lauryl ether (Brij 35) in 1-buthyl-3-methyl imidazolium chloride and hexafluorophosphate has been reported [26].

Evans et al. also showed that the 1 : 1 mixture of EAN and β, γ-distearoyl-phosphotidylcholine (DSPC) gives a smectic A texture in the temperature range of 57.3 to 100°C [27]. This is the first notice of lyotropic lamellar liquid crystals formed in the ionic medium. Additionally, Seddon et al. [28] and Neve et al. [29] have described the long-chained *N*-alkylpyridinium or 1-methyl-3-alkylimidazolium ions to display smectic liquid-crystalline phases above their melting points, when Cl^- or tetrachloro-metal anions like $CoCl_4^{2-}$ and $CuCl_4^{2-}$ are used as the counter ions. Lin et al. have also noted the liquid crystal behavior of 1-alkylimidazolium salts and the effect on the stereoselectivity of Diels-Alder reactions [30]. However, liquid crystals are classified as ionic liquid crystals (ILCs), and they are distinguished from liquid crystals that are dispersed in ionic liquids. Although the formation of micelles and liquid crystal phases in ionic liquids have been thus reported, there has been no mention of the self-assembly of developed nano-assemblies that are stably dispersed in ionic liquids. In the next section the formation of bilayer membranes and vesicles in ionic liquids is discussed.

11.3 SUGAR-PHILIC IONIC LIQUIDS: DISSOLUTION OF CARBOHYDRATES AND FORMATION OF GLYCOLIPID BILAYER MEMBRANES, IONOGELS

We developed ionic liquids that molecularly dissolve biopolymers by using ether-containing *N*, *N'*-dialkylimizalolium derivatives, Schemes 11.1 and 11.2 [6]. The

1 (*n* = 2)
2 (*n* = 1)

Scheme 11.1

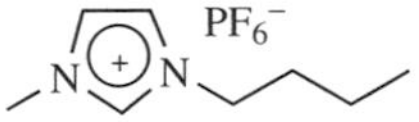

3

Scheme 11.2

ether linkages were introduced in the alkyl chain moieties in anticipation of their role as a hydrogen bond acceptor for hydrodxyl or ammonium groups. These compounds become solid upon freeze-drying, but they will liquify at room temperature when exposed to air, probably due to the absorption of water from the atmosphere. These ionic liquids were miscible with water and the content of water in these ionic liquids was 2.6 wt% for **1** and 2.5 wt% for **2**, as determined by the Karl Fischer method. To compare the solvent properties of these ionic liquids with conventional ionic liquids, the compound of Scheme 11.2 was saturated with water (water

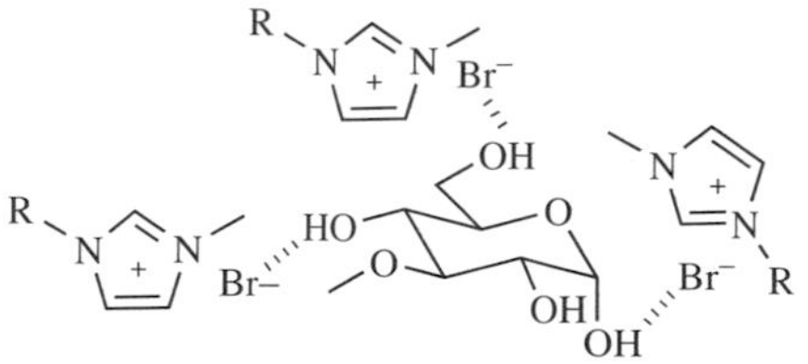

Figure 11.2 *Schematic illustration of carbohydrates solvated by Br$^-$ ions. Hydrogen bonding between hydroxyl groups and Br$^-$ ions causes the solvation.*

content, 2.5 wt%) and used as a reference. Interestingly, when heated, the ether-containing ionic liquids **1** and **2** homogeneously dissolve carbohydrates such as β-D-glucose (solubility, 450 mg mL^{-1}), α-cyclodextrin (350 mg mL^{-1}), amylose (30 mg mL^{-1}), and agarose (10–20 mg mL^{-1}). The observed solubility is not ascribed to the water molecules slightly coexisting in these ionic liquids, since amylose, which is a coiled polysaccharide of α-1,4-glucosyl units and only slightly soluble in pure water (water solubility below 0.5 mg ml^{-1})[31], can be dissolved. A highly glycosylated protein, glucose oxidase (GOD), was also soluble in these ionic liquids (concentration, 1 mg mL^{-1}), while other proteins such as cytochrome *c*, myoglobin, hemoglobin, and catalase were insoluble.

The sugar-philic property observed for the ether-containing ionic liquids **1** and **2** could be attributed to the presence of bromide anion because the solubility of sugar derivatives was diminished when the bromide was replaced by *bis*-trifluoromethane sulfonimide (TFSI). It appears that bromide ions effectively solvate the hydroxyl groups by hydrogen bonding (Figure 11.2). The dissolution of cellulose in chloride- or bromide-containing ionic liquids upon heating was reported by Rogers et al [32]. Interestingly, when the heated agarose solution of **1** was cooled to room temperature, the ionic liquid gelatinized. We have gave this physical state of gelation the name "ionogel," because the similiar terms hydrogel and orgenogel are commonly used to refer to gelatinized solvents [6]. This is the first case of ionogels observed to be formed by biopolymers, although gelation of ionic liquids has been known to occur in synthetic polymers [33].

To investigate the possibility of ordered molecular assemblies forming in ionic liquids, the solubilities of glycolipids, Schemes 11.3 and 11.4, were tested. L-Glutamate derivatives **4** (*n* = 12, 16) were employed because they are insoluble in water, even at a concentration of 1 mM. It was found that these water-insoluble

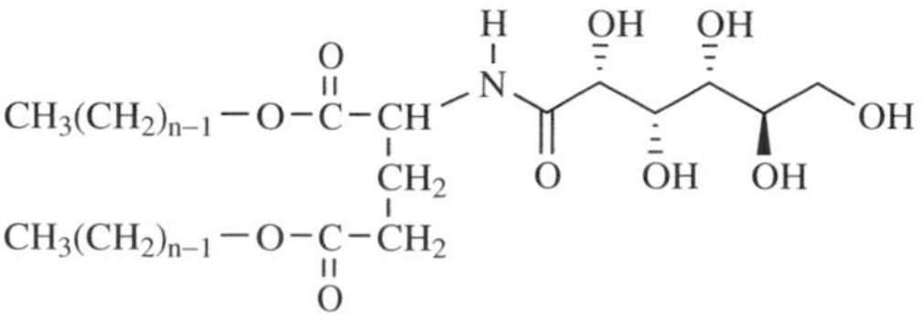

4 (*n* = 12, 16)

Scheme 11.3

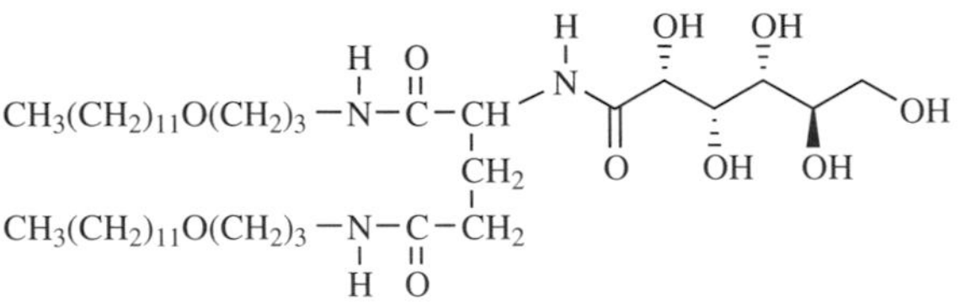

$$CH_3(CH_2)_{11}O(CH_2)_3 \overset{\overset{H}{|}}{-N} \overset{\overset{O}{||}}{-C} -CH \overset{H}{\underset{CH_2}{|}} \overset{OH \quad OH}{\underset{O \quad OH \quad OH}{N \bigwedge OH}}$$

$$CH_3(CH_2)_{11}O(CH_2)_3 \overset{\overset{H}{|}}{-N} \overset{\overset{O}{||}}{-C} -CH_2$$

5

Scheme 11.4

glycolipids can be dispersed in ionic liquids **1** and **2** as microcrystalline aggregates (Figure 11.3*a*). Gel-to-liquid crystalline phase transition is one of the basic aggregate characteristics of bilayer membranes and that of **4** ($n = 16$) was investigated by differential scanning calorimetry (DSC). The endothermic peaks (T_c) were observed at 47°C (ΔH, 40.6 kJ mol^{-1}; ΔS, 127.0 J K^{-1} mol^{-1}) and at 66°C (ΔH, 32.4 kJ mol^{-1}; ΔS, 95.7 J K^{-1} mol^{-1}) in ionic liquid **1**, and in ionic liquid **2** a peak was observed at 43°C (ΔH, 33.0 kJ mol^{-1}; ΔS, 104.3 J K^{-1} mol^{-1}). These ΔH and ΔS values are in the same order as those observed for the gel-to-liquid crystal phase transition of aqueous bilayer membranes, and therefore glycolipids **4** become dispersed in ionic liquids **1** and **2** with the basic structure of bilayer membrane.

Surprisingly, when glycolipid **5**, which contains ether-linkages in alkylchains and three amide bonds, was dissolved in ionic liquids in **1** and **2**, physical gelation occurred (concentrations above 10 mM, Figure 11.4). The developed fibrous structures whose length extends to several hundred micrometers could be abundantly seen in dark-field optical microscopy (Figure 11.3*b*). The DSC measurement of **5** in ionic liquid **1** showed an endothermic peak at 40°C, at which temperature the ionogel was dissolved. This is the first case of physical gelation occurring in ionic liquids because of the self-assembly of low-molecular weight compounds [6]. To our surprise this phase transition was accompanied by a reversible morphological transformation of the fibrous aggregate into vesicles (Figure 11.3*c*; with diameters of 3–5 μm) [6]. These findings indicate that the ether-containing ionic liquids behave as water-like solvents for carbohydrates and molecular assemblies. The ether-linkages appear to have enhanced the net polarity of the ionic liquids and thus to facilitate the self-assembly.

11.4 CHARGED BILAYER MEMBRANES IN IONIC LIQUIDS

Intrinsic to ionic surfactants and amphiphiles are solvophilic (or ionophilic) groups. We were therefore interested in learning whether the simple dialkylammonium bromides ($2C_nN^+$, $n = 12, 14$; Figure 11.1*d*) that belong to the original family of synthetic bilayer membranes [10] form a bilayer in ionic liquids. Amphiphiles $2C_nN^+$ were dispersed in three ionic liquids (Schemes 11.1–11.2) by ultrasonication (concentrations, 10 mM). While $2C_{12}N^+$ as homogeneously dispersed in the conventional

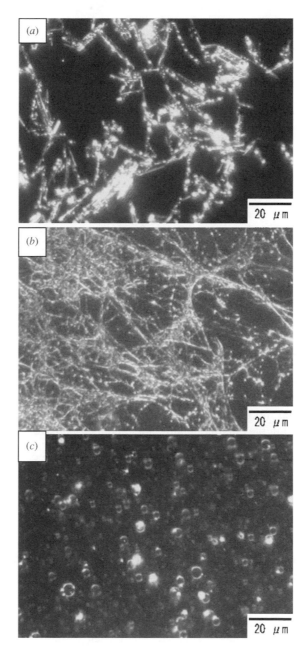

Figure 11.3 Dark-field optical micrographs of glycolipids dispersed in ionic liquid **1**: (a) **4** (n = 12), (b) **5** (20° C), and (c) **5** (50° C). [**4**] = [**5**] = 10 mM.

Figure 11.4   *Picture of self-assembling ionogel formed from* **5** *in* **1** *(10 mM).*

ionic liquid **3**, $2C_{14}N^+$ was insoluble even at the lower concentration of 1 mM. In the DSC measurement, the $2C_{12}N^+$ in ionic liquid **3** gave no endothermic peak (Figure 11.5c). Since no aggregate structure was observed under dark-field optical microscopy, it appears that $2C_{12}N^+$ must have dispersed without forming bilayer membranes.

Unlike the conventional ionic liquid **3**, both of the amphiphiles were homogeneously dispersed in the ether-containing ionic liquids and gave translucent dispersions. The $2C_{12}N^+$ showed an endothermic peak at 52°C (ΔH, 18.0 kJ mol^{-1}; ΔS, 55.3 J K^{-1} mol^{-1}, Figure 11.5a) in ionic liquid **1** and similarly at 51°C (ΔH, 12.5 kJ mol^{-1}; ΔS, 37.2 J K^{-1} mol^{-1}) in **2**. The longer alkyl-chained compound $2C_{14}N^+$ gave endothermic peaks at higher temperatures (72°C in **1**, ΔH, 23.8 kJ mol^{-1}, ΔS, 69.2 J K^{-1} mol^{-1}, Figure 11.5.b; 71°C in **2**, ΔH, 22.1 kJ mol^{-1}, ΔS, 64.4 J K^{-1} mol^{-1}). These endothermic peaks are considerably higher than those observed for the corresponding aqueous bilayers ($n = 12$; T_c, below 0°C, $n = 14$; T_c, 16°C) [34]. The gel-to-liquid crystalline phase transition temperature of charged bilayer membranes is affected by the balance of electrostatic repulsions and attractive interactions such as van der Waals forces. In ionic liquids it is expected that the electrostatic repulsion between ammonium head groups are highly screened. This leads to the increased association interactions in the totally ionized solvent that simultaneously enhances the thermal stability of gel(crystalline)-state bilayers. In dark-field optical microscopy, the $2C_{12}N^+$ that dispersed in **1** showed microcrystalline aggregates at room temperature, and this is consisted with its multilayered structure (Figure 11.6a). Remarkably these aggregates were reversibly transformed into vesicles upon heating above T_c (Figure 11.6b).

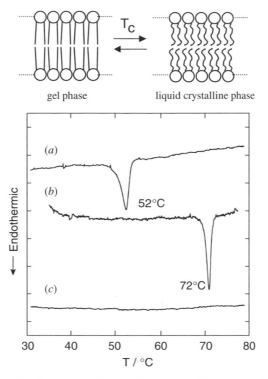

Figure 11.5 *Schematic illustration of gel-to-liquid crystalline phase transition and DSC thermograms: (a) 2C$_{12}$N$^+$ dispersed in* **1**, *(b) 2C$_{14}$N$^+$ dispersed in* **1**, *(c) 2C$_{12}$N$^+$ dispersed in* **3**. *Similar endothermic peaks were also observed for 2C$_n$N$^+$ (n = 12, 14) in* **2**. *[2C$_n$N$^+$] = 10 mM.*

Similar morphological changes were also observed for the 2C$_{12}$N$^+$ dispersed in ionic liquid **2** and for the 2C$_{14}$N$^+$ dispersed in ionic liquids **1** (Figure 11.6c) and **2**. These observations confirm that the ammonium amphiphiles 2C$_n$N$^+$ form bilayer membranes in ether-containing ionic liquids [35]. Since these bilayers did not form in ionic liquid **3**, the ether linkages and bromide counter ions in **1**, **2** again must play some important role. Since ether linkages are ionophilic [36], they should display high affinity to the ammonium groups. The bromide anion in ionic liquids **1**, **2** also secured stable solvation of the ion pairs unlike the hydrophobic PF$_6^-$ anions, which lowered their solubility.

11.5 CONTROL OF IONOPHILIC–IONOPHOBIC INTERACTIONS AND GENERALIZATION OF MOLECULAR SELF-ASSEMBLIES IN IONIC LIQUIDS

It should be now be clear that the molecular assembly in ionic liquids is governed by (1) the balance of ionophilicity and ionophobicity of the constituent molecules and (2) the chemical structure of the ionic liquids. It may well be that an increase in intermolecular interactions in the bilayer-forming amphiphiles makes their

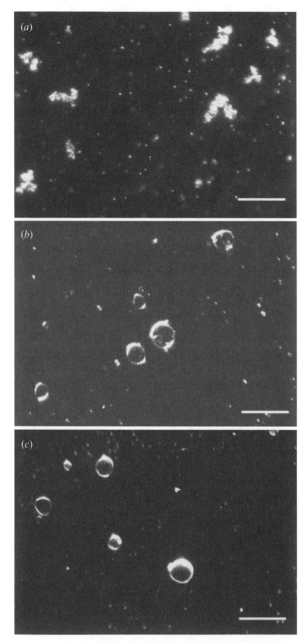

Figure 11.6 Dark-field optical micrographs of $2C_nN^+$ dispersed in ionic liquids: (a) $2C_{12}N^+$ in **1**, at room-temperature, (b) $2C_{12}N^+$ in **1**, at $60°C$, (c) $2C_{14}N^+$ in **1**, at $80°C$. Scale bar: $20\,\mu m$. $[2C_nN^+] = 10\ mM$.

self-assembly possible even in conventional ionic liquids. In aprotic organic media the hydrogen bonding occurs efficiently. The multiple-hydrogen bond-forming bilayers [37,38] and the mesoscopic-sized supramolecular assemblies [22] are stably dispersed in the organic media. The cationic glutamate amphiphile, Scheme 11.8, has three amide linkages, and it forms fibrous nanoassemblies when dispersed in chloroform [38]. A diester-type of amphiphile, Scheme 11.9, is used as a reference. The amphiphiles, **8** and **9** (Schemes 7, 8), are dispersed in ionic liquids **6** and **7** (Schemes 5, 6) [39], which contain TFSI anions. Although the diester-type amphiphile **9** did not form aggregates in ionic liquid **7**, vesicles were formed in the ether-containing ionic liquid **6** (above T_c). This is consistent with the molecular assembly-supporting property of ether-containing ionic liquids. In contrast, the amide-enriched compound **8** afforded self-assembling ionogels in both ionic liquids **6** and **7** (concentrations, 10 mM) [40].

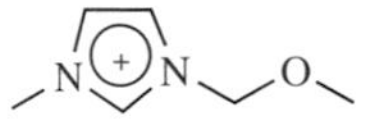

TFSI⁻

6

TFSI⁻ = $N(SO_2CF_3)_2^-$

Scheme 11.5

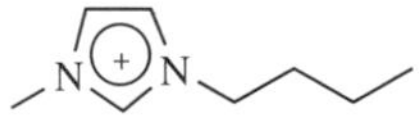

TFSI⁻

7

TFSI⁻ = $N(SO_2CF_3)_2^-$

Scheme 11.6

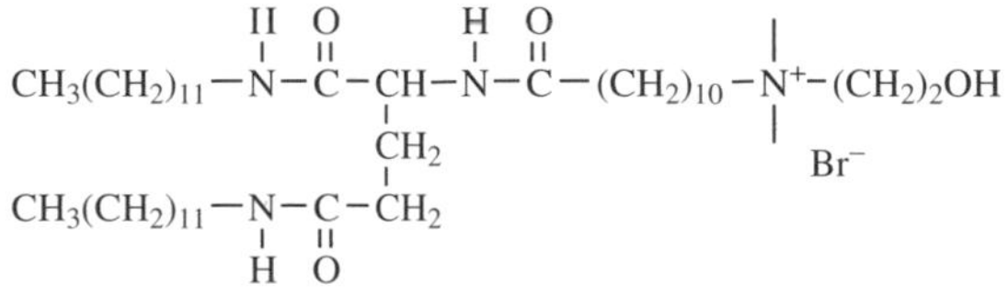

8

Scheme 11.7

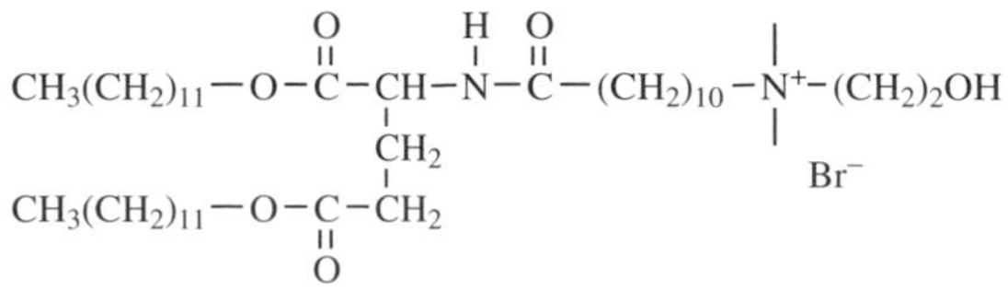

9

Scheme 11.8

Apparently, also in conventional ionic liquids, increases in intermolecular interaction due to hydrogen bonding drive the self-assembly. It is noteworthy that amphiphile **8** does not form hydrogels when dispersed in pure water. Physical gelation of liquids requires cross-linking of the nanofibrous assemblies dispersed in the solvents, and cross-linking is suppressed in water because of the electrostatic repulsive forces operating among the aggregates [41]. The ionogels observed to form from ionic liquid **8** must be as a consequence of the suppressed electrostatic repulsive forces in highly ionic environments (the concentration of ionic liquids is ca. 7 M).

In summary, the formation of self-assemblies in ionic liquids is a general phenomenon that depends on the molecular structures of amphiphiles, ionic liquids, and their combinations. The ether-introduced ionic liquids facilitate the formation of molecular self-assemblies. However, stable self-assemblies can form also in conventional ionic liquids if the intermolecular interactions are suitably tuned.

11.6 SUMMARY, UPDATES, AND OUTLOOK

In this chapter we introduced molecular self-assembly, which is becoming growing research area in the chemistry of ionic liquids. Our own earlier reports [6, 35] had touched off a burst of research on molecular assembly in ionic liquids and ionogel formation. A cholesterol-based organogelator, Scheme 11.9, was shown to induce

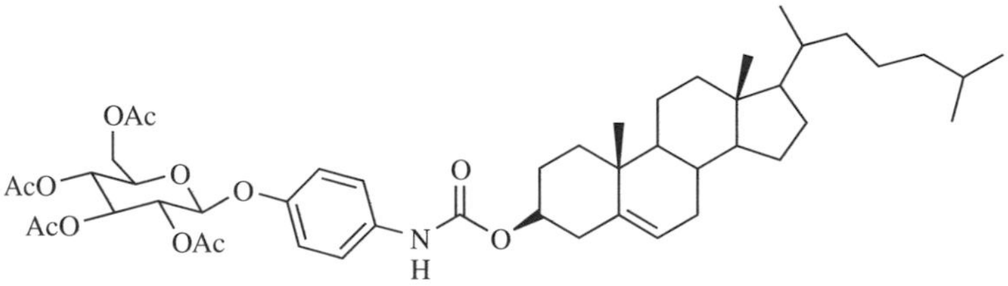

10

Scheme 11.9

gelation of various N, N′-dialkylimidazolium and N-alkylpyridinium salts [42]. Ionogels consisting of amino acid–derived low molecular weight gelators and 1-hexyl-3-methylimidazolium iodide were used for the fabrication of a dye-sensitized solar cell [43]. Hydrated samples of 1-decyl-3-methylimidazolium bromide were found to form liquid-crystalline ionogels [44].

A hydroxyl-terminated mesogenic compound, Scheme 11.10, showed limited miscibility with 1-ethyl-3-methylimidazolium tetrafluorobrate, but anisotropic ion

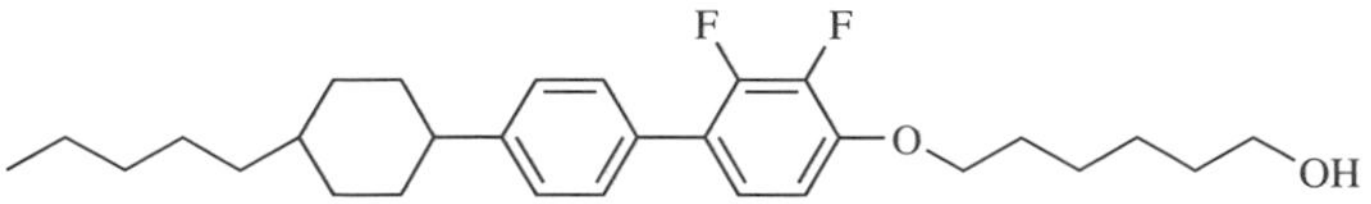

11

Scheme 11.10

conduction was observed in the layered liquid-crystalline phase [45]. Observations of anisotropic conductivity were also made in macroscopically oriented molecular assemblies such as multibilayer cast films [46]. The formation of molecular assemblies is found to simultaneously introduce supramolecular interfaces in ionic liquids. The design of such molecular interfaces should lead to the development of important functional materials. The introduction of microscopic interfaces also expands the area of ionic liquid research. For example, in a recent study we reported on the sol-gel synthesis of hollow TiO_2 microspheres at the surface of toluene microdroplets dispersed in ionic liquids [47]. Ionic liquids act not only as solvents but also as stabilizers of the produced inorganic materials. Other recent reports on microporous lamellar silica [48] and TiO_2 nanocrystals [49] confirm the potential for research of ionic liquids in inorganic nanochemistry.

The strategy of self-assembly could be applied to open up developments in molecular-based nanomaterials. We believe that the combination of ionic liquids and biomolecules, organic molecular self-assemblies and inorganic nanomaterials, can lead to new dimensions in materials science.

REFERENCES

1. T. Welton, *Chem. Rev.*, **99**, 2071 (1999).

2. J. G. Huddleston, H. D. Willauer, R. P. Swatloski, A. E. Visser, R. D. Rogers, *Chem. Commun.*, 1765 (1998).

3. (a) W. Xu, A. Angell, *Science*, **302**, 422 (2003). (b) F. Endres, *Chem. Phys. Chem.*, **3**, 144 (2002).

4. P. Wang, S. M. Zakeeruddin, P. Comte, I. Exnar, M. Grätzel, *J. Am. Chem. Soc.*, **125**, 1166 (2003).

5. K. R. Seddon, A. Stark, M.-J. Torres, *Pure Appl. Chem.*, **72**, 2275 (2000).

6. N. Kimizuka, T. Nakashima, *Langmuir*, **17**, 6759 (2001).

7. A. M. Leone, S. C. Weatherly, M. E. Williams, H. H. Thorp, R. W. Murray, *J. Am. Chem. Soc.*, **123**, 218 (2001).

8. (a) R. M. Lau, F. van Rantwijk, K. R. Seddon, R. A. Seddon, *Org. Lett.*, **2**, 4189 (2000). (b) T. Itoh, E. Akasaki, K. Kudo, S. Shirakami, *Chem. Lett.*, 262 (2001). (c) M. Erbeldinger, A. J. Mesiano, A. J. Russell, *J. Biotechnol. Prog.*, **16**, 1129 (2000).

9. (a) N. Kimizuka, *Curr. Opin. Chem. Biol.*, **7**, 702 (2003). (b) N. Kimizuka, in *Supramolecular Design for Biological Applications*, N. Yui, ed., CRC Press, 2002, p. 373.

10. T. Kunitake, *Angew. Chem. Int. Ed.*, **31**, 709 (1992).

11. H. Hoffmann, C. Elbert, *Angew. Chem. Int. Ed. Engl.*, **27**, 902 (1988).

12. (a) A. D. Miller, *Angew. Chem. Int. Ed.*, **37**, 1768 (1998). (b) I. S. Zuhorn, D. Hoekstra, *J. Membrane Biol.*, **189**, 167 (2002).

13. S. Okada, S. Peng, W. Spevak, D. Charych, *Acc. Chem. Res.*, **31**, 229 (1998).

14. (a) K. Sakata, T. Kunitake, *Chem. Commun.*, 504 (1990). (b) S. Asakuma, H. Okada, T. Kunitake, *J. Am. Chem. Soc.*, **113**, 1749 (1991).

15. D. F. Evans, *Langmuir*, **4**, 3 (1988).

16. I. Rico, A. Lattes, *J. Phys. Chem.*, **90**, 5870 (1986).

17. A. Ray, *Nature*, **231**, 313 (1971).

18. M. P. Turberg, J. E. Brady, *J. Am. Chem. Soc.*, **110**, 7797 (1988).

19. Y. Ishikawa, H. Kuwahara, T. Kunitake, *J. Am. Chem. Soc.*, **116**, 5579 (1994).

20. N. Kimizuka, T. Wakiyama, H. Miyauchi, T. Yoshimi. M. Tokuhiro, T. Kunitake, *J. Am. Chem. Soc.*, **118**, 5808 (1996).

21. N. Kimizuka, M. Tokuhiro, H. Miyauchi, T. Wakiyama, T. Kunitake, *Chem. Lett.*, 1049 (1997).

22. N. Kimizuka, T. Kawasaki, K. Hirata, T. Kunitake, *J. Am. Chem. Soc.*, **117**, 6360 (1995).

23. H. Bloom, V. C. Reinsborough, *Aust. J. Chem.*, **21**, 1525 (1968).

24. D. F. Evans, A. Yamaguchi, R. Roman, E. Z. Casassa, *J. Colloid Interface Sci.*, **88**, 89 (1982).

25. D. F. Evans, A. Yamaguchi, G. J. Wei, V. A. Bloomfield, *J. Phys. Chem.*, **87**, 3537 (1983).

26. J. L. Anderson, V. Pino, E. C. Hagberg, V. V. Sheares, D. W. Armstrong, *Chem. Commun.*, 2444 (2003).

27. D. F. Evans, E. W. Kaler, W. J. Benton, *J. Phys. Chem.*, **87**, 533 (1983).

28. C. J. Bowlas, D. W. Bruce, K. R. Seddon, *Chem. Commun.*, 1625 (1996).

29. F. Neve, O. Francescangeli, A. Crispini, J. Charmant, *Chem. Mater.*, **13**, 2032 (2001).

30. C. K. Lee, H. W. Huang, I. J. B. Lin, *Chem. Commun.*, 1911 (2000).

31. J. Putaux, A. Buléon, H. Chanzy, *Macromolecules*, **33**, 6416 (2000).

32. R. P. Swatloski, S. K. Spear, J. D. Holbrey, R. D. Rogers, *J. Am. Chem. Soc.*, **124**, 4974 (2002).

33. R. T. Carlin, J. Fuller, *Chem. Commun.*, 1345 (1997).

34. Y. Okahata, R. Ando, T. Kunitake, *Ber. Bunsen-Ges. Phys. Chem.*, **85**, 789 (1981).

35. T. Nakashima, N. Kimizuka, *Chem. Lett.*, 1018 (2002).

36. N. Kimizuka, T. Wakiyama, T. Kunitake, *Chem. Lett.*, 521 (1996).

37. (a) H. Ihara, H. Hachisako, C. Hirayama, K. Yamada, *J. Chem. Soc., Chem. Commun.*, 1244 (1992). (b) N. Yamada, E. Koyama, M. Kaneko, H. Seki, H. Ohtsu, T. Furuse, *Chem. Lett.*, 387 (1995).

38. N. Kimizuka, M. Shimizu, S. Fujikawa, K. Fujimura, M. Sano, T. Kunitake, *Chem. Lett.*, 967 (1998).

39. P. Bonhôte, A.-P. Dias, N. Papageorgiou, K. Kalyanasundaram, M. Grätzel, *Inorg. Chem.*, **35**, 1168 (1996).

40. T. Nakashima, Ph. D thesis, Graduate School of Engineering, Kyushu University, 2003.

41. T. Nakashima, N. Kimizuka, *Adv. Mater.*, **14**, 1113 (2002).

42. A. Ikeda, K. Sonoda, M. Ayabe, S. Tamaru, T. Nakashima, N. Kimizuka, S. Shinkai, *Chem. Lett.*, 1154 (2001).

43. W. Kubo, T. Kitamura, K. Hanabusa, Y. Wada, S. Yanagida, *Chem. Commun.*, 374 (2002).

44. M. A. Firestone, J. A. Dzielawa, P. Zapol, L. A. Curtiss, S. Seifert, M. L. Dietz, *Langmuir*, **18**, 7258 (2002).

45. M. Yoshio, T. Mukai, K. Kanie, M. Yoshizawa, H. Ohno, T. Kato, *Adv. Mater.*, **14**, 351 (2002).

46. R. Aoki, T. Kunitake, N. Kimizuka, M. Shimomura, *Synthetic Metals*, **18**, 861 (1987).

47. T. Nakashima, N. Kimizuka, *J. Am. Chem. Soc.*, **125**, 6386 (2003).

48. Y. Zhou, M. Antonietti, *Adv. Mater.*, **15**, 1452 (2003).

49. Y. Zhou, M. Antonietti, *J. Am. Chem. Soc.*, **125**, 14960 (2003).

Chapter 12

Solubilization of Biomaterials into Ionic Liquids

Kyoko Fujita, Yukinobu Fukaya, Naomi Nishimura, and Hiroyuki Ohno

Ionic liquids (ILs) are expected to be unique solvents for electrochemical reactions because of their extraordinary properties. ILs can dissolve a wide variety of molecules and materials. ILs are excellent solvents for the bioelectrochemistry field, but great care must be taken with the procedure of solubilization because most biomaterials such as proteins and enzymes are characterized by higher ordered structures that have an origin in their numerous functions. Therefore consideration must be given to the structural changes of biomaterials when they are dissolved in ILs.

Not all ILs are good solvents for proteins, however. There is the interesting example of lipase. Lipase is soluble in both aqueous and organic solvents, so it can be easily solubilized in ILs. Certain lipases even become dispersed or dissolved in some ILs. Since lipase is a very stable enzyme, it catalyzes the hydrolysis of lipids. Enzymatic activity is reported to be maintained in ILs [1]. There is not much published on the solubilization of biomaterials in ILs. In the present chapter we introduce a procedure to use in solubilizing biomaterials in ILs. First we consider the preparation of the IL, and then the chemical modification of biomaterials suitable for dissolution. We have found this procedure helpful when we tried to use electrochemically active biomaterials in ILs.

Electrochemical Aspects of Ionic Liquids   Edited by Hiroyuki Ohno
ISBN 0-471-64851-5   Copyright © 2005 John Wiley & Sons, Inc.

12.1 DESIGN OF ILS AS SOLVENTS FOR DNA

DNA is said to be the molecular memory of living things. DNA is expected as electro-conductive [2] and ion conductive [3] materials, base materials for electro-luminescence [4], and so on. There is wide number of trials of DNA reported in electrochemistry. We have been trying to solubilize DNAs in several organic solvents including ILs [5]. J. Davis and coworkers have also pursued this subject, and both groups have determined some ILs to be good solvents for DNA [6].

We studied the relation of an IL structure to the solubilization of DNA. To prepare the ILs or their models (see Chapter 19), one must neutralize the amines with an acid. We tried four different acids to prepare 19 different salts as solvents for DNA (see Table 12.1). To these organic salts (1.00 g) we added 10 mg DNA, and heated this mixture at a rate of $1°C \cdot min^{-1}$. No water or other solvents were added to this mixture, and the lowest temperature to solubilize 10 mg DNAs was recorded. Table 12.1 shows the lowest temperature we used to solubilize 10 mg DNA into 1.00 g salts. We call this temperature the "solubility temperature." Although the relation between the solubility temperature and the cation structure was not clear, could not be determined, there appeared to be a dependence of anion species on the temperature. We found that DNA cannot be dissolved in ionic liquids containing the TFSI anion in this concentration. When 1-methylpyrazolium BF_4 was used as a solvent, the solubility temperature was 100°C, which was the lowest of the ILs. The water content of all ILs was checked by Karl Fischer moisture titration and was confirmed to be less than 0.15 wt%. However, since 0.15 wt% of water in 1 g of ILs corresponds to be about 1.5 mg water content, the 10 mg of DNA could not be dissolved in the water. The ILs become contaminated with the 1.5 mg of water and could not solubilize the DNA. Therefore the DNA solubilized ILs had to be vacuum dried at 80°C for several hours to prevent any phase separation of the DNA specimen.

Our study of the relation between the solubility temperature and anion structure showed that the 1-hexyl-3-methylimidazolium cation coupled with the anions, namely Cl^-, Br^-, BF_4^-, and ClO_4^-. Roughly, the ILs having smaller anions showed

TABLE 12.1 Lowest Temperature (°C) to Solubilize 1 wt% DNA Into Ionic Liquids Prepared by the Neutralization of Corresponding Bases and Acids

| Bases | Solubility Temperatures (°C) | | | |
| | Acids | | | |
	HBF$_4$	HClO$_4$	HTFSI	HCl
1-Methylimidazole	130	—	X	120
1-Ethylimidazole	102	120	X	130
1-Butylimidazole	109	130	X	115
1-Ethyl-2-methylimidazole	137	107	X	—
Triethanolamine	130	110	X	—
1-Methylpyrazole	100	—	X	—

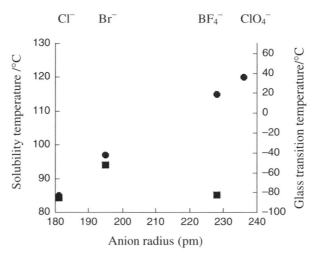

Figure 12.1 *Solubility temperature (●) of DNA into HMIM salts having different anions and T_g of ILs (■).*

lower solubility temperatures (Figure 12.1). There was strong hydrogen bond between Cl^- and the imidazolium cation [7], which indicated the extent to which hydrogen bonding of ILs works to solubilize DNA, just as is the case for cellulose [8]. Among only the above-mentioned anions, the chloride anion has the smallest anion radius and is also the strongest hydrogen bonding acceptor. So chloride appears to be the best anion to solubilize DNA. There appeared to be no relation between the melting point of the ILs and the solubility temperature of the DNA (Figure 12.1). Although the presence of chloride anions is important for solubilizing biomolecules, there are few studies on the development of low-viscosity ILs containing chloride anions.

We have already reported on low molecular compounds having more than one OH groups, such as triethanolamine and glycerin, being good solvents for DNA [9]. ILs with the OH groups appear to be excellent solvents for DNA. More on these ILs will be published elsewhere. Further the mixture of DNA with some ILs has produced excellent ion conductive films (see Chapter 28).

12.2 SOLUBILIZATION OF PROTEINS IN IL: IMPORTANCE OF POLYETHER MODIFICATION

The construction of a biosensor, a biochip, or a bioreactor frequently requires the fixation of corresponding biomacromolecules, especially proteins, on an electrode without denaturation. Protein functions have been studied only in an aqueous medium. However, the aqueous medium is not the best matrix from in practical applications. The available temperature range of an aqueous system is not sufficiently wide, and it is scarcely possible to maintain long-term stability of the proteins in an aqueous medium. It is therefore important to find a nonaqueous medium in

PEO

$\left(CH_2CH_2-O\right)_n$

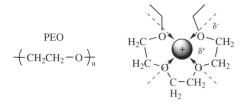

Figure 12.2   *Scheme of solvated ion by poly(ethylene oxide) (PEO).*

which these biomolecules can play their respective roles without denaturation. We have already reported that salt-containing polyethers are excellent nonaqueous media to use in creating an electrochemical reaction field for biomolecules [10]. Polyethers, especially poly(ethylene oxide) (PEO), are a linear and polar polymer having a repeating unit similar to that of a water molecule (Figure 12.2). In our studies redox reactions of different molecules were conducted in liquid and solid polyethers without using any water at all [11, 12].

We observed the redox reactions of different PEO-modified proteins dissolved in other organic solvents as well as in PEO [11–13]. PEO modification had already proved to be an excellent way to make proteins soluble in nonaqueous solvents [14]. There are nevertheless some reports of enzymatic activities [15, 16] and structure [17] determination of PEO-modified enzymes in homogeneous solutions of organic solvents. Although native heme proteins are insoluble and redox inactive in dehydrated PEO oligomers, these heme proteins have proved to be soluble and show redox activity even in PEO oligomers when they are chemically modified with PEO [18]. Furthermore it has been revealed that PEO-modified proteins, dissolved in solvent PEO, continue to function over a wide temperature range [19] and show long-term stability [20]; these characteristics are not possible to obtain in an aqueous medium. Since there is an effective interaction between the cation and ether oxygen unit of PEO through the ion–dipole interaction, many different salts are soluble in PEO [21]. The affinity between PEO and ions is a strong indication that PEOs are soluble in ILs, as they indeed are soluble in a number of ILs. Table 12.2 shows the solubility of PEOs in 1-ethyl-3-methylimidazolium *bis*-(trifluoromethanesulfonyl)imide, which is a typical ionic liquid. PEO was mixed with this ionic liquid, and stirred for 12 h. It is an IL in which most PEOs with different average molecular weight are soluble. This is because low molecular weight PEO is liquid, so it is freely miscible with this IL. Even for a solid PEO, after the dissolution into the IL there are no phase separation and re-liquation of the PEO. This compelled us to examine the solubilization of PEO-modified heme-proteins in ILs for which we expected no denaturation.

TABLE 12.2   Solubility of PEO into EMIm TFSI

Average molecular weight of PEO	550	2000	5000	10,000
State of PEO	Viscous liquid	Waxy solid	Waxy solid	White powder
Solubility (mg/ml)	Freely miscible	500	300	200

12.3   PEO-MODIFIED CYTOCHROME *c* IN IL

We modified cytochrome *c*, which is a typical heme-protein, with PEO chains
(PEO-cyt.*c*). We synthesized the PEO-cyt.*c* the same way as in our previous paper
[10]. The terminal-activated PEO (a-PEO) was prepared from these PEO mono-
methyl ethers. The ratio of a-PEO to cyt.*c* was changed in the feed in order to prepare the PEO-cyt.*c* with different degrees of PEO modification. The modification
degree was determined by titration of the remaining amino groups of cyt.*c* with
2,4,6-trinitrobenzene sulfonic acid [22]. The cyt.*c* was modified with 16.7 chains
of PEO_{150} (PEO_{150}-cyt.*c*(16.7)), 14.3 chains of PEO_{550} (PEO_{550}-cyt.*c*(14.3)),
15.4 chains of PEO_{1000} (PEO_{1000}-cyt.*c*(15.4)), 6.1 or 13.5 chains of PEO_{2000}
(PEO_{2000}-cyt.*c*(6.1 or 13.5)), and 9.7 chains of PEO_{5000} (PEO_{5000}-cyt.*c*(9.7)) to
compare the effects of both the PEO chain length and the degree of modification
on solubilization in 1-ethyl-3-methylimidazolium bis(trifluoromethanesulfonyl)-
imide, **1** which is a typical ionic liquid (shown in Figure 12.3).

Native cyt.c, or the prepared PEO-cyt.c, was mixed with **1**, and stirred slowly to
reach the final concentration of 0.1 mM. Native cyt.c was insoluble in **1**, but cyt.cs
modified with PEO_{150}(16.7), PEO_{550}(14.3), PEO_{1000}(15.4), or PEO_{2000}(6.1) was
partly soluble. However, when cyt.cs was modified with PEO_{2000}(13.5) or
PEO_{5000}(9.7), it dissolved well in **1**. Figure 12.4 shows the photographs after

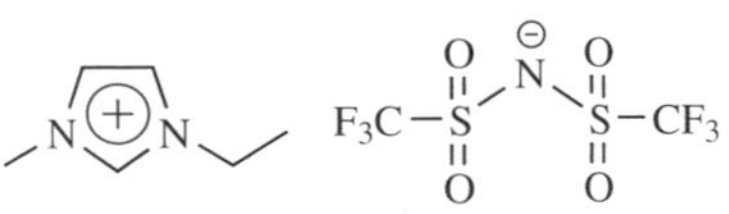

Figure 12.3   *1-Ethyl-3-methylimidazolium TFSI.*

Figure 12.4   *Solubility of native cyt.c (a) and PEO_{2000} modified cyt.c (b) into EMIm TFSI.*

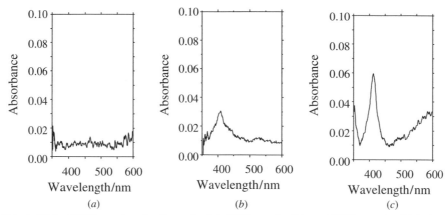

Figure 12.5 OWG spectra of native cyt.c (a), PEO₁₅₀-cyt.c (b), and PEO₂₀₀₀-cyt.c (c) in **1**.

mixing. Figure 12.5 shows the OWG spectra of the IL containing native cyt.c (*a*), PEO$_{150}$-cyt.c(16.7) (*b*), and PEO$_{2000}$-cyt.c(13.5) (*c*) [23]. (See Chapter 5 for more on the OWG spectroscopy.) Since native cyt.c was insoluble in **1**, no absorption was observed. However, a Soret band was clearly observed at 408 nm in the case of PEO$_{2000}$-cyt.c(13.5) in **1**. The appearance of this band varied depending on the IL species. It turned out that at least 10 chains of PEOs with averaged molecular weight of over 2000 were required for the dissolution of cyt.c in **1**. From the absorption spectrum we learned that the dissolved PEO-cyt.c contained heme with a ferric ion (Fe^{3+}) and that the vicinity of the heme remained unchanged. Our results showed that modified of PEO chains can be used to dissolve proteins into ILs.

PEO modification was thus confirmed to be an effective way to dissolve biomolecules in ILs. There are some other factors that served to regulate the solubility of the prepared PEO-modified molecules in out series of IL studies. The details will be reported elsewhere.

REFERENCES

1. J. L. Kaar, A. M. Jesionowski, J. A. Berberich, R. Moulton, A. J. Russell, *J. Am. Chem. Soc.*, **125**, 4125 (2003).

2. C. J. Murphy, M. R. Arkin, Y. Jenkins, N. D. Ghatlia, S. H. Bossmann, N. J. Turro, J. K. Barton, *Science*, **262**, 1025 (1993).

3. N. Nishimura, H. Ohno, *J. Mater. Chem.*, **12**, 2299 (2002).

4. L. Wang, J. Yoshida, N. Ogata, *Chem. Mater.*, **13**, 1273 (2001).

5. N. Nishimura, H. Ohno, *Proc. Polymer Preprints, Japan*, II Pd114 (2001).

6. A. C. Robinson, J. H. Davis Jr., N. F. Campbell, *225th ACS Nat. Meeting*, CHED-891, (2003).

7. A. Elaiwi, P. B. Hitchcock, K. R. Seddon, N. Srinivasan, Yu-May Tan, T. Welton, J. A. Zora, *J. Chem. Soc., Dalton Trans.*, 3467 (1995).

8. R. P. Swatloski, S. K. Spear, J. D. Holbrey, R. D. Rogers, *J. Am. Chem. Soc.*, **124**, 4974 (2002).

9. Y. Fukaya, N. Nishimura, H. Ohno, *Proc. Chemical Society of Japan*, 4 G8-01 (2003).

10. H. Ohno, T. Tsukuda, *J. Electroanal.Chem.*, **341**, 137 (1992).

11. (a) G. Shi, H. Ohno, *J. Electroanal. Chem.*, **314**, 59 (1991). (b) H. Ohno, H. Yoshihara, *Solid State Ionics*, **80**, 251 (1995).

12. H. Ohno, T. Tsukuda, *J. Electroanal. Chem.*, **367**, 189 (1994).

13. (a) H. Ohno, F. Kurusu, *Chem. Lett.*, 693 (1996), (b) N. Nakamura, Y. Nakamura, R. Tanimura, N. Y. Kawahara, H. Ohno, Deligeer, S. Suzuki, *Electrochim. Acta*, **46**, 1605 (2001).

14. G.-A. Humberto, V. Brenda, S.-R. Gloria, V.-D. Rafael, *Bioconjugate Chem.*, **13**, 1336 (2002).

15. K. Takahashi, T. Yoshimoto, A. Ajima, Y. Tamaura, Y. Inada, *Enzyme*, **32**, 235 (1984).

16. A. Matsushima, M. Okada, Y. Inada, *FEBS Lett.*, **178**, 275 (1984).

17. P. A. Mabrouk, *J. Am. Chem. Soc.*, **117**, 2141 (1995).

18. H. Ohno, N. Yamaguchi, M. Watanabe, *Polym. Adv. Technol.*, **4**, 133 (1993).

19. (a) H. Ohno, N. Yamaguchi, *Bioconjugate Chem.*, **5**, 379 (1994). (b) N. Y. Kawahara, H. Ohno, *Bioconjugate Chem.*, **8**, 643 (1997).

20. F. Kurusu, H. Ohno, *Electrochim. Acta*, **45**, 2911 (2000).

21. H. Ohno, *Electrochim. Acta.*, **37**, 1649, (1992).

22. A. F. S. A. Habeeb, *Anal. Biochem.*, **14**, 328, (1966).

23. H. Ohno, C. Suzuki, K. Fukumoto, M. Yoshizawa, K. Fujita, *Chem. Lett.*, 450 (2003).

Chapter *13*

Redox Reaction of Proteins

Kyoko Fujita and Hiroyuki Ohno

There are some studies on the direct electron transfer between an electrode and heme proteins that suggest the applications of functional proteins in vitro [1]. Almost all of these studies have been investigated in aqueous media [2, 3]. However, the electrochemistry of proteins can be captured in a variety of media, including organic solvents, polymers, and even ionic liquids. The redox reaction of molecules in ILs is usually analyzed by CV measurement [4, 5]. However, the redox response of PEO_{2000}-cyt.c in ILs cannot be detected with ordinary CV measurements. The serious drawback has been that the large background current of the ILs masks the faradaic component. Also the high viscosity of these media has been shown to induce low diffusion coefficients and low electrochemical currents [6, 7]. Therefore, it is necessary to carry out the voltammetric experiments at relatively high concentrations of the electroactive substance, typically 10 mM, to get sufficient signal-to-noise ratios [8]. Spectroelectrochemistry is useful in such cases (see Chapter 7), because data can be obtained on the redox processes in the absence of background current [9–12]. Since the absorption spectrum is the result of the integrated changes, it is easy to detect reactions that are difficult to analyze with ordinary electrochemical techniques. Specroelectrochemistry, OWG spectroscopy (see Chapter 7), has proved to be particularly useful for sensitive and simple spectral analysis of redox reactions of molecules in a small amount of solution on opaque electrodes [13]. The available μL range analysis, makes it further suitable for expensive and highly valuable samples such as ILs. In this chapter the redox reactions of proteins dissolved in ILs are analyzed with the introduction of OWG spectroscopy.

Electrochemical Aspects of Ionic Liquids   Edited by Hiroyuki Ohno
ISBN 0-471-64851-5   Copyright © 2005 John Wiley & Sons, Inc.

As was shown in Chapter 12, PEO modification to cytochrome c (cyt.c) can effectively solubilize proteins without changing the heme vicinity in 1-ethyl-3-methylimidazolium bis(trifluoromethanesulfonyl)imide, **1**, one of the typical ionic liquids. In particular, cyt.cs modified with $PEO_{2000}(13.5)$ or $PEO_{5000}(9.7)$ dissolved well in **1**. The redox activity of dissolved PEO_{2000}-cyt.c(13.5) in **1** was analyzed by OWG spectroelectrochemistry [10]. The electrochemical cell system was constructed using the waveguide as shown previously in Chapter 7. Carbon plate (3 × 40 mm) was used as the working electrode, and Pt and Ag wires (0.5 mm in diameter) were used as the counter and reference electrode, respectively. PEO_{2000}-cyt.c(13.5) was dissolved in **1** (0.1 mM), and 150 µl of this solution was directly introduced into the cell system. The OWG spectra were analyzed by applying the potential ($-800 \sim +800$ mV vs. Ag, sweep rate: 5 mV/s^{-1}). However, the spectral change due to the redox reaction of PEO-cyt.c in **1** was scarcely evident. Since the resistance of the solution was very small due to the excellent ionic conductivity of the IL as a solvent, this could not be due to the solvent's resistance. This was thought to be caused by the lack of small ions, despite the extremely high density of ions in ILs. During the redox reaction, globular proteins containing active enter into the inner domain, where small and suitably sized ion species are essential for the electron transfer activity in the ILs. Since KCl is known to be a good electrolyte for the electrochemical reaction of heme-proteins in PEOs [14], we added KCl as the supporting electrolyte into **1**. The solubility of KCl in **1** was, however, low (i.e., less than 0.1 M). PEO_{2000}-cyt.c (13.5) (0.1 mM) was then dissolved in the KCl saturated **1** in a similar analysis to that mentioned above. The spectrum was the same as that without KCl (Figure 13.1, solid line), indicating an oxidized form. When the

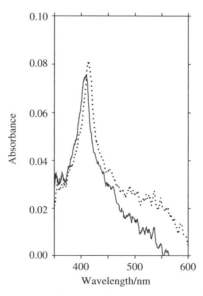

Figure 13.1 OWG spectra of oxidized (solid) and reduced (dotted) PEO_{2000}-cyt.c in KCl-saturated **1**.

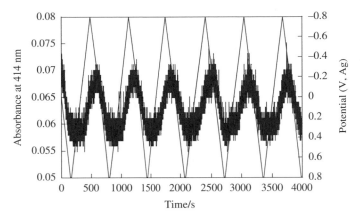

Figure 13.2 *Spectral change at 414 nm of PEO$_{2000}$-cyt.c due to the potential apply in* **1** *containing 0.02 M KCl.*

negative potential (-0.8 V vs. Ag) was applied to the spectrum, the Soret band peak showed a red shift from 408 to 414 nm based on the reduction (Figure 13.1, dotted line). After that, the λ_{max} returned to 408nm by applying positive potential ($+0.5$V vs. Ag). This spectral change due to the redox reaction of PEO-cyt.c in **1** was repeatedly observed in situ and corresponded to the potential switching.

From these absorbance changes, the repetition of the redox response of PEO-cyt.c was also observed in situ followed by the potential sweep. Figure 13.2 shows the absorbance change at 414 nm induced by the potential sweeping (sweep rate; 5 mV/s, potential sweep range; between -0.8 and $+0.8$ V vs. Ag). The stable absorbance change due to the redox reaction of PEO-cyt.c was observed without attenuation.

Figure 13.3 shows the effect of the given potential on the absorbance change at 414 nm of PEO-cyt.c dissolved in **1** (*a*) and in KCl saturated **1** (*b*). The KCl addition clearly affects the redox reaction of PEO-cyt.c in **1**. No absorbance change was confirmed along with the potential sweep, when PEO-cyt.c was dissolved in **1** without salts addition (Figure 13.3*a*). Nevertheless, the redox reaction of PEO-cyt.c was improved by the addition of KCl in **1** (Figure 13.3*a*). Thus PEO modification makes it possible to introduce cyt.c into **1**, which is composed of only oversized ions, but it is difficult to realize the electron transfer reaction of cyt.c under this condition where the cyt.c are surrounded by relatively huge ions. It became evident that in the huge ionic system the presence of micro ions is essential to the electron transfer reaction of proteins with the mounted active center. Through the analyses of various salts, the Cl$^-$ ion was determined to be the most effective ion to use in improving the redox reaction of PEO-cyt.c dissolved in **1**.

From the absorbance change at certain wavelengths as the function of the given potential, we could obtain the electrochemical characteristics such as the redox potential. From the absorbance change as the function of the applied potential in Figure 13.3*b*, the oxidation-reduction potential of PEO-cyt.c dissolved in **1**/KCl

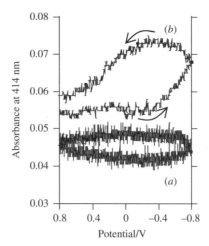

Figure 13.3   *Effect of the given potential on the absorbance change at 414 nm of PEO$_{2000}$-cyt.c dissolved in (a)* **1** *only, and (b) KCl-saturated* **1***.*

mixed electrolyte solution was estimated to be about -0.3V vs. Ag. This redox reaction was the same even after we kept the solution in a dry desiccator for several months. The extreme stability is understood to be due to the lack of water molecules.

Our study confirmed that a suitable sized electrolyte is essential for the smooth electron transfer reaction of heme-proteins in ILs, as shown in Figure 13.4.

Figure 13.4   *Scheme of solvated PEO-modified protein in ionic liquid containing added salts.*

REFERENCES

1. F. A. Armstrong, H. A. O. Hill, N. J. Walton, *Acc. Chem. Res.*, **21**, 407 (1988).
2. P. Yeh, T. Kuwana, *Chem. Lett.*, 1145 (1977).
3. K. Niki, T. Yagi, H. Inokuchi, K. Kimura, *J. Am. Chem. Soc.*, **101**, 3335 (1979).
4. D. L. Compton, J. A. Laszlo, *J. Electroanal. Chem.*, **520**, 71 (2002).

5. J. D. Wadhawan, A. J. Wain, A. N. Kirkham, D. J. Walton, B. Wood, R. R. France, S. D. Bull, R. G. Compton, *J. Am. Chem. Soc.*, **125**, 11418 (2003).

6. T. Welton, *Chem. Rev.*, **99**, 2071 (1999).

7. P. Wasserscheid, W. Keim, *Angew. Chem. Int. Ed.*, **39**, 3772 (2000).

8. C. Lagrost, D. Carrié, M. Vaultier, P. Hapiot, *J. Phys. Chem.* A, **107**, 745 (2003).

9. H. Ohno, C. Suzuki, K. Fukumoto, M. Yoshizawa, K. Fujita, *Chem. Lett.*, 450 (2003).

10. K. Fujita, H. Ohno, *Polym. Adv. Technol.*, **14**, 486 (2003).

11. N. Matsuda, A. Takatsu, K. Kato, Y. Shigesato, *Chem. Lett.*, 125 (1998).

12. J. T. Bradshaw, S. B. Mendes, S. S. Saavedra, *Anal. Chem.*, **74**, 1751 (2002).

13. K. Fujita, C. Suzuki, H. Ohno, *Electrochem. Comm.*, **5**, 47 (2003).

14. (a) H. Ohno, N. Yamaguchi, *Bioconjugate Chem.*, **5**, 379 (1994). (b) F. Kurusu, H. Ohno, *Electrochim. Acta*, **45**, 2911 (2000).

Part III

Ionic Devices

Chapter *14*

Application of Ionic Liquids to Li Batteries

Hikari Sakaebe and Hajime Matsumoto

14.1 INTRODUCTION

An "ionic liquid (IL)", or classically "a room-temperature molten salt", is an interesting series of materials being investigated in a drive to find a novel electrolyte system for electrochemical devices. ILs contain anions and cations, and they show a liquid nature at room temperature without the use of any solvents. The combination of anionic and cationic species in ILs gives them a lot of variations in properties, such as viscosity, conductivity, and electrochemical stability. These properties, along with the nonvolatile and flame-resistant nature of ILs, makes this material especially desirable for lithium-ion batteries, whose thermal instability has not yet been resolved despite investigations for a long time. In this chapter we discuss the efforts made for battery application of ILs.

14.2 SAFETY ASPECTS OF THE LI-ION BATTERY AND THE ADVANTAGES OF USE OF IONIC LIQUIDS

The exothermic reaction at the first stage of thermal runaway of a Li-ion battery with an organic electrolyte involves electrolyte decomposition and an electrolyte–electrode reaction which accelerates the rising heat [1]. Because the organic electrolyte

Electrochemical Aspects of Ionic Liquids   Edited by Hiroyuki Ohno
ISBN 0-471-64851-5   Copyright © 2005 John Wiley & Sons, Inc.

contains volatile solvents, such as propylenecarbonate, dimethylcarbonate, and dimethoxyethane, the inner pressure of the battery can be pushed up by these solvents and cause the battery to explode. Once the battery is opened to air, these organic electrolytes will work as a fuel because they are flammable. Even if the battery could stay unopened in the elevated temperature, the positive electrode material at the charged state will decompose at around 200°C and provide oxygen inside the battery in the presence of organic solvent. Thus it will be important to replace the volatile and flammable electrolyte with a different system. The solid-state electrolyte will ultimately fit this purpose. However, the manufacture of a solid-state battery requires a big scientific breakthrough. This is because it is extremely difficult to keep the electrode and electrolye solid in close contact through the entire charge–discharge process and because the conductivity is still unsufficient, though great efforts are being made. It will take a long time before the solid-state battery is commercialized.

As we mentioned earlier, ILs are the ideal material for use as an electrolyte because they are nonvolatile and flame resistive. These properties will make the battery safer. We will not discuss here in detail the basic and general properties of ILs, but rather point out the more interesting properties that pertain to battery application. These are listed briefly below:

1. Liquidity in a wide temperature range, including ambient temperature.
2. Thermal stability, nonvolatility, flame-resistance.
3. Wide electrochemical window.
4. Higher conductivity among the novel electrolyte.
5. Chemical stability.

14.3 SOME EXAMPLES FOR APPLICATIONS TO LI-ION BATTERIES

Various ionic liquids have been investigated as possible candidates for the electrolyte of the Li-ion battery. The experiments tapped into much human ingenuity, depending both drawbacks and advantages for each investigated ionic liquid. The application techniques that have been reported can be roughly classified into three categories:

1. IL + supporting electrolyte + additives
2. IL + supporting electrolyte + alternative negative electrode with higher potential
3. IL (neat) + supporting electrolyte

We will discuss the results using this classification in the following way:

   IL with chloroaluminate anions

IL with alkylimidazolium cation–fluorinated anion combination

IL with non-imidazolium cation–fluorinated anion combination

14.3.1 IL with Chloroaluminate Anions

An IL was first reported as a new bath for electrolysis about 50 years ago [2]. A mixture of organic bromide and aluminum chloride was found to have a melting point lower than room temperature, and this led to the discovery that a mixture of organic chloride and aluminum chloride stays liquid in a wide temperature range. More researchers became attracted to this ILs because this series of ILs have lower viscosity and higher conductivity as well as a relatively wide electrochemical window. However, the early series of ILs was extremely sensitive to water and corrosive to certain metals.

There are many examples of battery applications of chloroaluminate-containing ILs because of the long history of use. The corrosive nature of this IL limited its application, and it could not be applied to conventional batteries. The examples of their use include novel battery systems that turned the difficulties to advantages, among these the chloroaluminate acid–base concentration cell [3], the Al–chloride cell [4], the Cd–bromide cell [5], the Al–polyaniline cell [6], and the Mg–metal vanadate cell [7]. The application of ILs to Li-ion batteries remains problematic mainly because the cathodic stability of the chloroaluminate-containing electrolyte against lithium has not yet been made sufficient. At the present time it appears that the use of alternative negative electrodes (e.g., Li–Al alloy-layered oxides cell [8–10], Li–Sn alloy negative electrode [11]) or additives (e.g., $SOCl_2$ [12], H_2O [13]) can enhance the potential of ILs of the chloroaluminate family. For details, the reader is referred to the article by Webber and Blomgren [14]. Here we will review two relatively recent successful applications to lithium batteries from the literature.

Li/AlCl₃–EMIC–LiCl–SOCl₂/Graphite [12]. EMIC:1-ethyl-3-methylimidazolium chloride

(IL+ supporting electrolyte + additives)

There is not much literature that gives successful results for Li^+ intercalation-deintercalation into/from the graphitized carbon electrode. One reason may be that the alkylimidazolium-containing melt decomposes at 1 V above Li/Li^+, and another that these melts do not readily form an SEI layer on the carbon electrode surface the way the organic electrolyte will. Moreover, the imidazolium cation intercalates (probably irreversibly) into the interlayer of carbon [15, 16], which leads to complicated reactions in the reduction of the melts on the graphitized carbon electrode. However, in the presence of $SOCl_2$ in the chloroaluminate-containing IL, the charge–discharge properties of the graphitized carbon electrode become improved. A discharge capacity of 300 mAh g^{-1} has been demonstrated as long as

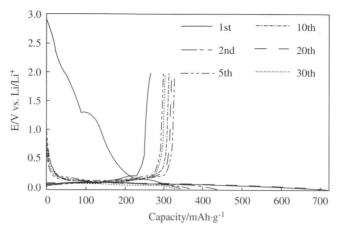

Figure 14.1 *Reduction/oxidation cycles obtained with the binder-free graphite electrode (artificial graphite, KS-25, 1.8 mg on Mo) in the Li–Cl saturated AlCl$_3$ (60 mol%)–EMIC (40 mol%) melt containing 0.11 mol dm^{-3} SOCl$_2$. First cycle at 0.5 C (0.28 mA cm^{-2}); remaining cycles at 0.2 C. (Adapted from Koura et al. [12])*

the electrode is fabricated without a binder (Figure 14.1). The coulombic efficiency of each cycle is reported to be more than 90% except for the first cycle of nearly 40%. The irreversible capacity observed in the initial charge is thought to be used for the cathodic decomposition of SOCl$_2$ and lead to the SEI formation. This research group has more recently fabricated a binder-free electrode by electrophoretic deposition, and the electrode showed higher efficiency over all the cycles [17].

Li–Al/MeEtImCl–AlCl$_3$–LiCl–C$_6$H$_5$SO$_2$Cl/LiCoO$_2$ [9]

MeEtImCl: 1-methyl-3-ethylimidazolium chloride

(IL+ supporting electrolyte + alternative negative electrode with higher potential, also with additives)

Li–Al, whose electrode potential is nearly 0.3 V higher than Li/Li$^+$, was applied as a negative electrode in a melt MeEtImCl–AlCl$_3$–LiCl containing C$_6$H$_5$SO$_2$Cl as an additive. The lithium metal deposited on the Al was left shiny, thanks to the additive's effect. The authors concluded that Li–Al could be stably used in this cell configuration. The charge–discharge curves of the cell in the first cycle are shown in Figure 14.2. During the first charge a plateau was observed around 3.5 V with excess charge capacity. The succeeding discharge curve lost the plateau, with a coulombic efficiency of less than 50%. This is similar to the case of Li/AlCl$_3$–EMIC–LiCl–SOCl$_2$/graphite [12]. The irreversible capacity at the initial charge was used to form the SEI layer on the surface of negative electrode.

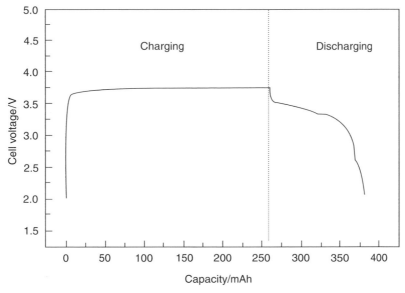

Figure 14.2 *Variation in cell voltage of the LiAl/LiCoO$_2$ battery using the RTMS as the electrolyte media with the addition of C$_6$H$_5$SOCl$_2$ (0.05 mol kg^{-1}) at the first cycle. Positive electrode: LiCoO$_2$ 85 wt.%, graphite 10 wt.%, Teflon 5 wt.%. RTMS: MeEtImCl/AlCl$_3$/LiCl = 1.0/ 1.2/0.15. Current density: 1.0 mA cm^{-2}. (Adapted from Fung and Zhou [9]).*

14.3.2 IL with Alkylimidazolium Cation–Fluorinated Anion Combination

A certain kind of fluorinated anion was found to form an IL having a melting point lower than room temperature and having sufficient anodic stability. Moreover its sensitivity toward water was low [2b]. It was evident that the imide anion had greatly improved the chemical stability in air, which allowed easy handling of the IL. The EMI cation is prevalent in the IL of the nonchloroaluminate family and mostly it imparts to the electrolyte the lower viscosity that makes it so manageable. Despite the kinetic advantage, the EMI-containing IL usually falls short of cathodic stability. As shown in Figure 14.3 [18], the cathodic decomposition of the neat IL breaks away in the positive potential region toward Li/Li$^+$. This indicates that its application to Li-battery systems will be essentially difficult. As is the case with the chloroaluminate-containing IL, the electrolyte design or negative electrode material needs to be reconsidered for lithium battery applications. An example for a successful battery application showing good cycleability and long life is outlined next.

Li$_4$Ti$_5$O$_{12}$/EMIBF$_4$–LiBF$_4$/LiCoO$_2$ *[19].* EMIBF$_4$: 1-ethyl-3-methylimidazolium tetrafluoroborate

(IL+ supporting electrolyte + alternative anode with higher potential)

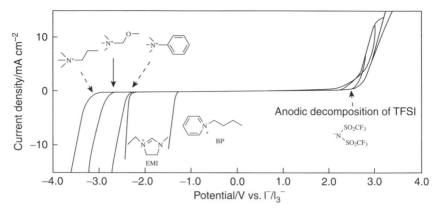

Figure 14.3 *Linear sweep voltammograms of Glassy Carbon in various ionic liquids with a TFSI anion at 25°C. Sweep rate: 50 mv/s. (Adapted from Matsumoto et al. [18]).*

To be obtain an effective application of the EMI-containing IL to a lithium battery system, there must used an alternative negative electrode material. Several kinds of Li–alloys and metal oxides are found to work over 0.2 to 2.0 V vs. Li/Li$^+$ and help overcome the drawbacks of cathodic stability. In this case the spinel phase of Li$_4$Ti$_5$O$_{12}$ working at 1.5 V vs. Li/Li$^+$ was used as a negative electrode. The structural change of Li$_4$Ti$_5$O$_{12}$ was intensively studied by Ohzuku et al. They called Li$_4$Ti$_5$O$_{12}$ the "zero-strain insertion material" because it appeared to have an almost negligible volume change of the unit cell of the crystal structure during the entire charge–discharge process [20]. These crystallographic properties could keep the electrode composite from degrading during the cycling, so the Li$_4$Ti$_5$O$_{12}$ could give the battery cycle long life. Figure 14.4 shows the initial charge–discharge curves of the Li$_4$Ti$_5$O$_{12}$/EMIBF$_4$–LiBF$_4$/LiCoO$_2$ cell. The cell voltage that was decreased around 2 V reduced the energy density of the battery, but the capacity decay became extremely small after 50 cycles.

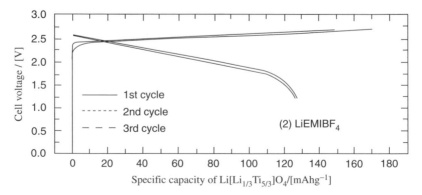

Figure 14.4 *Discharge curves of the cell Li$_4$Ti$_5$O$_{12}$/EMIBF$_4$–LiBF$_4$/LiCoO$_2$. (Adapted from Nakagawa et al. [19]).*

14.3.3 IL with Non-imidazolium Cation–Fluorinated Anion Combination

It is difficult to find "neat" ILs that work in Li-batteries as well as the ILs introduced in Sections 14.3.1 and 14.3.2. By "neat" it is meant that the IL contains only the supporting electrolyte and no additive. The non-imidazolium system, the quaternary asymmetric ammonium system, and the pyrazolium system can be regarded as "neat" examples of ILs that are well-suited to Li-batteries.

Li/DMFP(or EMP)BF₄–LiBF₄ or LiAsF₆/LiMn₂O₄ [21, 22]

$DMFPBF_4$:1,2-dimethyl-4-fluoropyrazolium tetrafluoroborate

$EMPBF_4$:1-ethyl-2-methylpyrazolium tetrafluoroborate

$DMFPBF_4$ and $EMPBF_4$ are known to have excellent thermal stability. This property makes them the ideal electrolyte to introduce more safety to batteries. However, the cathodic decomposition of these salts sets in at 1.0 to 1.3 V above Li/Li^+. When an electrolyte of $DMFPBF_4$–$LiBF_4$ was used, 75% of the theoretical capacity was obtained in the initial cycles, and that capacity quickly faded. Replacing $LiBF_4$ with $LiAsF_6$ made the battery's utilization very low (ca. 30%), but its coulombic efficiency improved to 100%. The authors concluded that SEI layer on Li metal could be made to show good conductivity by adding an arsenate compound [21]. Figure 14.5 gives the charge–discharge curves for the $Li/EMPBF_4$–$LiAsF_6/LiMn_2O_4$ cell during the cycling. In this configured cell, the coulombic efficiency is close to 100%, but the cathode is utilized at less than 60% [22]. The authors

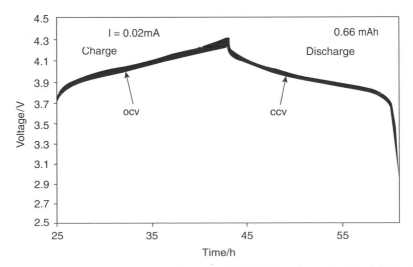

Figure 14.5 *Charge–discharge curves for the cell $Li/EMPBF_4 + 0.8$ molar $LiAsF_6/LiMn_2O_4$ at room temperature. (Adapted from Caja, Dunstan, and Katovic [22]).*

attributed this mainly to the cell's large resistance, combined of the high resistance of the electrolyte and the resistive product on Li.

Li/LiTFSI + PP13–TFSI/LiCoO₂ [23]. (PP13:N-methyl-N-propylpiperidinium; Figure 14.6a)

The authors believe that the simple composition can be used in battery materials to increase the battery's reliability. For this reason they have turned their interest to the so-called neat IL, which excludes volatile additives. Some years ago when a certain kind of ammonium cation was first known to form an IL, one of the authors, Matsumoto, created a novel IL with aliphatic and asymmetric quaternary ammonium cations. This series of ILs was found to be the most stable cathodically, and reversible plating/stripping of the Li metal was observed in that media [24]. Recall from Figure 14.3 that the linear sweep voltammograms of several classes of ILs with the *bis*(trifluoromethane sulfonyl)imide anion (TFSI, Figure 14.6c), showed the trimethylpropylammonium cation (TMPA)–TFSI, to be stable around the Li/Li⁺ potential. This led the authors to apply several ILs, including EMI-TFSI, to the electrolyte of Li/LiCoO₂ cell. Then the charge–discharge properties were evaluated [23, 25]. The result showed that sufficient cathodic stability should be the minimum requirement for the acceptable battery's performance. In addition it was evident that PP13-TFSI, which is one of the cathodically most stable ILs, could give an excellent performance as an electrolyte in the Li/LiCoO₂ cell [23, 25]. Figure 14.7 gives the charge–discharge curves of the Li/LiTFSI + PP13–TFSI/LiCoO₂ cell. The obtained capacity was very close to the theoretical one (Figure 14.7a) and the coulombic efficiency of each cycle was more than 97% over 30 cycles (Figure 14.7b). For the first cycle, the efficiency was a little bit lower, but 93%, which implied the no significant decomposition occurred.

The physical properties of the ILs studied in this work were listed in Table 14.1 [13, 18, 25, 26]. The ammonium series of ILs has less advantageous kinetic properties than EMI-containing ILs. The conductivity of PP13–TFSI turned out to be 1.5 mS cm⁻¹ at 25°C, and it decreased to 0.5 mS cm⁻¹ when 0.4 mol dm⁻³ LiTFSI was dissolved into the PP13–TFSI. Further work to greatly improve the conductivity and viscosity is necessary before the ammonium system can be commercialized. Despite the drawback in conductivity, however, the thinner electrode did raise the cycling to a higher rate. Figure 14.8 gives the result of the charge–discharge cycling test of the Li/LiTFSI + PP13–TFSI/LiCoO₂ cell at the 0.5 C current rate. The

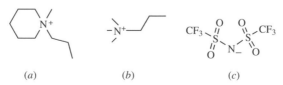

(a)   (b)   (c)

Figure 14.6 *Typical cations and anion of an ionic liquid with an ammonium cation. (a) N-Methyl-N-propylpiperidinium cation (PP13); (b) trimethylpropylammonium cation (TMPA); (c) bis(trifluoromethane sulfonyl)imide anion (TFSI).*

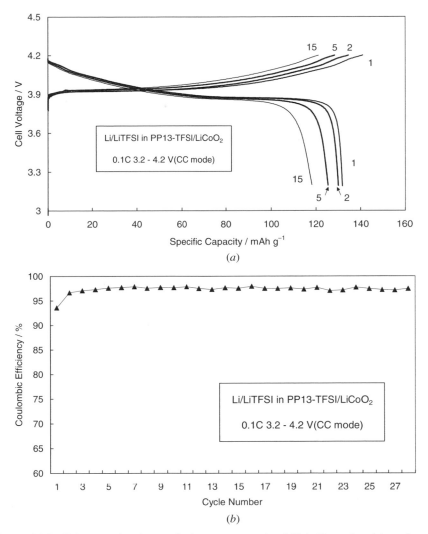

Figure 14.7 *Galvanostatic charge–discharge curves for initial 25 cycles (a) and cycle properties (b) of Li/0.4 mol dm^{-3} LiTFSI in PP13–TFSI/LiCoO$_2$ cell at C/10 current rate, 3.2 to 4.2 V. (Adapted from Sakaebe and Matsumoto [25])*

TABLE 14.1 Physical Properties of the ILs

ILs	Melting Point (°C)	Density at 20°C (g cm^{-1})	Viscosity at 25°C (mPa s)	Conductivity at 25°C (mS cm^{-1})	Reference
EMI–TFSI	−15	1.52	34	10.8	13
TMPA–TFSI	22	1.44	72	3.27	18
P13–TFSI	12	1.45	63	1.40	26
PP13–TFSI	8.7	—	151	1.51	25

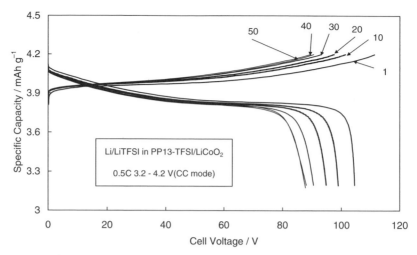

Figure 14.8 Galvanostatic charge–discharge curves for initial 50 cycles of Li/0.4 mol dm^{-3} LiTFSI in PP13–TFSI/LiCoO$_2$ cell at C/2 current rate, 3.2 to 4.2 V. (Adapted from Sakaebe and Matsumoto [25])

electrode loading was about 1 mg for the cell in Figure 14.7 and 14.8. The retention of capacity at 0.5 C compared to that at 0.1 C ([first discharge capacity at 0.5 C]*100/[first discharge capacity at 0.1 C]) was 80%. Since the cell's construction affects the cycling test results, it is extremely difficult to compare the absolute values of the PP13-TFSI and other ILs. Nevertheless, the cell using PP13–TFSI as an electrolyte base showed well-balanced performance in cell-voltage, capacity utilization, coulombic efficiency of each cycle, cycling properties, and so on.

It is known that the Al current collector undergoes corrosion in an organic electrolyte containing the LiTFSI supporting electrolyte [27, 28]. The TFSI anion is actually stable, releasing very little free fluoride anion. When LiPF$_6$ is used instead, more F$^-$ is present in the organic electrolyte, which helps to form the protective surface film on Al. Still there can be found in the literature an indication that Al dissolves into an organic electrolyte to form a species [Al(TFSI)$_x$]$^{3-x}$ [28]. Because a number of TFSI anions are contained in the PP13-TFSI electrolyte, there was concern that the Al current collector can also corrode in the IL electrolyte. For this reason the electrochemical measurements of the Li/0.4 mol dm^{-3} LiTFSI + PP13–TFSI/Al cell and the Li/0.4 mol dm^{-3} LiTFSI + EC–DMC/Al cell (EC: ethylenecarbonate; DMC: dimethylcarbonate), were compared, as were the surface morphologies [29]. The cyclicvoltammogram between O.C.V. and 5.5 V showed the current density that passed through the cell containing IL to be two figures smaller than that containing EC-DMC. This result indicated that Al corrosion in an IL containing electrolyte would not be significant, and this result was confirmed by an SEM image. On the Al oxidized in the organic electrolyte (Figure 14.9a) a lot

(a)

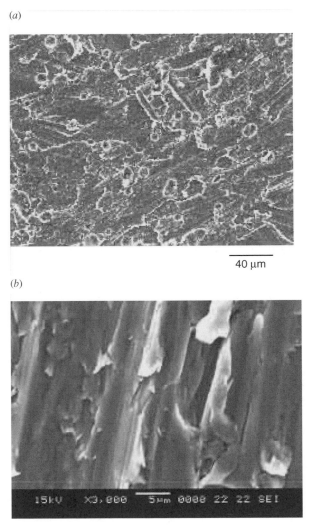

40 µm

(b)

Figure 14.9 *Scanning electron micrographs for the surface of Al oxydized (a) up to 5.5 V in 0.4 mol dm^{-3} LiTFSI/EC-DMC and (b) up to 6.5 V in 0.4 mol dm^{-3} LiTFSI/PP13–TFSI. The Al surface was scratched in an Ar atmosphere by emery paper before the electrochemical reaction. (Reproduced from Sakaebe and Matsumoto [29])*

of pit formed as a result of the corrosion, but no corrosion appeared to take place on the Al oxidized in the PP13–TFSI electrolyte. An experiment using severe conditions was tried, with oxidation at 5.5 V and storage for one month at 60°C, and also oxidation at 6.5 V (Figure 14.9b), no pit was observed on the Al surface [29]. The Al current collector, which is cheap, easy to handle, light-weight, and stable in the higher voltage region, is the key component for the commercial Li-ion battery, so the study on corrosion in ILs should be continued.

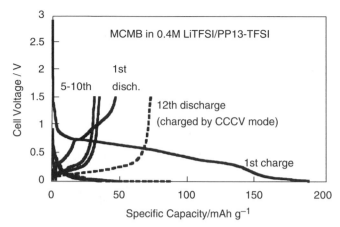

Figure 14.10 *Galvanostatic charge–discharge curves for the initial 12 cycles of Li/0.4 mol dm⁻³ LiTFSI in PP13-TFSI/ MCMB2800 cell at C/10 current rate (CC charge for initial 11 cycles and CCCV charge at 5 mV for the twelfth cycle), 0 to 1.5 V. (Adapted from Sakaebe, Matsumoto, and Tatsumi [30])*

The availability of ILs seems to depend on the kinetic nature and on electrode compatibility. Several electrode materials have been cycled in the PP13–TFSI based electrolyte. For the carbon electrode, which was in practical use, full capacity cannot yet be easily obtained using the PP13–TFSI based "neat" electrolyte. Some examples of the charge–discharge properties of MCMB2800 are shown in Figure 14.10 [30]. The constant current (CC) charging has provided only a 10% capacity of the theoretical one. The constant current–constant voltage (CCCV) charging at 5 mV increase the capacity to 20% of the theoretical capacity. This is partly because the inner resistance of the cell is very high ($\sim$20 Ω), and wettability of the electrode against IL may be responsible for the elevated resistance. Moreover it is evident that a favorable SEI for Li$^+$ intercalation cannot be formed properly. Nevertheless, the so-called hard carbons show more capacity (more than 50% of the capacity obtained in organic electrolyte), and this tells us that an alternative electrode material could leads us to a suitable cell configuration using ILs.

14.4 SUMMARY AND FUTURE VIEW

Although at the present time the asymmetric quaternary ammonium-based IL is left with its kinetic properties unsolved, because of their cathodic stability it remains worthwhile to continue the investigation of their potential as an electrolyte for Li-ion batteries. And at the same time, including other class of ILs, it seems important to make the best use of them by considering the additives and alternative electrode materials. A lot of work still remains to be done to obtain both a general and a unified view of the basic behavior of ILs. For instance, there is nothing known about the surroundings of the Li$^+$ ion in the electrolyte within the battery, and how they move between electrode. Any possible impurities, and how they affect the electrochemical properties of the battery, should also be considered for battery application. Then there should be atteation given to battery design in research on design compatibility with different ILs and battery components like the separator and binder.

The exploration of novel types of ILs has involved important and intensive work. For the battery use, higher concentration of Li^+ in the electrolyte is required and for this purpose the lithium salt itself hopefully could be melted at room temperature. Fujinami et al. have designed a novel bulky anion and demonstrated that certain kinds of Li salts can become viscous fluids at ambient temperature with a conductivity of between 10^{-4} and 10^{-5} S cm^{-1} [31]. This result is still far from practical application and the electrochemical window is still unknown, but the work has shown unique progress for ionic liquids.

Summarizing this chapter, it can be said that the several interesting findings have brought ILs closer to a practical application for the lithium-ion battery, and have attracted research interest. At this point we don't yet have the IL in which both electrochemical stability and conductivity are compatible to the organic electrolyte. It seems that improvements are possible by researching additives, but no big change is yet evident. It is nevertheless important to find a favorable way to optimize the configuration of the IL-containing battery, and to promote work on "ionic-liquid batteries."

REFERENCES

1. J. Yamaki, Y. Baba, N. Katayama, H. Takatsuji, M. Egashira, S. Okada, *J. Power Sources*, **119–121**, 789 (2003), and references therein.

2. (a) M. Matsunaga, *Electrochemistry*, **70**, 126 (2002). (b) R. Hagiwara, *Electrochemistry*, **70**, 130 (2002).

3. C. J. Dymek Jr., J. L. Williams, D. J. Groeger, *J. Electrochem. Soc.*, **131**, 2887 (1984).

4. P. R. Gifford, J. B. Palmisano, *J. Electrochem. Soc.*, **135**, 650 (1988).

5. C. J. Dymek, G. F. Reynolds, J. S. Wilkes, *J. Electrochem. Soc.*, **134**, 1658 (1987).

6. N. Koura, H. Ejiri, K. Takeishi, *J. Electrochem. Soc.*, **140**, 602 (1993).

7. P. Novak, W. Scheifele, F. Joho, O. Haas, *J. Electrochem. Soc.*, **142**, 2544 (1995).

8. J. Dvynck, R. Messina, J. Pingarron, *J. Electrochem. Soc.*, **131**, 2274 (1984).

9. Y. S. Fung, R. Q. Zhou, *J. Power Sources*, **81**, 891 (1999).

10. K. Ui, N. Koura, Y. Idemoto, K. Iizuka, *Denki Kagaku.*, **65**, 161 (1997).

11. Y. S. Fung, D. R. Zhu, *J. Electrochem. Soc.*, **149**, A319 (2002).

12. N. Koura, K. Etoh, Y. Idemoto, F. Matsumoto, *Chem. Lett.*, 1320 (2001).

13. J. Fuller, R. T. Carlin, and R. A. Osteryoung, *J. Electrochem. Soc.*, **144**, 3881 (1997).

14. A. Webber, G. E. Blomgren, in *Advances in Lithium-Ion Battery*, W. van Schlkwijk, B. Scrosati, eds., Kluwer Academic, 2002, p. 185.

15. R. T. Carlin, J. Fuller, W. K. Kuhn, M. J. Lysaght, P. C. Trulove, *J. Appl. Electrochem.*, **26**, 1147 (1996).

16. T. E. Sutto, D. M. Fox, P. C. Trulove, H. C. De Long, Meeting Abstract of 203rd ECS Meeting, Paris, Vol. 2003-1, Abstract No. 1161 (2003).

17. K. Ui, T. Minami, Y. Idemoto, N. Koura, *Meeting Abstract of the 44th Battery Symp. in Japan*, 3D08 (2003)

18. H. Matsumoto, M. Yanagida, K. Tanimoto, T. Kojima, Y. Tamiya, Y. Miyazaki, in *Molten Salts XII*, P. C.Trulove et al., eds., Electrochemistry Society, 2000, p. 186.

19. H. Nakagawa, S. Izuchi, S. Sano, K. Takeuchi, K. Yamamoto, H. Arai, *Meeting Abstract of the 41th Battery Symp. in Japan*, 3C18 (2000). H. Nakagawa, S. Izuchi, K. Kuwana, T. Nukuda, Y. Aihara,. *J. Electrochem. Soc.*, **150**, A695 (2003).

20. T. Ohzuku, A. Ueda, N. Yamamoto, *J. Electrochem. Soc.*, **142**, 1430 (1995).

21. J. Caja, T. D. Dunstan, D. M. Ryan, V. Katovic, *Molten Salts XII*, P. C. Trulove et al., eds., Electrochemistry Society, 2000, p. 150.

22. J. Caja, T. D. Dunstan, V. Katovic, *Molten Salts XIII*, edited by H. C. Delong, R. W. Bradshaw, M. Matsunaga, G. R. Stafford, P. C. Trulove, eds., Electrochemistry Society, 2002, p. 1014.

23. H. Sakaebe, H. Matsumoto, H. Kobayashi, Y. Miyazaki, *Meeting Abstract of the 42th Battery Symp. in Japan*, 2B04 (2001).

24. H. Matsumoto, Y. Miyazaki, *Chem. Lett.*, 922 (2000).

25. H. Sakaebe, H. Matsumoto, *Electrochem. Commun.*, **5**, 509 (2003).

26. D. R. MacFarlane, P. Meakin, J. Sun, N. Amini, M. Forsyth, *J. Phys. Chem. B*, **103** 4164 (1999).

27. K. Kanamura, T. Umegaki, S. Shiraishi, M. Ohashi, Z. Takehara, *J. Electrochem. Soc.*, **149**, A185 (2002).

28. X. Wang, E. Yasukawa, S. Mori, *Electrochimica Acta*, **45**, 2677 (2000).

29. H. Sakaebe, H. Matsumoto, *Meeting Abstract of the 43th Battery Symp. in Japan*, 1B16 (2002).

30. H. Sakaebe, H. Matsumoto, K. Tatsumi, *Meeting Abstract of 203rd ECS Meeting*, Paris, Vol. 2003-1, Abstract No.1159 (2003); in *New Trends in Intercalation Compounds for Energy Storage and Conversion*, C. Julien, ed., Electrochemical Society, 2003.

31. T. Fujinami, Y. Buzoujima, *J. Power Sources*, **119–121**, 438 (2003).

Chapter *15*

Application of Ionic Liquids to Photoelectrochemical Cells

Hajime Matsumoto

15.1 INTRODUCTION

Photoelectrochemical solar energy conversions have been a subject of intensive study since the discovery of the Honda-Fujishima effect [1]. In particular, the photoelectrochemical cell, denoted as the PEC cell, for solar energy conversion to electricity has attracted much attention because of the possibility of reducing the fabrication cost compared with conventional solid-state semiconductor devices. The PEC cells investigated so far are divided into two systems based on the kind of adopted photoelectrode. One is a system that uses the n- or p-type semiconductor electrode as the photoanode or photocathode [2], respectively (Figure 15.1, right). The other is a system that uses a dye-loaded semiconductor (Figure 15.1, left) [3]. The reason the cost of the PEC cell is lower than that of a solid-state solar cell is that a much higher quality of semiconductor material is not needed for the PEC cell's construction [4]. For solid-state systems, both p-type and n-type semiconductors are needed to form a Schottky-junction, which is necessary to separate the photogenerated electrons and holes. Both types of semiconductors are prepared from highly pure semiconductors by a severely controlled doping process. For PEC

Electrochemical Aspects of Ionic Liquids   Edited by Hiroyuki Ohno
ISBN 0-471-64851-5   Copyright © 2005 John Wiley & Sons, Inc.

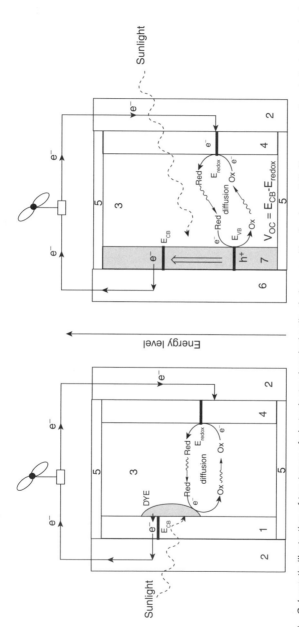

Figure 15.1 Schematic illustration of two types of photoelectrochemical cells. A dye-sensitized solar cell (left); a conventional electrochemical solar cell (right). (1) A dye-loaded porous wide gap semiconductor; (2) transparent conducting grass; (3) an electrolyte containing a redox couple, (4) Pt or carbon counter electrode, (5) a sealing material, (6) back contact, (7) a semiconductor electrode (n-type). Photogeneration of electron has occurred at the gray zone.

cells, a similiar junction can be achieved by a semiconductor electrode immersed in an electrolyte with a redox compound [4].

Although PEC cells can reduce fabrication costs, the usual problems encountered in a practical application of electrochemical devices, which are derived from the use of volatile solvents, must be resolved. The use of a liquid electrolyte containing organic solvents for such devices not only can achieve fast mobility of the active materials but also maintain good contact between the electrolyte and an electrode with a high specific surface area. The room-temperature ionic liquid, denoted as RTIL, is one of the possible candidates to resolve the problems with the use of conventional solvents due to its unique properties, such as nonvolatility and noncombustibility as described elsewhere in this book. The major problem with the use of RTILs in electrochemical energy devices is their relatively high viscosity, which limits the mobility of the electrochemically active material in RTILs such as Li^+. This problem might also limit the performance of PEC cells using ILs. Schematic illustrations of two typical PEC systems are shown in Figure 15.1. A redox compound such as ferrocene/ferricinium, iodide/triiodide, is contained in an electrolyte to provide carrier transport (Figure 15.1). The mobility of these materials limits the current density output of the PEC cells.

PEC cells constructed using a dye-loaded semiconductor electrode are called "dye-sensitized solar cells" (Figure 15.1 left). These cells have been intensively investigated since the high conversion efficiency of over 7% was achieved in 1991 by Prof. Grätzel et al. [5]. The key for such a high efficiency is the development of a novel dye that can efficiently absorb sunlight and the mesoporous titanium oxide electrode. This type of cell is called the "Grätzel cell." The highest efficiency of the systems reaches over 10 % at AM 1.5 [6]. Grätzel's colleague also discovered moisture stable ionic liquids based on the *bis*(trifluormethylsulfonyl)imide anion [7] and a low-temperature melting iodide [8]. However, a conventional PEC cell using a semiconductor electrode with a chloroaluminate system was reported in 1980 by Singh et al. [9]. Based on these reports, the problem of the relatively highly viscous RTILs seems to have been overcome by using redox hopping, or the so-called Grotthus mechanism [8].

In this chapter, PEC systems using RTILs will be introduced. The discussion will focus on the relationship between the viscosity of the RTILs and the short-circuit photocurrent and show how the problem of the viscosity of RTILs in an electrochemical device application has been resolved. Then the chapter will review recent reports on quasi-solid-state DSSCs using RTILs [23, 25, 27–33].

15.2 PARAMETERS FOR THE PERFORMANCE EVALUATION OF PEC CELLS

Figure 15.2 shows a typically observed current-voltage characteristic curve for PEC cells [4]. This curve was obtained by linear sweep voltammetry using a conventional electrochemical apparatus. The two major parameters used to estimate the PEC cell are as follows: (1) the short-circuit photocurrent, which is denoted as

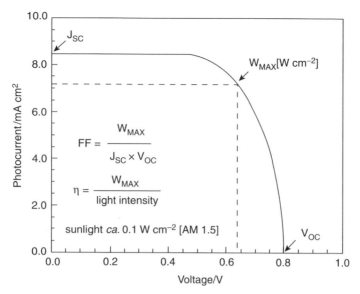

Figure 15.2 *Typically observed voltage–current characteristic curve for PEC cell. J_{SC}: a short-circuit photocurrent; V_{OC}: a open circuit photovoltage; FF: fill factor; η: conversion efficiency; W_{MAX}: a maximum output.*

J_{SC}, and (2) the open-circuit photovoltage, which is denoted as V_{OC}. J_{SC} is generally related to the mobility of the redox couple in an electrolyte. V_{OC} is basically the energy difference between the energy level of the semiconductor electrode and the redox potential using a redox couple, as shown in Figure 15.2. A fill factor, which is denoted as FF, is reflected in the overall performance of a PEC cell. The conversion efficiency is the most important parameter in terms of the practicality of the PEC cells. This review focuses on J_{SC} in order to clarify the relationship between J_{SC} and the viscosity of the RTILs.

15.3 IONIC LIQUIDS USED AS AN IN-VOLATILE SOLVENT

15.3.1 *N*-butylpyridinium Chloroaluminate Systems

In 1980 Singh et al. reported the first example of a photoelectrochemical cell using an RTIL. They employed a chloroaluminate melt based on the *N*-butylpyridinium cation, denoted as BP, as a solvent and III-V semiconductors, namely n-GaAs [9–12] or n-InP [13], as the photoanode. Ferrocene and ferricinium chloride were used as the redox couple. The chloroaluminate-based RTIL used here was prepared by mixing equimolar amounts of $AlCl_3$ and BP–Cl. The viscosity of the RTIL is 21 mPas at 25°C [14].

The PEC cell's current voltage characteristic, which indicates cell performance, is shown in Figure 15.3 [10]. Twenty years ago the short lifetime of the PEC cells

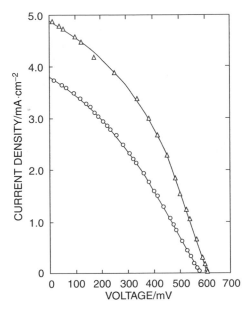

Figure 15.3 *Current–voltage characteristics of n-GaAs in BP–AlCl₄ containing 0.2 M ferrocene and 0.02 M ferricinium chloride: Single crystal n-GaAs (triangle); polycrystalline n-GaAs (circle). The light intensity in these experiments was 100 mW cm⁻² [10]. (Reprinted by permission of the Publisher, The Electrochemical Society).*

due to the photocorrosion of the semiconductor electrode was a big problem [15]. Although the performance of the PEC cell in RTILs under 1 sun (100 mW cm^{-2}) shown in Figure 15.3 was not very high ($\eta = 1.7\%$) compared with that of organic solvents, the photocorrosion of n-GaAs could be suppressed by using chloroaluminate-based RTILs for 720 hours of irradiation [10]. The addition of benzene (50 vol%), which increased J_{SC} about 1.3 times, indicated that the higher viscosity of the RTIL compared with molecular liquids limited cell performance. However, the current increase has not proved to be as high as was expected from the decrease in viscosity ($21 \rightarrow 5 \text{ mPa s}$). The chloride anion contained in the chloroaluminate melt became attached to the n-GaAs surface and acted as a surface recombination center, which is the main cause behind the decreased conversion efficiency [11].

15.3.2 Grätzel Cell with an Ionic Liquid Based on Iodide

Papageorgiou et al. reported the first example of an RTIL-based dye-sensitized solar cell system (Grätzel cell). They found that 1-hexyl-3-methylimidazolium iodide, denoted as HMI-I, melts at room temperature [8]. However, the viscosity of HMI-I is very high, at over 1000 mPa s at room temperature. Therefore the J_{SC} of DSSC when HMI-I is used (0.75 mA cm^{-2} at the irradiation of 120,000 Lux [= 1 sun]) is much lower than that when an organic solvent is used (over 15 mA cm⁻² at 1 sun). This indicates that the slow diffusion of the iodide/triiodide redox

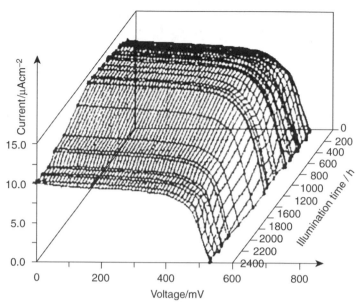

Figure 15.4 *Current–voltage characteristics of quasi-transparent, nanocrystalline, dye-sensitized solar cells operating with molten salts as the electrolyte and redox mediator (90% 1-ethyl-3-methylimidazolioumtriflate, 10% 1-hexyl-3-methylimidazoliumidodide, 5 mM I₂) measured under 0.020 AM1. The irradiation was aged under AM1 as a function of illumination time [8]. (Reprinted by permission of the Publisher, The Electrochemical Society)*

couple in HMI-I limits the overall photocurrent. However, as shown below, the J_{SC} of 0.75 mA cm^{-2} is unexpectedly high despite its high viscosity. Papageorgiou et al. also investigated the diffusion coefficient of triiodide in pure HMI-I and in a mixture of HMI-I and a relatively low RTIL such as 1-butyl-3-methyl-imidazolium triflate and an organic solvent. They applied both the Walden product and the Einstein-Stokes equation to the experimentally obtained diffusion coefficient and the dynamic viscosity, and concluded that a Grotthus-type charge carrier transfer mechanism between triiodide and iodide exists in pure HMI-I; the charge transfer is the reason for the unexpectedly high J_{SC} even in a highly viscous RTIL. They also showed the long-term stability of the DSSC system using an RTIL as shown in Figure 15.4. Over 2000 hour exposure in light (AM 1) did not seriously damage the DSSC systems. Furthermore, at under 5000 lx (= 0.04 sun), the DSSC systems using RTILs continuously operated for over 230 days. This result first demonstrated that the nonvolatile nature of RTILs can indeed undergo long-term durability for use in electrochemical devices [8].

15.3.3 Grätzel Cell Containing a Low Viscous Ionic Liquid

The hydrofluoric acid based molten salts such as quaternary ammonium cations and F(HF)$_n^-$ anions have a low viscosity and are used as an electrochemical fluorination bath [17]. Hagiwara et al. reported new RTILs with a low viscosity based on the

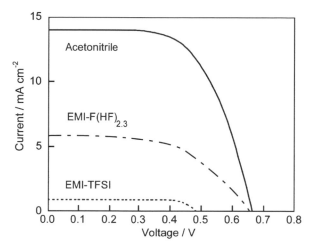

Figure 15.5 *Photocurrent–voltage characteristics of a dye-sensitized solar cell. Electrolyte composition: (ionic liquids) 0.9 M of DMHI-I and 30 mM of I_2, (acetonitrile) 0.8 M of DMHI-I, 0.1 M LiI, 50 mM I_2, 0.2 M t-butylpyridine. Temperature: 25°C. Area: 1.0 cm². AM 1.5 (100 mW cm⁻²)* [21]. *(Reprinted by permission of the Publisher, The Electrochemical Society of Japan).*

1-ethyl-3-methylimidazolium cation (EMI) and the $F(HF)_{2.3}$ anion, whose viscosity is 4.8 mPas at 25°C [18–19] and on their application in the Grätzel cell [20].

The current-voltage characteristics of DSSC using the low viscous RTILs are shown in Figure 15.5 [20–21]. Although the additives such as t-butylpyridine, which improves the performance of the DSSC cells [20], were not contained for the RTILs, this figure clearly shows that J_{SC} significantly depends on the viscosity of liquid media. Table 15.1 shows the parameters taken from Figure 15.5 and the diffusion coefficient roughly estimated by the cyclic voltammogram of the iodide redox in each liquid. As is clear from these results, the J_{SC} decreased with the decreasing diffusion coefficient of the iodide redox in the liquid media. To improve J_{SC}, which is almost the same as increasing the conversion efficiency, much less viscous RTILs than EMI–$F(HF)_{2.3}$ must be prepared. However, the preparation of much less viscous RTILs as low as acetonitrile (0.4 mPas at 25°C) may be very difficult because of the strong coulombic interaction existing between a cation and an anion.

15.3.4 Relationships between J_{SC} and Viscosity of Various Liquid Media

Figure 15.6 shows the relationship between J_{SC} of the Grätzel cell and the viscosity of various liquid media under the same conditions [21]. The J_{SC} of the DSSC system depends simply on the viscosity of the liquid media. However, as described above, the J_{SC} of iodide-based RTILs is much higher even where viscosity is higher than 200 mPas at 25°C. As shown in the figure, the amount of iodine (I_2) affects the improvement in J_{SC}. This implies that the diffusion coefficient of the iodide

TABLE 15.1 Photovoltaic Characteristics of Dye-sensitized Solar Cell in Various Ionic Liquids at 25°C (AM 1.5, 100 mW cm^{-2})

	Viscosity (mPas)	[DMHI-I][a] (M)	[I_2] (mM)	J_{sc} (mAcm^{-2})	V_{oc} (V)	Fill Factor	η[b] (%)	D(I_3^-)[c] (10^{-7} cm^2 s^{-1})	Reference
HMI-I	1800	(4.6)[d]	50	0.8	—	—	—	0.9	8
EMI-TFSI	34	0.9	30	0.9	0.50	0.80	0.36	7.6	20
EMI-F(HF)$_{2.3}$	4.8	0.9	30	5.8	0.65	0.56	2.1	43	20
Acetonitrile[e]	0.4	0.9	50	13.7	0.70	0.59	5.7	110	20

[a] 1,2-dimethyl-3-hexylimidazoliumiodide.
[b] Conversion efficiency.
[c] Diffusion coefficient of triiodide ion calculated from the reduction peak current of a cyclic voltammogram.
[d] Concentration of 1-hexyl-3-methylimidazoliumiodide.
[e] The electrolyte contained 0.1 M of LiI, and 0.2 M of t-butylpyridine.

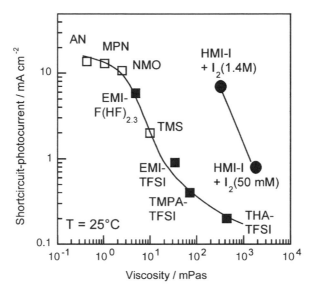

Figure 15.6 *Relationship between J_{SC} and viscosity of the redox medium containing 0.9 M DMHI-I, 50 mM iodine. Organic solvents: AN: acetonitrile; MPN: methoxypropionitrile; NMO: N-methyloxazolidinone; TMS: sulfolane. Ionic liquids: EMI: 1-ethyl-3-methylimidazolium; TMPA: trimethylpropylammonium, THA: tetrahexylammonium, HMI-I: 1-hexyl-3-methyl-imidazolium-iodide, TFSI: bis(trifluoromethyl-sulfonyl)imide [21]. (Reprinted by permission of the Publisher, The Electrochemical Society of Japan).*

redox increases with the increasing amount of triiodide even for the high viscosity case.

Kawano et al. used a microelectrode technique and the Dahms-Ruff equation to explore the diffusion mechanism of the iodide/triiodide redox couple and explain the physical diffusion accompanying an exchange reaction [22]. They showed that when high concentrations of iodide and triiodide are added to a RTIL as the solvent, the diffusion coefficient derived from the exchange reaction, expressed by $I^- + I_3^- \rightarrow I_3^- + I^-$, become significant and superior to the simple physical diffusion mechanism. This is due to the operation of a Grotthus-like mechanism that appears to enhance the diffusion coefficient of the iodide redox in RTILs based on an iodide. From these finding the iodide-based RTILs, it appears that the problematic high viscosity of the RTILs can be reduced in actual application by adoption of a Grotthus-like hopping mechanism at least for the output current density such as J_{SC}.

However, the conversion efficiency of a DSSC using RTILs does not reach the level of an organic solvent-based system. Kubo et al. studied the effects of iodide-based RTILs on the mechanism of the charge transfer involving the photocurrent generation process of a DSSC system such as electron diffusion in the TiO_2 electrode, which is a charge recombination process [23]. The existence of a very high concentration of cationic species, which is inevitable with the use of ILs, decreases the diffusion constant of the electrons in TiO_2, and the relatively high concentration of triiodide that results, which is needed to increase J_{SC}, increases the recombination loss between the photogenerated electrons and triiodide. Furthermore the triiodide acts as a filter because of its relatively high absorption coefficient in the visible light region [23—24]. All these phenomena are causes for the relatively low conversion efficiency of a DSSC using iodide-based RTILs compared with that using an organic solvent electrolyte (>10%).

15.4 QUASI-SOLID-STATE DSSC SYSTEM USING RTILs BASED ON IODIDE

The solidification of a liquid electrolyte that maintains its performance is one of the important objectives sought in producing a DSSC system for actual devices. Recently various studies on the solidification of a liquid electrolyte using iodide-based RTILs and a gelator [23, 25], cross-linked gel [26–28], fluoropolymer-based gel [29], and silica nanoparticles [30–31] have been reported. Remarkably the conversion efficiency of these cells stayed almost the same before and after the solidification of the electrolyte. Especially noteworthy is the conversion efficiency of 7.0% that was achieved with the use of silicon nanoparticles as a gelator instead of an organic gelator [30]. All these results indicate that iodide RTILs are a key material to continue to investigate for the solidification of an electrolyte in

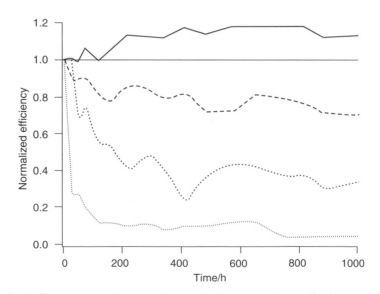

Figure 15.7 *Time-courses change under a dry heat test of normalized photoconversion efficiency of solar cells stored at 85°C, with a gelled molten salt electrolyte (solid line), a molten salt electrolyte (long dashed line), a gelled organic solvent electrolyte (short dashed line), and a organic solvent electrolyte (dotted line) [25]. (Reprinted by permission of the Publisher, The Royal Society of Chemistry).*

DSSC systems, and that through novel techniques the shortcomings in the use of RTILs can be overcome. Kubo et al. reported that the quasi-solidification with a gelator is valid for long-term stability at 85°C as shown in Figure 15.7. The decrease in the normalized efficiency of the iodide-based RTIL without a gelator was found to be due to the decrease in the triiodide and adsorbed dye onto the TiO_2 [23, 25].

Additionally CuI, a p-type semiconductor, is being investigated for use as a hole transport layer for DSSC systems. The performance of a DSSC with the use of microcrystalline CuI has been shown to be much improved after the CuI is combined with ILs [32–33]. These studies report that the ILs serve as a protecting reagent for CuI, keeping it from degradation and maintaining a particle size small enough to allow effective contact with a dye-loaded oxide electrode.

15.5 SUMMARY

In this chapter, PEC systems using RTILs were introduced. The discussion focused on the relationship between the viscosity of the RTILs and the short-circuit photocurrent. It was shown that a Grotthus-like mechanism observed in a high concentration of iodide/triiodide redox in RTILs compensates for the shortcomings of the relatively high viscosity of RTILs. This is a critical fact for the application of RTILs in actual electrochemical power devices.

REFERENCES

1. K. Honda, A. Fujishima, *Nature*, **238**, 37 (1972).
2. (a) H. Gerischer, in *Solar Energy Conversion*, B. O. Seraphin, ed., Springer, 1979, p. 115. (b) R. Memming, *Semiconductor Electrochemistry*, Wiley-VCH, 2001.
3. H. Tsubomura, M. Matsumura, Y. Nomura, T. Amamiya, *Nature*, **261**, 402 (1977).
4. Y. V. Pleskov, Y. Y. Gurevich, *Semiconductor Photoelectrochemistry*, Consultants Bureau, 1986, ch. 8.
5. B. O'Regan, M. Grätzel, *Nature*, **353**, 737 (1991).
6. M. Grätzel, *J. Photochem. Photobio. C*, **145** 4 (2003), and references therein.
7. P. Bonhôte, A. P. Dias, N. Papageorgiou, K. Kalyanasundaram, M. Grätzel, *Inorg. Chem.*, **35**, 1168 (1996).
8. N. Papageorgiou, Y. Athanassov, M. Armand, P. Bonhôte, H. Pettersson, A. Azam, M. Grätzel, *J. Electrochem Soc.*, **143**, 3099 (1996).
9. P. Singh, K. Rajeshwar, J. DuBow, R. Job, *J. Am. Chem. Soc.*, **102**, 4676 (1980).
10. P. Singh, R. Singh, K. Rajeshwar, J. DuBow, *J. Electrochem. Soc.*, **128**, 1145 (1981).
11. P. Singh, K. Rajeshwar, *J. Electrochem. Soc.*, **128**, 1724 (1981).
12. K. Rajeshwar, P. Singh, R. Thapar, *J. Electrochem. Soc.*, **128**, 1750 (1981).
13. R. Thapar, J. Du Bow, K. Rajeshwar, *J. Electrochem. Soc.*, **129**, 1145 (1982).

14. J. Robinson, R. C. Bugle, H. L. Chum, D. Koran, R. A. Osteryoung, *J. Am. Chem. Soc.*, **101**, 3776 (1979).

15. Y. V. Pleskov, Y. Y. Gurevich, *Semiconductor Photoelectrochemistry*, Consultants Bureau, 1986, p. 243.

16. M. K. Nazeeruddin, A. Kay, I. Rodico, R. Humphry-Baker, E. Muller, P. Liska, N. Vlachopoulos, M. Grätzel, *J. Am. Chem. Soc.*, **115**, 6382 (1993).

17. A. Tasaka, T. Yachi, T. Makino, K. Hamano, T. Kimura, K. Momota, *J. Fluorine Chem.*, **97**, 253 (1999), and references there in.

18. R. Hagiwara, K. Matsumoto, Y. Nakamori, T. Tsuda, Y. Ito, H. Matsumoto, K. Momota, *J.Electrochem. Soc.*, **150**, D195 (2003).

19. R. Hagiwara, T. Hirashige, T. Tsuda, Y. Ito, *J. Fluorine Chem.*, **99**, 1 (1999).

20. H. Matsumoto, T. Matsuda, T. Tsuda, R. Hagiwara, Y. Ito, Y. Miyazaki, *Chem. Lett.*, 26 (2001).

21. H. Matsumoto, T. Matsuda, *Electrochemistry*, **70**, 190 (2002).

22. R. Kawano, M. Watanabe, *Chem. Commun.*, 330 (2003).

23. W. Kubo, S. Kambe, S. Nakade, T. Kitamura, K. Hanabusa, Y. Wada, S. Yanagida, *J. Phys. Chem. B*, **107**, 4374 (2003).

24. T. Matsuda, H. Matsumoto, *Electrochemistry*, **70**, 446 (2002).

25. W. Kubo, T. Kitamura, K. Hanabusa, Y. Wada, S. Yanagida, *Chem. Commun.*, 374 (2002).

26. Y. Shibata, T. Kato, T. Kado, R. Shiratuchi, W. Takashima, K. Kaneto, S. Hayase, *Chem. Commun.*, 2730 (2003).

27. S. Murai, S. Mikoshiba, H. Sumino, S. Hayase, *J. Photochem. Photobio. A*, **148**, 33 (2002).

28. S. Murai, S. Mikoshiba, H. Sumino, T. Kato, S. Hayase, *Chem. Commun.*, 1534 (2003).

29. P. Wang, S. M. Zakeeruddin, I. Exnar, M. Gräzel, *Chem. Commun.*, 2972 (2002).

30. P. Wang, S. M. Zakeeruddin, P. Comte, I. Exnar, M. Grätzel, *J. Am. Chem. Soc.*,**125**, 1166 (2003).

31. E. Stathatos, P. Lianos, S. M. Zakeeruddin, P. Liska, M. Grätzel, *Chem. Mater.*, **15**, 1825 (2003).

32. A. Konno, G. R. A. Kumara, R. Hata, K. Tennakone, *Electrochemistry*, **70**, 432 (2002).

33. Q.-B. Meng, A. Konno A. K. Takahashi, X.-T. Zhang, I. Sutanto, T. N. Rao, O. Sato, A. Fujishima, *Langmuir*, **19**, 3572 (2003).

Chapter *16*

Fuel Cell

Masahiro Yoshizawa and Hiroyuki Ohno

We face in the multiple crises including burgeoning energy needs and continuing discharge of global warming gases. Technologies for improving these environmental problems are being studied. In particular, there is the need to develop eco-friendly energy sources that is receiving much attention worldwide. The focus is on a device that produces electricity through the redox reaction of molecules, called a fuel cell. From such a fuel cell there would be zero emission in terms of greenhouse gases. A number of motor vehicle manufacturers have undertaken the development of proton membrane fuel cells to power electric vehicle [1–4]. However, improved electrolyte materials have to be developed to obtain high-performance fuel cells. There are many problems with the present proton-transport materials in the emerging field of electric vehicle fuel cell applications. A high proton conductivity and high thermal stability are two of the critical requirements for fuel cell electrolytes. A major drawback is that in the proton membrane fuel cell the upper limit of available temperature is around 130°C because the system needs to be hydrated in order to transport protons.

Ionic liquids (ILs) are being considered more and more as alternatives for conventional electrolyte materials [5–7]. ILs offer the unique features of nonvolatility and nonflammability even in a liquid state. Systems that show ionic conductivity of over 10^{-2} S cm^{-1} at room temperature have been reported close to the level required for fuel cell applications [8–10]. However, this value is based on the IL itself, and they do not include target ions such as the proton. This is a critical subject of research on making the present system viable.

Electrochemical Aspects of Ionic Liquids   Edited by Hiroyuki Ohno
ISBN 0-471-64851-5   Copyright © 2005 John Wiley & Sons, Inc.

Proton-conducting membranes based on perfluorinated ionomer and IL were prepared by Doyle et al. [11]. Traditional proton-conducting membranes such as Nafion® suffer from the volatility of water at over 100°C. Proton conductivity therefore decreases immediately above 130°C. It was thought that ILs work as a thermal stable solvent in transporting proton because ILs are nonvolatile liquids. Indeed, the membranes containing 1-butyl-3-methylimidazolium trifluoromethane sulfonate have been demonstrated to have an ionic conductivity of above 0.1 S cm^{-1} at 180°C. Although the primary charge carrier in these membranes is not known, the high ionic conductivity has been demonstrated to be in the temperature range of 100 to 200°C when using ILs.

Recently ILs have been vigorously studied for proton transfer toward fuel cell electrolytes under water-free conditions [12–17]. These ILs are also called "Brønsted acid–base ILs" [14, 15]. Brønsted acid-base ILs can be obtained by simple coupling of a wide variety of tertiary amines with various acids. Table 16.1 gives a summary of the structures of tertiary amines and acids, thermal properties, and ionic conductivity for Brønsted acid-base ILs [12, 13, 15a]. Note that there are many systems showing T_m below 100°C and high ionic conductivity over 10^{-2} S cm^{-1} at 130°C. These have activated protons, so they can be used as proton conductors for the fuel cell electrolyte in a wide temperature range. Angell et al. found a simple relation between the boiling point (the temperature at which the total vapor pressure reaches one atm) and the difference in pKa value (ΔpKa) for the acid and base determined in dilute aqueous solutions [13]. For ΔpKa values above 10, the boiling point elevation becomes so high (>300°C) that preemptive decomposition prevents any accurate measurement. The completeness of proton transfer in such cases is suggested by the ionic liquid-like values of the Walden products.

Imidazole receives interesting interests as a proton conductive material. Imidazole, which is a self-dissociation compound, shows high proton conductivity over its T_m without any acid doping [18]. This behavior is due to the fact that imidazole has both proton donor and proton accepter groups within one molecule [19]. Recently a simple combination of imidazole and various acids has been reported as a novel proton conductive material [14, 20]. For the specific mixing ratio of imidazole and acid, this combination showed higher ionic conductivity than that of pure imidazole and of pure acid. It was thought that the maximum conductivity would be obtained when the acid-base mixing ratio is 1: 1. However, the imidazole–HTFSI mixture showed the maximum conductivity at the imidazole excess region. There are two typical mechanisms of proton conduction—namely proton hopping (Grotthuss mechanism) and matrix transport (vehicle mechanism)—in such a system. Figure 16.1 shows models of proton transfer process via the Grotthuss mechanism and the vehicle mechanism. It appears that because the Grotthuss mechanism is dominant during the proton conduction as the imidazole ratio is increased, proton conductivity should also increase in the imidazole–HTFSI mixed system.

The evaluation of ILs as electrolytes for fuel cell has already been done. Watanabe et al. tested the power generation of the hydrogen/oxygen fuel cell by using a

TABLE 16.1 Structures of tertiary amines, melting points, and ionic conductivities for tertiary amines neutralized with an equimolar amount of HTFSI

Structure of Amine	T_m (°C)	$\sigma_i \times 10^2$ (S cm^{-1}) at 130°C
(pyrrolidine structure)	35.0	3.96
(piperidine structure)	37.9	2.35
(pyridine structure)	60.3	3.04
(propylamine, NH$_2$ structure)	16.2	1.04
(dibutylamine, H–N structure)	42.6	1.26
(triethylamine-type structure)	3.5	3.23
(imidazole, N NH structure)	73.0	2.71
(pyrazole, N–N–H structure)	58.9	2.65
(benzimidazole structure)	101.9	1.31
(triazole structure)	22.8	2.20

platinum electrode for the imidazole–HTFSI mixture system, as shown in Figure 16.2 [15a, 21]. Because the oxygen reduction at the cathode was slow, early on the polarization became very large, and also in the Nafion® system. This appears to have been the first attempt to apply ILs for power generation in the absence of water molecules.

Moreover, polymer film electrolytes have been prepared by polymerization of the imidazole–acid mixtures. The copolymer was synthesized by imidazole

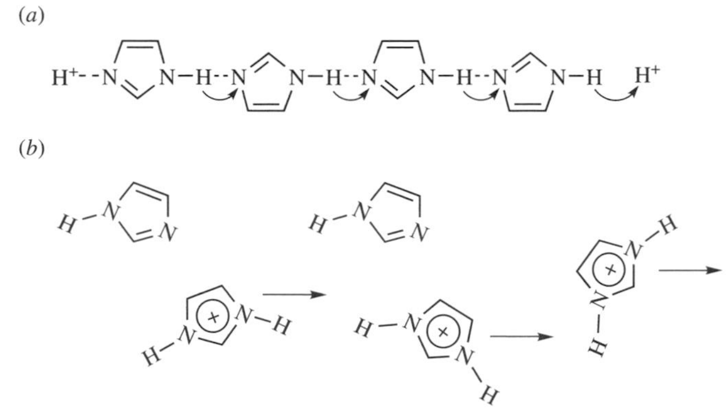

(a)

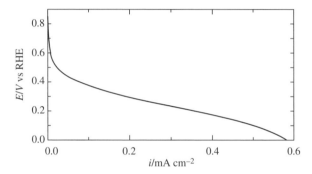

(b)

Figure 16.1 *Models of proton transfer process via (a) Grotthuss mechanism and (b) Vehicle mechanism.*

Figure 16.2 *Fuel cell characteristics for imidazole/HTFSI neutral salt at 130°C. Scan rate is 1 mV s^{-1}. W.E. is a Pt-wire in O$_2$ atmosphere, C.E. is a Pt-wire in H$_2$ atmosphere and R.E. is a Pt-wire in H$_2$ atmosphere. (Reprinted by permission of the Publisher, The Royal Society of Chemistry)*

derivatives and phosphoric acid derivatives having a polymerizable group [22]. Their ionic conductivity was very low, 10^{-8}–10^{-9} S cm^{-1} at even 100°C, which is even lower than that of monomer systems. This effect can be explained as due to the suppression of molecular motion. To overcome this effect, a flexible spacer was tethered between the imidazole and polymerizable group [23–26]. Both groups showed ionic conductivity of over 10^{-5} S cm^{-1} at 100°C, indicating that such a system can effectively maintain high proton conductivity. This result is consistent with that of IL-type polymer brushes tethering the imidazolium salt with a flexible spacer [27]. The long flexible spacer, that is, to maintain low T_g, appears to be crucial in maintaining high ionic conductivity for polymer systems and to be even more influential than the density of imidazole moieties. Other approaches for proton conductive ILs mentioned in this book should be helpful for this research too. Excellent proton conductive films can be obtained using these strategies.

REFERENCES

1. B. C. H. Steele, A. Heinzel, *Nature* **414**, 345 (2001).

2. K. D. Kreuer, *Chemphyschem*, **3**, 771 (2002).

3. K. D. Kreuer, *J. Membrane Sci.*, **185**, 29 (2001).

4. K. D. Kreuer, *Chem. Mater.*, **8**, 610 (1996).

5. H. Ohno, Ed., *Ionic Liquids: The Front and Future of Material Developments*, CMC, Tokyo, 2003.

6. V. R. Koch, C. Nanjundiah, G. B. Appetecchi, B. Scrosati, *J. Electrochem. Soc*, **142**, L116 (1995).

7. D. R. MacFarlane, J. Huang, M. Forsyth, *Nature*, **402**, 792 (1999).

8. P. Bonhôte, A.-P. Dias, M. Armand, N. Papageorgiou, K. Kalyanasundaram, M. Grätzel, *Inorg. Chem.*, **35**, 1168 (1996).

9. R. Hagiwara, Y. Ito, *J. Fluorine Chem.*, **105**, 221 (2000).

10. A. B. McEwen, H. L. Ngo, K. LeCompte, J. L. Goldman, *J. Electrochem. Soc.*, **146**, 1687 (1999).

11. M. Doyle, S. K. Choi, G. Proulx, *J. Electrochem. Soc.*, **147**, 34 (2000).

12. (a) M. Hirao, H. Sugimoto, H. Ohno, *J. Electrochem. Soc.*, **147**, 4168 (2000). (b) M. Yoshizawa, W. Ogihara, H. Ohno, *Electrochem. Solid-State Lett.*, **4**, E25 (2001). (c) H. Ohno, M. Yoshizawa, *Solid State Ionics*, **154–155**, 303 (2003).

13. M. Yoshizawa, W. Xu, C. A. Angell, *J. Am. Chem. Soc.*, **125**, 15411 (2003).

14. A. Noda, M. A. B. H. Susan, K. Kudo, S. Mitsushima, K. Hayamizu, M. Watanabe, *J. Phys. Chem. B*, **107**, 4024 (2003).

15. (a) M. A. B. H. Susan, A. Noda, S. Mitsushima, M. Watanabe, *Chem. Commun.*, 938 (2003). (b) M. A. B. H. Susan, M. Yoo, H. Nakamoto, M. Watanabe, *Chem. Lett.*, **32**, 836 (2003).

16. M. Yamada, I. Honma, *Electrochim. Acta*, **48**, 2411 (2003).

17. J. Sun, L. R. Jordan, M. Forsyth, D. R. MacFarlane, *Electrochim. Acta*, **46**, 1703 (2001).

18. K. D. Kreuer, *Solid State Ionics*, **94**, 55 (1997).

19. W. Munch, K. D. Kreuer, W. Silvestri, J. Maier, G. Seifert, *Solid State Ionics*, **145**, 437 (2001).

20. K. D. Kreuer, A. Fuchs, M. Ise, M. Spaeth, J. Maier, *Electrochim. Acta*, **43**, 1281 (1998).

21. *Chem. & Eng. News*, April 21, p. 48 (2003).

22. A. Bozkurt, W. H. Meyer, J. Gutmann, G. Wegner, *Solid State Ionics*, **164**, 169 (2003).

23. M. Schuster, W. H. Meyer, G. Wegner, H. G. Herz, M. Ise, M. Schuster, K. D. Kreuer, J. Maier, *Solid State Ionics*, **145**, 85 (2001).

24. J. C. Persson, P. Jannasch, *Chem. Mater.*, **15**, 3044 (2003).

25. H. G. Herz, K. D. Kreuer, J. Maier, G. Scharfenberger, M. F. H. Schuster, W. H. Meyer, *Electrochim. Acta*, **48**, 2165 (2003).

26. M. F. H. Schuster, W. H. Meyer, M. Schuster, K. D. Kreuer, *Chem. Mater.*, **16**, 329 (2004).

27. (a) M. Yoshizawa, H. Ohno, *Chem. Lett.*, 889 (1999). (b) M. Yoshizawa, H. Ohno, *Electrochim. Acta*, **46**, 1723 (2001). (c) S. Washiro, M. Yoshizawa, H. Nakajima, H. Ohno, *Polymer*, **45**, 1577 (2004).

Chapter *17*

Application of Ionic Liquids to Double-Layer Capacitors

Makoto Ue

17.1 INTRODUCTION

Ionic liquids are being considered for use as liquid electrolytes for electrochemical energy storage devices because they can increase device safety by their nonvolatile and nonflammable properties [1–4]. The double-layer capacitor (DLC) has shown the most promise as an electrochemical energy storage device for hybrid electric vehicles and hybrid fuel cell vehicles. This is because it has higher pulse power capability than conventional rechargeable batteries [5–7]. In this chapter the technology of applying ionic liquids to the double-layer capacitor as a liquid electrolyte is reviewed based mainly on the results of our laboratory.

17.2 OUTLINE OF DOUBLE-LAYER CAPACITORS [8]

When an electrode (electronic conductor) is contacted with an electrolyte (ionic conductor), it shows some potential and attracts ions with opposite sign, forming "electrical double-layer" at the electrode/electrolyte interface, as shown in Figure 17.1a. Increasing its electrode potential causes further adsorption of ions

Electrochemical Aspects of Ionic Liquids   Edited by Hiroyuki Ohno
ISBN 0-471-64851-5   Copyright © 2005 John Wiley & Sons, Inc.

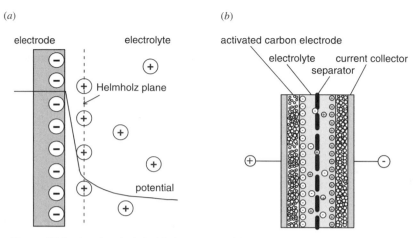

Figure 17.1 An electrical double-layer model (a) and a double-layer capacitor (b).

(charging current), and eventually an electrode reaction (electrolytic current, Faradaic current) occurs. The "polarizable electrode" is the one where charge transfer does not occur when the electrode potential is varied.

A typical example of the ideal polarizable electrode is Hg, which shows a "double-layer capacitance" of $10 \sim 20$ μF cm^{-2} in aqueous electrolyte solutions. Since the double-layer capacitance is dependent on electrode potential V, minimum values of differential double-layer capacitance C_d were adopted, as defined below, where Q is the charge:

$$C_d = \frac{dQ}{dV}. \tag{17.1}$$

Formally the capacitance C of a plane capacitor having plates of equal area S in parallel configuration, separated by a distance d, is given by equation (17.2), where ε_0 and ε_r are the permittivity of the vacuum and the relative permittivity of the dielectric material, respectively:

$$C = \frac{\varepsilon_r \varepsilon_0 S}{d}. \tag{17.2}$$

The relative permittivity ε_r of the double layer is calculated to be $3 \sim 7$, if Helmholtz model is considered, where the concept of a double-layer corresponds to a model consisting of two array layers of opposite charges separated by a small distance d (e.g., 0.3 nm).

Therefore the utilization of porous materials with high surface areas serving as electrodes enables capacitors to be made with high capacitance. For example, the use of an activated carbon (10 μF cm^{-2}) with the specific surface area of 1000 m^2 g^{-1} gives a capacitance as high as 100 F g^{-1}. It should be noted that the capacitance measured in a unit cell corresponds to a quarter of the capacitance per unit weight or volume of the single electrode (F g^{-1} or F cm^{-3}), because the cell consists of the two single electrodes (half capacitance and double weight), as shown in Figure 17.1b [9, 10]. In this case the capacitance of the unit cell becomes around 25 F g^{-1} if the weight of other components is neglected.

The double-layer capacitor is one of the "electrochemical capacitors" showing intermediate performances between conventional capacitors and rechargeable batteries from the viewpoint of energy and power densities. Although the terms "supercapacitor" and "ultracapacitor" are often used for double-layer capacitors, in a sense that they have higher capacitance than conventional capacitors (ceramic, film, aluminum electrolytic, or tantalum electrolytic capacitors), these terms are not to be used because they are the trademarks of certain companies' products.

17.3 REQUIREMENTS FOR ELECTROLYTE MATERIALS [8]

The selection of the electrolyte material is dependent on the electrode type. In principle, any ionic conductors can be used as the electrolyte material of a double-layer capacitor as long as the current due to the electrochemical reactions (Faradaic current) does not flow in the desired potential region. The electrolyte merely works as an ion source to form a double-layer.

The energy W and power P of the electrochemical capacitor when it is discharged at a constant current I from initial voltage V_i to final voltage V_f are written as

$$W = \tfrac{1}{2}C(V_i^2 - V_f^2) = \tfrac{1}{2}C[(V_0 - IR)^2 - V_f^2] \tag{17.3}$$

$$P = \tfrac{1}{2}I(V_i + V_f) = \tfrac{1}{2}I(V_0 + V_f - IR) \tag{17.4}$$

where C, V_0, and R are the capacitance, the open circuit voltage, and the internal resistance, respectively.

Both C and V_0 must be large and R must be small for the capacitors to have high energy and high power. The following performances are required for the electrolyte materials:

1. *High double-layer capacitance.* The capacitance C of a capacitor is proportional to the capacitance of an electrode, which is dependent on the type of electrolyte material chosen. An electrolyte material showing a high double-layer capacitance C_d for a given electrode is desired.

2. *High decomposition voltage.* The maximum operational voltage V_0 of a capacitor is governed by the decomposition voltage V_s of an electrolyte material. An electrolyte material having a wide stable potential region (electrochemical window) for a given electrode is desired.

3. *High electrolytic conductivity.* The internal resistance R of a capacitor leads to energy loss due to voltage drop during charge–discharge cycling. Although it includes not only the resistance of an electrolyte but also those from other

components, an electrolyte material with a high electrolytic conductivity κ is desired.

4. *Wide operational temperature range.* The electrochemical capacitor must operate in the temperature range ΔT of at least $-25°$ to $70°C$, and its performances are closely related to the temperature characteristics of the electrolyte material. An electrolyte material satisfying the above requirements 1 through 3 at a wide temperature range is desired.

5. *High safety.* One advantage of the double-layer capacitor is that it uses environmentally friendly materials in contrast to the rechargeable batteries using heavy metals. An electrolyte material with high safety and low environmental impact is desired.

For commercial double-layer capacitors using activated carbon electrodes, a nonaqueous solution such as 0.5 to 1 mol dm^{-3} ($\equiv$M) Et$_4$NBF$_4$ in propylene carbonate (PC) or an aqueous solution such as 3.7 to 4.5 M (30 $\sim$ 35 wt%) H$_2$SO$_4$ is used. The advantages of the nonaqueous liquid electrolytes are as follows:

- High decomposition voltage
- Wide operational temperature range
- Noncorrosive property being able to use components of low cost metals such as Al. (The coin or cylindrical type cell can be fabricated.)

The disadvantages are as follows:

- Low electrolytic conductivity
- Need of tight closure to isolate from atmospheric moisture
- High environmental impact
- High cost

Based on these properties, the double-layer capacitor comprising from a pair of the activated carbon electrodes and the nonaqueous liquid electrolyte is the most favorable one from the viewpoint of energy density. Ionic liquids are a kind of nonaqueous liquid electrolytes.

17.4 FUNDAMENTAL PROPERTIES OF IONIC LIQUIDS

Typical ionic liquids consist of a cation and an anion represented in Figure 17.2 [11, 12]. However, as described in the previous section, a fundamental electrochemical property such as high electrolytic conductivity is required for ionic liquids in their use as a liquid electrolyte. Many combinations of heterocyclic cations and various anions have been examined, and the ionic compounds containing the 1-ethyl-3-methylimidazolium cation (EMI$^+$) have generally showed the highest conductivity. No cation better than EMI$^+$ has been found.

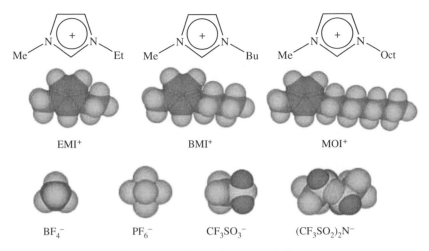

Figure 17.2 *Popular ions for ionic liquids.*

To apply ionic liquids as an electrolyte for double-layer capacitors, they should have wide operational temperature range and high safety. Although ionic liquids have favorable properties such as nonvolatility and nonflammability from the viewpoint of thermal stability and safety, they have the disadvantage that because of their higher viscosity, they rapidly lose their electrolytic conductivity as the temperature decreases.

The fundamental properties of EMI salts include the melting point T_{mp}, density d, viscosity η, electrolytic conductivity κ, and electrochemical window $E_{red}–E_{ox}$; these are collected in Table 17.1. Since the electrochemical windows were measured by different working and reference electrodes, and different sweep rates, it is not easy to compare each other. Therefore the original data obtained by platinum or glassy carbon (GC) electrodes were converted to values based on Li/Li$^+$ in PC by taking, for convenience, into account the effect of the sweep rate and the cutoff current density [35] and using the reduction potential of EMI$^+$ as a probe. The converted values, using a relation depicted in Figure 17.3, include some deviations ($\sim$0.2 V) because most reference electrodes suffer from the "solvent effect" of ionic liquids and each EMI salt shows a different equilibrium potential. The number of ionic liquids having lower melting point than 25°C is very limited.

17.4.1 Electrolytic Conductivity

The temperature dependences of electrolytic conductivity of several EMI salts are given in Figure 17.4 and compared with those of an aqueous solution (4.5 M H$_2$SO$_4$/H$_2$O) and a nonaqueous propylene carbonate solution (1 M Et$_3$MeNBF$_4$/ PC) [36–38]. Note that most ionic liquids (Figure 17.4a) show inferior conductivity compared to their aqueous and nonaqueous counterparts (Figure 17.4b) at low temperatures. This is because of their higher viscosities, and their conductivities at −20°C are less than 1 mS cm^{-1}. EMIF·2.3HF is the only one exception. It shows

TABLE 17.1   Physicochemical Properties of EMI Salts

Salt	T_{mp} (°C)	d (g cm^{-3})	η (mPa s)	κ (mS cm^{-1})	E_{red}	E_{ox} (V vs. Li/Li$^+$ in PC[c])	References
EMIAlCl$_4$	8	1.29	18	22.6	1.0[d]	5.5[d]	13–15
EMIAl$_2$Cl$_7$	nc	1.39	14	14.5	3.0[d]	5.5[d]	13–15
EMIF·HF	51	1.26					16
EMIF·2.3HF	−90	1.14	5	100	1.5	5.5	17,18
EMINO$_2$	55	1.27[a]					19
EMINO$_3$	38	1.28[a]					19
EMIBF$_4$	15	1.28	32	13.6	1.0	5.5	19–21
EMIAlF$_4$	45						22
EMIPF$_6$	62	1.56					23,24
EMIAsF$_6$	53	1.78					21,24
EMISbF$_6$	10	1.85	67	6.2	1.8	5.8	21
EMINbF$_6$	−1	1.67	49	8.5	2.1	5.7	21,25
EMITaF$_6$	2	2.17	51	7.1	1.1	5.7	21,25
EMIWF$_7$	−15	2.27	171	3.2	3.2[d]	5.7[d]	21
EMICH$_3$CO$_2$	−45		162[a]	2.8[a]			19,26
EMICF$_3$CO$_2$	−14	1.29[b]	35[a]	9.6[a]	1.1	4.6	26
EMIC$_3$F$_7$CO$_2$	nc	1.45[b]	105[a]	2.7[a]			26
EMICH$_3$SO$_3$	39	1.25	160	2.7	1.2	4.8	13
EMICF$_3$SO$_3$	−10	1.38	43	9.3	1.0	5.3	13
EMIC$_4$F$_9$SO$_3$	28						26
EMI(CF$_3$CO)(CF$_3$SO$_2$)N	−2	1.46	25	9.8	1.1[d]	5.2[d]	27,28
EMI(CF$_3$SO$_2$)$_2$N	−15	1.52	28	8.4	1.0	5.6	24,26
EMI(CF$_3$SO$_2$)(C$_2$F$_5$SO$_2$)N	nc	1.55	49	4.2			29
EMI(C$_2$F$_5$SO$_2$)$_2$N	−1		61	3.4	1.0	5.7	24,30
EMI(CF$_3$SO$_2$)$_3$C	39		181	1.7	1.0[d]	5.9[d]	31,32
EMI(CN)$_2$N	−12	1.08[a]	17[b]	27[a]	1.0	4.9	33,34
EMI(CN)$_3$C	−11	1.11[a]	18[b]	18[a]	1.0	4.4	34

Note: nc: not crystallized, 25°C ([a]20°C, [b]22°C), [c]Pt, 1 mA cm^{-2}, 50 mV s^{-1} ([d]GC).

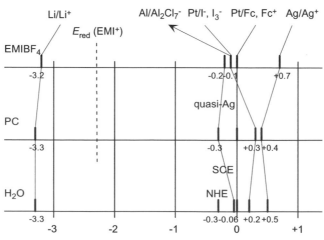

Figure 17.3   Equilibrium potentials of reference electrodes in EMIBF$_4$, PC, and H$_2$O.

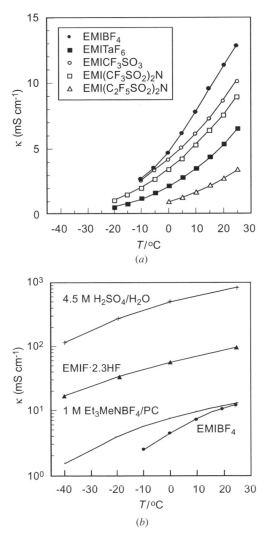

Figure 17.4 *Temperature dependences of electrolytic conductivities of ionic liquids (a) compared with conventional aqueous and nonaqueous electrolyte solutions (b). (Reproduced by permission of The Electrochemical Society, Inc.)*

very high conductivity at 25°C that is comparable to that of a neutral aqueous solution (1 M KCl/H$_2$O), and superior conductivity even at −40°C to the nonaqueous counterpart (Figure 17.4*b*) [17, 18].

17.4.2 Decomposition Voltage

It has been proved that EMI$^+$ has not only inferior cathodic stability to Et$_3$MeN$^+$ and but also inferior anodic stability to BF$_4$$^−$, from the comparison between the

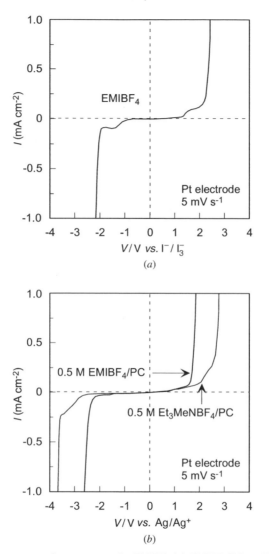

Figure 17.5   Linear sweep voltammograms for EMIBF₄ (a), EMIBF₄/PC and Et₃MeNBF₄/PC (b).

electrochemical window of EMIBF₄ (Figure 17.5a) and those of 0.5 M EMIBF₄/PC and 0.5 M Et₃MeNBF₄/PC (Figure 17.5b) [36]. This narrower electrochemical window of EMI⁺ indicates that the EMI⁺ is weak for not only reduction but also oxidation due to the existence of a π-electron conjugated system. Although most ionic liquids showed the same cathodic limit governed by the reduction of EMI⁺, EMIF·2.3HF showed a bit higher reduction potential, which is probably due to the existence of acidic protons. EMIAl₂Cl₇, EMISbF₆, EMINbF₆, and EMIWF₇ showed higher reduction potentials because of the reduction of their center atoms.

Anodic limits were usually governed by the oxidation of EMI^+; however, some organic anion salts such as $EMIRCO_2$ ($R = CH_3$, CF_3, etc.), $EMICH_3SO_3$, $EMI(CF_3CO)(CF_3SO_2)N$, and $EMI(CN)_nX$ ($X = N$, C) showed lower anodic limits because of the oxidation of their anions. It was also shown that the ionization energies of the MF_n^- type anion (inorganic fluoroanions) are larger than those of the $(CF_3SO_2)_nX^-$ anion (organic fluoroanions) by *ab initio* molecular orbital and density functional theories [35]. The following anodic stability order observed in PC solutions [39] was found to be applicable to ionic liquid systems:

$$SbF_6^- > AsF_6^- \geq PF_6^- > BF_4^- > (CF_3SO_2)_3C^- \geq (CF_3SO_2)_2N^- > (CF_3SO_2)O^-.$$

$$(17.5)$$

17.4.3 Double-Layer Capacitance

There are two reports that determined the double-layer capacitance of ionic liquids [31, 40]. By an electrocapillary curve measurement using dropping mercury electrode (DME), the integral double-layer capacitances C_d^{int} of ionic liquids were shown to be smaller than those of aqueous solutions and larger than those of non-aqueous solutions, as summarized in Table 17.2 [31]. This behavior can be explained by the thinner double-layer being due to the higher ionic concentration than that of nonaqueous solutions. However, the correlation between the double-layer capacitance and anion size [41] observed in PC solutions [8] is not clear. It was further shown that the double-layer capacitance of the ionic liquid was not dependent on the choice of electrode from among DME, GC, and activated carbon fiber [31].

TABLE 17.2 Double-Layer Capacitance on Hg

Electrolyte	C_d^{int} ($\mu F\ cm^{-2}$)
$EMIBF_4$	10.6
$EMICF_3SO_3$	12.4
$EMI(CF_3SO_2)_3C$	10.6
$EMI(CF_3SO_2)_2N$	11.7
$EMI(CF_3SO_2)_2N$	12.0^a
$EMI(CF_3SO_2)_2N$	11.4^b
1.5 M $EMI(CF_3SO_2)_2N$/PC	9.1
1M Et_4NBF_4/PC	7.0
0.1 M KCl/H_2O	15.1
3 M H_2SO_4/H_2O	14.6

[a]GC.
[b]SpectraCarb 2220 yarn.

17.5 PERFORMANCES OF IONIC LIQUIDS IN DOUBLE-LAYER CAPACITORS

A double-layer capacitor consisting of activated carbon cloth electrodes (Spectra-Carb 2220 yarn) and an ionic liquid ($EMI(CF_3SO_2)_2N$) was proposed by Covalent Associates, Inc. [42]. A moderate capacitance around 25 F g^{-1} was obtained for 1000 charge–discharge cycles at 2.0 V, but its capacitance faded rapidly at 3.0 V down to 40% after 3500 cycles, as shown in the top panel of Table 17.3. After this study the research group changed to use ionic liquids as the supporting electrolyte salt in the organic solvents, and this was probably due to the bad low temperature characteristics [24, 43–48]. $EMIAlCl_4$ was also tested for double-layer capacitor using carbon black, but the charge–discharge efficiency was very low (52 ~ 32%) because the electrolyte decomposed at a high cutoff voltage (3.5 ~ 4.0 V) and the capacity was also low, as shown in the lower panel of Table 17.3, due to inappropriate selection of carbon material [49].

To learn about the potentiality of ionic liquids, we have examined the performances of double-layer capacitors composed of a pair of activated carbon electrodes and each ionic liquid listed in Table 17.1, which we compared with the conventional nonaqueous electrolyte (1 M Et_3MeNBF_4/PC [8]) and the aqueous electrolyte (4.5 M H_2SO_4/H_2O) [36–38]. $EMIF \cdot 2.3HF$, $EMIBF_4$, $EMITaF_6$, $EMICF_3SO_3$, $EMI(CF_3SO_2)_2N$, and $EMI(C_2F_5SO_2)_2N$ were selected, because they have melting points below 25°C and enough electrolytic conductivity and electrochemical window. $EMIAlCl_4$ was excluded because of its moisture sensitivity.

17.5.1 Preparation of Ionic Liquids

$EMIBF_4$, $EMICF_3SO_3$, $EMI(CF_3SO_2)_2N$, and $EMI(C_2F_5SO_2)_2N$ were prepared by the neutralization of $EMIOCO_2CH_3$ (1-ethyl-3-methylimidazolium methylcarbonate)

TABLE 17.3 Preceding Examples of Double-Layer Capacitors Using Ionic Liquids

System	ΔV	$i(mA\,cm^{-2})$	$C(F\,g^{-1})$	
(a)			*Cycle#*	
Carbon fiber				
(BET surface area = 2000 cm² g⁻¹)	2.0	3.0	5	25.0
		(3C rate)	1000	25.0
EMI(CF₃SO₂)₂N			5000	20.8
39 mg for each electrode	3.0	3.0	5	30.5
		(1.6C rate)	1000	24.3
			3500	12.4
(b)			*T(°C)*	
Carbon black	4.0	7.9	25	3.8
(BET surface area > 2000 cm² g⁻¹)			52	6.3
			82	7.5
EMIAlCl₄ 136 mg for active components				

with the corresponding acid: HBF_4, CF_3SO_3H, $(CF_3SO_2)_2NH$, and $(C_2F_5SO_2)_2NH$, respectively. A methanol solution of $EMIOCO_2CH_3$ was prepared by the reaction between 1-ethylimidazole and dimethyl carbonate in methanol at 145°C for 13 hours [50]. $EMIF \cdot 2.3HF$ [17, 18] and $EMITaF_6$ [37] were prepared according to the methods described elsewhere.

17.5.2 Fabrication of Double-Layer Capacitors

Coconut shell charcoal (average pore diameter 2.0 nm, surface area 1700 $m^2\,g^{-1}$, average particle size 10 μm) 80 wt%, acetylene black conductor 10 wt%, and poly-tetrafluoroethylene binder 10 wt% were mixed, ground, and pressed at 6 MPa to form a disk composite 10 mm in diameter and 0.55 mm thick. A pair of these disk composite electrodes were dried at 300°C below 7.5×10^{-3} Pa for 3 hours. Afterward they were cooled in an argon atmosphere, and an electrolyte was immersed in them under reduced pressure. By sandwiching a nonwoven polypropylene separator with the two identical immersed electrodes, a 2032 coin cell was assembled with a stainless spacer, as depicted in Figure 17.6a.

The assembled cell was charged for $t = 50$ min by CC–CV mode from $V = 0$ to a given voltage (0.8, 2.0, or 2.8 V) at a constant current $I = 5$ mA, and discharged by CC mode to 0 V at a constant current $I = 5$ mA and a given temperature T as shown in Figure 17.6b. The capacitance C was calculated as $C = 2WV^{-2}$, where energy output W was calculated from the discharge curve as $W = \int IVdt$. The volumetric capacitance was obtained by dividing the capacitance above by the total volume of a pair of the disk electrodes. The internal resistance R was calculated from IR drop on the discharge curve. All data were the average values of the three same cells.

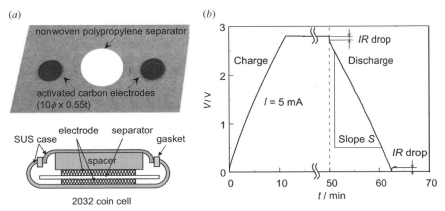

Figure 17.6 *Components and their configuration in a model cell (a) (Reproduced by permission of The Electrochemical Society, Inc.) and a charge–discharge curve (b).*

17.5.3 Initial Characteristics

Table 17.4 summarizes the performances of the double-layer capacitors using various electrolytes. To begin with, the charging voltage was set to 0.8 V in order to make a comparison with the aqueous system. The double-layer capacitor using EMIF·2.3HF showed an intermediate capacitance and internal resistance between the aqueous (4.5 M H_2SO_4/H_2O) and the nonaqueous (1 M Et_3MeNBF_4/PC) electrolytes. These results reflect the differences in double-layer capacitance and electrolytic conductivity of the electrolytes themselves. It is evident that low viscosity helps increase the capacitance and decrease the internal resistance, because the capacitance of $EMIBF_4$ was a bit smaller than that of the 1 M $Et_3MeNBF_4/$ PC and because the internal resistance of $EMIBF_4$ is larger than that of 1 M Et_3MeNBF_4/PC, even though $EMIBF_4$ has higher double-layer capacitance (Table 17.2, Et_3MeNBF_4 is equivalent to Et_4NBF_4) and the similar electrolytic conductivity compared with 1 M Et_3MeNBF_4/PC.

Then the charging voltage was raised to 2.8 V (2.0 V for EMIF·2.3HF as discussed later). The capacitances of the cells were around 13 F cm^{-3} or 19 F g^{-1}, based on the total volume or weight of the electrodes at 25°C, and no remarkable difference was observed between the ionic liquids and the 1 M Et_3MeNBF_4/PC except that EMIF·2.3HF showed almost two times capacitance of those of other ionic liquids (Table 17.4). All ionic liquids showed higher internal resistances than the 1 M Et_3MeNBF_4/PC. The capacitance and internal resistance generally increased and decreased respectively, when the temperature was raised to 75°C. The $EMICF_3SO_3$ showed the opposite tendency due to the electrolyte's decomposition. The high internal resistance of the EMIF·2.3HF system is also an indication of the electrolyte's decomposition.

It was reported by other groups that ionic liquids can show higher capacitance than Et_4NBF_4/PC solutions [51, 52]. This observation is attributed to the "ion-sieving effect" of the activated carbons because EMI^+ ($V_c = 116$ Å^3) is smaller than Et_4N^+ (159) [41]. The smaller anion also affords higher capacitance in the following order [51]: $EMIBF_4^-$ ($V_a = 49$ Å^3) > $CF_3SO_3^-$ (80) > $(CF_3SO_2)_2N^-$ (199).

TABLE 17.4 Capacitances and Internal Resistances of Double-Layer Capacitors Using Various Electrolytes

Electrolyte	κ(mS cm^{-1})	η(mPa s)	C(F cm^{-3})	R/Ω	C(F cm^{-3})	R/Ω	C(F cm^{-3})	R/Ω
		25°C	25°C	0.8 V	25°C	2.8 V	70°C	2.8 V
$EMIBF_4$	13.6	32	6.8	19.8	14.1	43	16.0	31
$EMITaF_6$	7.1	51			13.6	71	15.9	42
EMIF·2.3HF	100	4.9	11.1	6.6	24.7[a]	43[a]	36.2[a]	53[a]
$EMICF_3SO_3$	9.3	43			14.6	37	14.0	53
$EMI(CF_3SO_2)_2N$	8.4	28			13.6	39	16.1	19
$EMI(C_2F_5SO_2)_2N$	3.4	61			10.7	80	15.7	24
1M Et_3MeNBF_4/PC	13	3.5	8.1	11.8	13.1	24	13.3	18
4.5 M H_2SO_4/H_2O	848	2.5	23.2	4.6	—	—	—	—

[a]2.0 V.

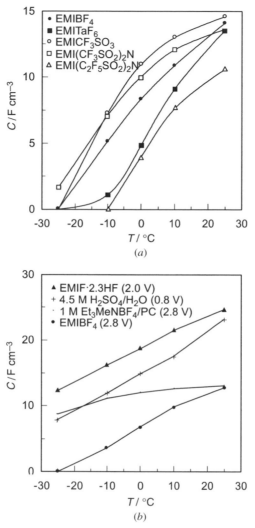

Figure 17.7 *Temperature dependences of the capacitances of double-layer capacitors using ionic liquids (a) compared with conventional aqueous and nonaqueous electrolyte solutions (b). (Reproduced by permission of The Electrochemical Society, Inc.)*

Figure 17.7 shows the change of the capacitances of the double-layer capacitors using various electrolytes, when operating temperature was varied from 25 to $-25°C$. The capacitance of all ionic liquids decreased rapidly compared with the 1 M Et_3MeNBF_4/PC due to the increase of internal resistance reflecting the temperature dependence of electrolytic conductivity in Figure 17.4. We have learned that this behavior is a fatal disadvantage of ionic liquids, and that EMIF·2.3HF is the one exception that affords enough capacitance even at $-25°C$, as reflects its high electrolytic conductivity.

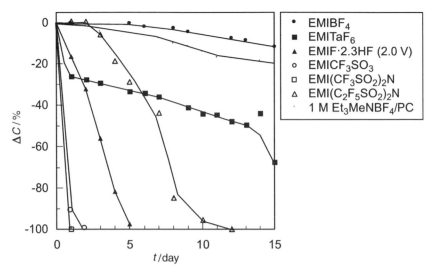

Figure 17.8 *Life test of double-layer capacitors using ionic liquids at 3 V, 70°C compared with a conventional nonaqueous electrolyte solution.*

17.5.4 Life Characteristics

Unlike for rechargeable batteries, a cycling test is not so important for double-layer capacitors, since the deterioration mostly occurs at maximum operating voltage. Figure 17.8 shows the deterioration over time when 3 V (or 2 V for EMIF· 2.3HF) is continuously applied to the cells at 70°C. The capacitance has generally deteriorated over time. The $EMIBF_4$ shows good life, similar to the 1 M Et_3MeNBF_4/PC. The $EMITaF_6$ is not as stable as the $EMIBF_4$ but more stable than any of the organic fluoroanion compounds, $EMICF_3SO_3$, $EMI(CF_3SO_2)_2N$, and $EMI(C_2F_5SO_2)_2N$. This behavior can be attributed to the inferior anodic stability of organic fluoranions [35]. The $EMIF·2.3HF$ did not have enough stability even at 2 V.

17.5.5 Voltage Dependence of Capacitance

The most important problem for the evaluation of new electrolytes is to determine the charging voltage, because a voltage that is too high will shorten the life and a voltage that is too low will not represent the maximum potential of the materials. The voltage behaviors of the double-layer capacitors using $EMIF·2.3HF$, $EMIBF_4$, and 1 M Et_3MeNBF_4/PC were examined by monitoring the charge–discharge curves at 70°C, as the applied voltages were increased from 0.8 V to a given voltage by a 0.1 V step after each charge–discharge cycling. When the voltage increase in the charge curves became sluggish, the result was an increase of internal resistance and a decrease of capacitance, as shown in Figure 17.9, accompanied by a decomposition of the electrolyte. The approximate decomposition voltages at 70°C were

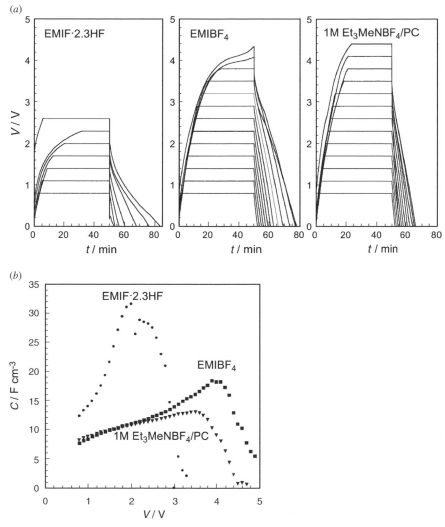

Figure 17.9 *Voltage dependences of charge-discharge curves (a) and capacitances of double-layer capacitors when applied voltages were increased from 0.8 V by a 0.1 V step at 70°C. (Reproduced by permission of The Electrochemical Society, Inc.)*

estimated to be 1.5, 3.0, and 3.2 V for EMIF·2.3HF, EMIBF$_4$, and 1 M Et$_3$MeNBF$_4$/PC, respectively. The capacitance of the double-layer capacitors was dependent on the applied voltage, particularly, for EMIF·2.3HF.

17.5.6 Energy Density

The capacitance density and energy density at 70°, 25°, and −25°C were estimated based on the experimental results where practical charge voltages were taken into

TABLE 17.5   Comparison of Energy Densities of Double-Layer Capacitors Using Various Electrolytes

Electrolyte	V/V	C/F cm^{-3}			W/Wh l^{-1}		
		70°C	25°C	−25°C	70°C	25°C	−25°C
EMIBF$_4$	2.8	16.0	14.1	0	17.4	15.3	0
1M Et$_3$MeNBF$_4$/PC	2.8	13.3	13.1	8.7	14.5	14.3	9.5
EMIF·2.3HF^a	1.5	27.2	18.5	9.2	8.5	5.8	2.9
4.5 M H$_2$SO$_4$/H$_2$O	0.8		23.2	7.9		2.1	0.7

^{a}assumed to be 75% capacitance at 2.0 V.

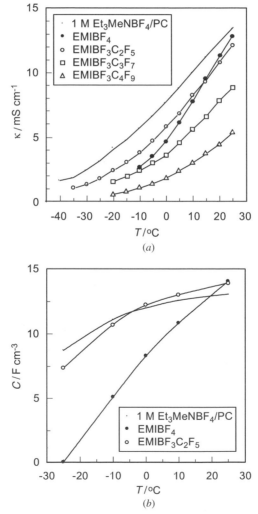

Figure 17.10   *Temperature dependences of electrolytic conductivity of a new ionic liquid (a) and capacitance of a double-layer capacitor using the ionic liquid (b).*

account. As listed in Table 17.5, $EMIBF_4$ affords higher energy density than the 1 M Et_3MeNBF_4/PC over 25°C, but it fails at −25°C. The energy density for $EMIF \cdot 2.3HF$ is situated between the aqueous and PC systems.

17.6 RECENT PROGRESS

New ionic liquids having a diethylmethyl(2-methoxyethyl)ammonium cation ($DEME^+$) have been successfully prepared to improve the intrinsic narrower electrochemical window of EMI salts. The double-layer capacitor using $DEMEBF_4$ (mp = 9°C, $\kappa = 4.8$ mS cm^{-1}) gave excellent thermal stability and worked at 150°C, where those using 1 M Et_4NBF_4/PC and $EMIBF_4$ had failed [53]. However, they still had bad low temperature characteristics, which is a common disadvantage of ionic liquids. We have succeeded in finding a new ionic liquid $EMIBF_3C_2F_5$ (mp = −1°C), whose electrolytic conductivity was comparable to the 1 M Et_3MeNBF_4/PC down to −35°C, as shown in Figure 17.10a. The double-layer capacitor using $EMIBF_3C_2F_5$ exhibited the excellent low temperature characteristics as shown in Figure 17.10b [54]. These inventions are technological breakthroughs, overcoming the low-temperature disadvantage of common ionic liquids and encourage the search of new chemical structures. Polymer electrolytes including ionic liquids were applied to the construction of the double-layer capacitors, which overcome the disadvantage of liquid electrolytes by immobilizing the ionic liquids by polymers (PAN and PEO) [55].

REFERENCES

1. J. S. Wilkes, in *Ionic Liquids in Synthesis*, P. Wasserscheid, T. Welton, eds., Wiley-VCH, 2003, p.1.

2. J. S. Wilkes, in *Green Industrial Applications of Ionic Liquids*, R. D. Rogers, K. R. Seddon, S. Volkov, eds., Kluwer Academic, 2002, p. 295.

3. A. Weber, G. E. Blomgren, in *Advances in Lithium-Ion Batteries*, W. A. van Schalkwijk, B. Scrosati, eds., Kluwer Academic 2002, p. 185.

4. H. Matsumoto, M. Ue, M. Watanabe, in *Ionic Liquids*, H. Ohno, ed., CMC, Tokyo, 2003, p. 208 (in Japanese).

5. B. E. Conway, *Electrochemical Supercapacitors: Scientific Fundamentals and Technological Applications*, Kluwer Academic, 1999.

6. A. Nishno, K. Naoi, eds., *Technologies and Materials for EDLC*, CMC, Tokyo, 1998 (in Japanese).

7. A. Nishino and K. Naoi, eds., *Technologies and Materials for EDLC II*, CMC, Tokyo, 2003 (in Japanese).

8. M. Ue, *Curr. Top. Electrochem.*, **7**, 49 (2000).

9. M. Ue, M. Takeda, *Electrochemistry*, **70**, 194 (2002) (in Japanese).

10. M. Ue, in Ref. 4, p. 232 and Ref. 7, p. 295 (in Japanese).

11. M. Ue, *Mater. Integration*, **16**, 40 (2003) (in Japanese).

12. M. Ue, *Mater. Stage*, **3** (7), 19 (2003) (in Japanese).

13. E. I. Cooper, E. J. M. O'Sullivan, in *Proc. 8th Int. Symp. Ionic Liquids*, R. J. Gale, G. Blomgren, H. Kojima, eds., PV92-16, Electrochemical Society, 1992, p. 386.

14. J. S. Wilkes, J. A. Levisky, R. A. Wilson, C. L. Hussey, *Inorg. Chem.*, **21**, 1263 (1982).

15. P. R. Gifford, J. B. Palmisano, *J. Electrochem. Soc.*, **134**, 610 (1987).

16. K. Matsumoto, T. Tsuda, R. Hagiwara, Y. Ito, O. Tamada, *Solid St. Sci.*, **4**, 23 (2002).

17. R. Hagiwara, T. Hirashige, T. Tsuda, Y. Ito, *J. Fluorine Chem.*, **99**, 1 (1999).

18. R. Hagiwara, T. Hirashige, T. Tsuda, Y. Ito, *J. Electrochem. Soc.*, **149**, D1 (2002).

19. J. S. Wilkes, M. J. Zaworotko, *J. Chem. Soc., Chem. Commun.*, **1992**, 965.

20. A. Noda, K. Hayamizu, M. Watanabe, *J. Phys. Chem. B*, **105**, 4603 (2001).

21. K. Matusmoto, R. Hagiwara, R. Yoshida, Y. Ito, Z. Mazej, P. Benkič, B. Žemva, O. Tamada, H. Yoshino, S. Matsubara, *J. Chem. Soc., Dalton Trans.*, **2004**, 144.

22. M. Ue, M. Takeda, T. Takahashi, M. Takehara, unpublished results (2001); M. Takehara, M. Ue, M. Takeda, *World Patent*, WO02101773 (2002).

23. J. Fuller, R. T. Carlin, H. C. De Long, D. Haworth, *J. Chem. Soc., Chem., Commun.*, **1994**, 299.

24. A. B. McEwen, H. L. Ngo, K. LeCompte, J. L. Goldman, *J. Electrochem. Soc.*, **146**, 1687 (1999).

25. K. Matsumoto, R. Hagiwara, Y. Ito, *J. Fluorine Chem.*, **115**, 133 (2002).

26. P. Bonhôte, A.-P. Dias, N. Papageorgiou, K. Kalyanasundaram, M. Grätzel, *Inorg. Chem.*, **35**, 1168 (1996).

27. H. Matsumoto, H. Kageyama, Y. Miyazaki, *J. Chem. Soc., Chem. Commun.*, **2002**, 1726.

28. H. Matsumoto, H. Kageyama, Y. Miyazaki, in *Molten Salts XIII*, H. C. Delong, R. W. Bradshaw, M. Matsunaga, G. R. Stafford, P. C. Trulove, eds., PV2002-19, Electrochemical Society, 2002, p. 1057.

29. H. Matusmoto, private communication (2003).

30. J. L. Goldman, A. B. McEwen, in *Selected Battery Topics*, W. R. Cieslak, ed., PV98-15, Electrochemical Society, 1999, p. 507.

31. C. Nanjundiah, S. F. McDevitt, V. R. Koch, *J. Electrochem. Soc.*, **144**, 3392 (1997).

32. A. B. McEwen, J. L. Goldman, D. Wasel, L. Hargens, in *Molten Salts XII*, H. C. De Long, S. Deki, G. R. Stafford, P. C. Trulove, eds., PV99-41, Electrochemical Society, 2000, p. 222.

33. D. R. MacFarlane, S. A. Forsyth, J. Golding, G. B. Deacon, *Green Chem.*, **4**, 444 (2002).

34. Y. Yoshida, K. Muroi, A. Otsuka, G. Saito, M. Takahashi, T. Yoko, *Inorg. Chem.*, **43**, 1458 (2004).

35. M. Ue, A. Murakami, S. Nakamura, *J. Electrochem. Soc.*, **149**, A1572 (2002).

36. M. Ue, M. Takeda, *J. Korean Electrochem. Soc.*, **5**, 192 (2002).

37. M. Ue, M. Takeda, T. Takahashi, M. Takehara, *Electrochem. Solid-State Lett.*, **5**, A119 (2002).

38. M. Ue, M. Takeda, A. Toriumi, A. Kominato, R. Hagiwara, Y. Ito, *J. Electrochem. Soc.*, **150**, A499 (2003).

39. M. Ue, M. Takeda, M. Takehara, S. Mori, *J. Electrochem. Soc.*, **144**, 2684 (1997).

40. R. J. Gale, R. A. Osteryoung, *Electrochim. Acta*, **25**, 1527 (1980).

41. M. Ue, A. Murakami, S. Nakamura, *J. Electrochem. Soc.*, **149**, A1385 (2002).

42. V. R. Koch, C. Nanjundiah, J. L. Goldman, in *Proc. 5th Int. Sem. Double Layer Capacitors and Similar Energy Storage Devices*, Florida Educational Seminars (1995); V. R. Koch, C. Nanjundiah, R. T. Carlin, *U.S. Patent*, 5827602 (1998).

43. C. Nanjundiah, J. L. Goldman, S. F. McDevitt, V. R. Koch, in *Electrochemical Capacitors II*, F. M. Delnick, D. Ingersoll, X. Andrieu, K. Naoi, eds., PV96-25, Electrochemical Society, 1997, p. 301.

44. A. B. McEwen, R. Chadha, T. Blakley, V. R. Koch, in *Electrochemical Capacitors II*, F. M. Delnick, D. Ingersoll, X. Andrieu, K. Naoi, eds., PV96-25, Electrochemical Society, 1997, p. 313.

45. A. B. McEwen, S. F. McDevitt, V. R. Koch, *J. Electrochem. Soc.*, **144**, L84 (1997).

46. A. B. McEwen, J. L. Goldman, T. Blakley, W. F. Averill, V. R. Koch, in *Batteries for Portable Applications and Electric Vehicles*, C. F. Holmes, A. R. Landgrebe, eds., PV97-18, Electrochemical Society, 1997, p. 602.

47. H. L. Ngo, A. B. McEwen, in *Selected Battery Topics*, W. R. Cieslak, ed., PV98-15, Electrochemical Society, 1999, p. 683.

48. J. L. Goldman, A. B. McEwen, *Electrochem. Solid-State Lett.*, **2**, 501 (1999).

49. J. Caja, T. D. J. Dunstan, G. Mamantov, in *Molten Salts XI*, P. C. Trulove, H. C. De Long, G. R. Stafford, S. Deki, eds., PV98-11, Electrochemical Society, 1998, p. 333.

50. M. Ue, in *Ionic Liquids* H. Ohno, ed., CMC, Tokyo, 2003, p. 9 (in Japanese); S. Mori, K. Ida, M. Ue, *U.S. Patent*, 4892944 (1998).

51. T. Kaneko, M. Watanabe, Abstracts of 43rd Battery Symp. in Japan, Fukuoka, Oct. 12 14, 2002, p. 412.

52. N. Nishina, S. Shiraishi, A. Oya, Abstracts of 2003 Fall Meeting of Electrochemical Society of Japan, Sapporo, Sept. 11–12, 2003, p. 235.

53. T. Sato, G. Masuda, K. Takagi, *Electrochim. Acta*, **49**, 3603 (2004).

54. Z.-B. Zhou, M. Takeda, M. Ue, *J. Fluorine Chem.*, **125**, 471 (2004).

55. A. Lewandowski, A. Świderska, *Solid State Ionics*, **161**, 243 (2003).

Part IV

Functional Design

Chapter *18*

Novel Fluoroanion Salts

Rika Hagiwara and Kazuhiko Matsumoto

In 1999 we developed a new room-temperature ionic liquid (RTIL), 1-ethyl-3-methylimidazolium fluorohydrogenate, $EMIm(HF)_{2.3}F$ (Figure 18.1), having a high ionic conductivity ($10^2\,mScm^{-1}$ at 298 K) [1]. Since then, we have synthesized a series of RTILs containing fluorohydrogenate ions as anionic components, and reported some fundamental studies on their physical and chemical properties [1–6] as well as applications for electrolytes of some electrochemical systems [7–9]. Recent progress in chemistry and electrochemistry of some RTIL fluorohydrogenates will be briefly reviewed in this section.

RTIL, 1-alkyl-3-methylimidazolium fluorohydrogenate, $RMIm(HF)_{2.3}F$ (R = n-alkyl group), is prepared by metathesis of starting chloride and anhydrous hydrogen fluoride:

$$R^{-N\overset{+}{\bigodot}N_{-}Me}\;Cl^{-}\;\xrightarrow[-HCl]{HF(large\;excess)\;r.t.}\;R^{-N\overset{+}{\bigodot}N_{-}Me}\;(HF)_{2.3}F^{-} \qquad (18.1)$$

The RTIL with high purity is obtained by elimination of volatile byproduct HCl with excess HF by evacuation using a rotary pump. The noninteger figure of 2.3 is observed for the HF composition of vacuum stable dialkylimidazolium fluorohydrogenates at ambient condition regardless of the type of the alkyl side-chains (n-alkyl groups) on the cations. Vibrational spectroscopies revealed that all the RTILs contain two kinds of anionic species, $(HF)_2F^-$ and $(HF)_3F^-$ shown in Figure 18.1. Since the ^1H-NMR and ^{19}F-NMR give single signals and do not distinguish the hydrogen and fluorine atoms in these two anions even at 233 K, HF

Electrochemical Aspects of Ionic Liquids Edited by Hiroyuki Ohno
ISBN 0-471-64851-5 Copyright © 2005 John Wiley & Sons, Inc.

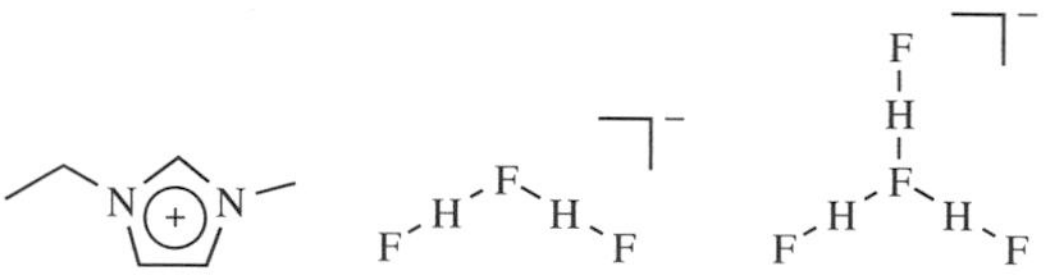

Figure 18.1 *Structure of EMIm(HF)$_{2.3}$F; EMIm$^+$, (HF)$_2$F$^-$, and (HF)$_3$F$^-$.*

exchange is considered to occur among the anions in shorter periods than the NMR resolution time scale. From the composition, the abundance ratio of these anions is roughly estimated to be 7 to 3. No free neutral HF is detectable in the liquid by spectroscopic methods. Thus EMIm(HF)$_{2.3}$F is considered to be a mixed salt composed of EMIm$^+$(HF)$_2$F$^-$ and EMIm$^+$(HF)$_3$F$^-$. There is still no clear explanation for the "magic number" 2.3. Because of the absence of free HF, these RTILs do not etch borosilicate glass container if they are thoroughly dried (Figure 18.2).

EMIm(HF)F (EMIm$^+$FHF$^-$), 1-ethyl-3-methylimidazolium bifluoride, prepared by thermal decomposition of EMImF(HF)$_{2.3}$F is a colorless crystalline solid at room temperature (Figure 18.3) [4], and it melts at 324 K. In this salt, cations form pillars by making hydrogen bonds between the ring proton and π electrons of the adjacent cation rings, as well as with the coplanar bifluoride ions. The bifluoride ions also form a pillar along the same direction as that of the cations sandwiching the alkyl side-chains of the cations as dielectric spacers. As a result these cations and anions form an interesting layered structure in which each sheet is two-dimensionally connected by hydrogen bond between the cations and anions. Judging from the separation that is close to the van der Waals distance, the interaction between the layers is weak due to the cancellation of the attractive interaction via hydrogen bond and electrostatic repulsion between the ions of the same

Figure 18.2 *EMIm(HF)$_{2.3}$F in a borosilicate glass beaker.*

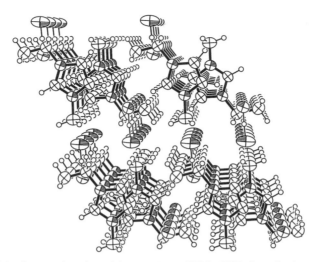

Figure 18.3 *Perspective view of the structure of EMImFHF along the b-axis direction.*

charges. High-energy X-ray diffraction studies of the liquid state fluorohydrogen-ates revealed such an ordering of the flat molecular cations and anions are partially preserved even in the liquid state [5, 6].

The reaction of EMIm(HF)F with stoichiometric amount of aHF gives EMIm(HF)$_n$F $(1 < n \leq 2.3)$ with negligibly small HF dissociation pressures at room temperature. A phase diagram of EMImF-HF has been constructed in this composition range [10]. A liquid phase is observed at higher than 298 K for the salts with $n > 1.3$. EMIm(HF)$_n$Fs with $n > 1.7$ exhibit melting points of lower than 223 K. The viscosity, and therefore conductivity, are significantly affected by the HF composition. The conductivity of EMIm(HF)$_n$F monotonously decreases with decreasing n. EMIm(HF)$_{1.3}$F exhibits approximately half the conductivity of the original EMmI(HF)$_{2.3}$F. On the other hand, the electrochemical window is about 3 V, not sensitive to the composition.

Figure 18.4 shows the temperature dependence of the conductivity of RMIm(HF)$_{2.3}$F (R = methyl, ethyl, n-propyl, n-butyl, n-pentyl, n-hexyl). In this ser-ies 1,3-dimethylimidazolium fluorohydrogenate, DMIm(HF)$_{2.3}$F, has been found to exhibit the highest conductivity (110 mScm^{-1}) at 298 K (Table 18.1). The conduc-tivities of these salts are higher than those found for the other RTILs by one order of magnitude. Extension of the alkyl side-chain of the cation increases the viscosity and decreases the conductivity. Figure 18.5 shows a Walden plot (plot of reciprocal molar conductivity against viscosity) for the alkylimidazolium RTILs including fluorohydrogenates (hatched points). Logarithmic scales are used for both the axes because the values are distributed over two orders of magnitude. The linear relation observed between these parameters suggests ionic conduction obeys Walden's rule, the conductivity being directly governed by the mobilities of the component ions in the system. Namely exceptionally high conductivities of these RTILs are explained by their low viscosities. Since protons are strongly bound to fluorine atoms in the

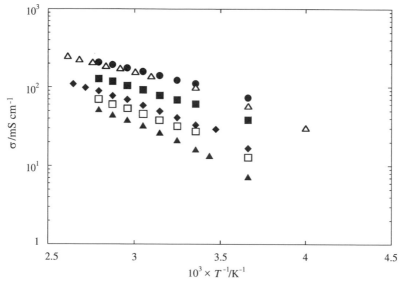

Figure 18.4 *Arrhenius plots of the conductivities of RMIm(HF)$_{2.3}$F.* ●: *DMIm(HF)$_{2.3}$F,* △: *EMIm(HF)$_{2.3}$F,* ■: *PrMIm(HF)$_{2.3}$F,* ◆: *BMIm(HF)$_{2.3}$F,* □: *PeMIm(HF)$_{2.3}$F,* ▲: *HMIm(HF)$_{2.3}$F.*

anions, the proton exchange (not HF exchange described above) between the anions which facilitates the conduction by Grottuss mechanism is ruled out in this system.

The electrochemical windows of these RTILs are 3 to 3.5 V, extension of the alkyl side-chain of the cation increasing the window mainly by the shift of cathodic limit toward negative potentials (Figure 18.6) [3]. However, anodic limits are roughly the same, fairly lower than those for the RTILs of aliphatic cations.

We have recently developed some new RTILs by combining fluorohydrogenate anions and some nonaromatic heterocyclic cations such as *N*-alkyl-*N*-methylpyrrolidinium (RMPyr) and *N*-alkyl-*N*-methylpiperidinium (RMPip) ions [10, 11].

TABLE 18.1 Some Physical Properties of RMIm(HF)$_{2.3}$F at 298 K

Salts	Molecular Weight	T_m (K)	T_g (K)	Density (g cm^{-3})	Viscosity (cP)	Conductivity (mS cm^{-1})
DMIm(HF)$_{2.3}$F	162	272	—	1.17	5.1	110
EMIm(HF)$_{2.3}$F	176	208	148	1.13	4.9	100
PrMIm(HF)$_{2.3}$F	190	—	152	1.11	7.0	61
BMIm(HF)$_{2.3}$F	204	—	154	1.08	19.6	33
PeMIrn(HF)$_{2.3}$F	218	—	158	1.05	26.7	27
HMIm(HF)$_{2.3}$F	232	—	157	1.00	25.8	16

Note: DMIm: 1,3-dimethylimidazolium; EMIm: 1-ethyl-3-methylimidazolium; PrMIm: 1-methyl-3-propylimidazolium; BMIm: 1-butyl-3-methylimidazolium; PeMIm: 1-pentyl-3-methylimidazolium; and HMIm: 1-hexyl-3-methylimidazolium.

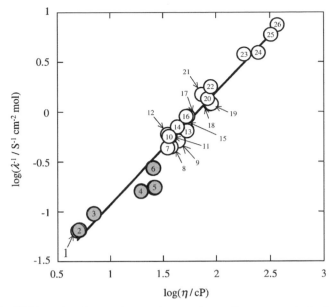

Figure 18.5 *Walden plot of room-temperature molten salts composed of alkylimidazolium cation and fluoroanions including the present salts. (1) EMIm(HF)$_{2.3}$F; (2) DMIm(HF)$_{2.3}$F; (3) PrMIm(HF)$_{2.3}$F; (4) BMIm(HF)$_{2.3}$F; (5) PeMIm(HF)$_{2.3}$F; (6) HMIm(HF)$_{2.3}$F; (7) 1,3-diethylimidazolium bis(trifluoromethylsulfonyl)amide; (8) EMImBF$_4$; (9) DMImN(SO$_2$CF$_3$)$_2$; (10) 1-ethyl-3,4-dimethylimidazolium bis(trifluoromethylsulfonyl)amide; (11) 1,3-dimethyl-4-methylimidazolium bis(trifluoromethylsulfonyl)amide; (12) EMImOCOCF$_3$; (13) 1,3-diethylimidazolium triflate; (14) 1,3-diethylimidazolium trifluoromethylcarboxylate; (15) 1-ethyl-3,4-dimethylimidazolium triflate; (16) BMImN(SO$_2$CF$_3$)$_2$; (17) 1-etoxymethyl-3-methylimidazolium bis(trifluoromethylsulfonyl)-amide; (18) 1-ethyl-2,3-dimethylimidazolium bis(trifluoromethylsulfonyl)amide; (19) BMImOSO$_2$CF$_3$; (20) 1-isobutyl-3-methylimidazolium bis(trifluoromethylsulfonyl)amide; (21) BMImOCOCF$_3$; (22) 1-butyl-3-ethylimidazolium trifluoromethylcarboxylate; (23) BMImOCOCF$_2$CF$_2$CF$_3$; (24) 1-(2,2,2-trifluoromethyl)-3-methylimidazolium bis(trifluoromethylsulfonyl)amide; (25) 1-butyl-3-ethylimidazolium perfluorobutylsulfate; (26) BMImOSO$_2$CF$_2$CF$_2$CF$_2$CF$_3$. Viscosities and conductivities of the salts are obtained in the present study or summarized in the reference [14].*

Table 18.2 shows some selected physical and electrochemical properties of them. The HF composition of vacuum stable RTILs is again 2.3, the same as in the case of the imidazolium salts described above. However, in the case of vacuum stable dimethylpyrrolidinium and dimethylpiperidinium salts, the RTILs become colorless gummy solids with the HF composition $n = 2$. The specific conductivities of these RTILs are a little lower than that for the imidazolium salts with the same alkyl side-chains on the cations, although they are still high compared to those of the RTILs combined with the other anions. Especially *N*-ethyl-*N*-methylpyrrolidinium fluorohydrogenate, EMPyr(HF)$_{2.3}$F, exhibits a remarkably high conductivity

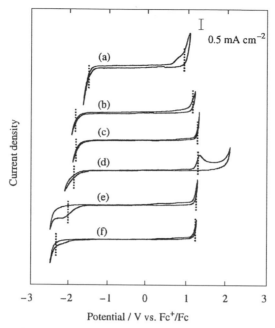

Figure 18.6 *Comparison of the electrochemical windows of RMIm(HF)$_{2.3}$F. W.E.: GC disk; C.E.: Pt plate; scanning rate: 10 mV s^{-1}. The potential is referred to the potential of ferrocene/ferrocenium redox couple in each salt. (a) DMIm(HF)$_{2.3}$F; (b) EMIm(HF)$_{2.3}$F; (c) PrMIm(HF)$_{2.3}$F; (d) BMIm(HF)$_{2.3}$F; (e) PeMIm(HF)$_{2.3}$F; (f) HMIm(HF)$_{2.3}$F. Vertical dotted lines denote cathode and anode limits.*

of 74.6 mS cm^{-1}. Figure 18.7 shows the cyclic voltammograms of these RTILs in comparison with that of EMIm(HF)$_{2.3}$Fs. Compared to the cases of RMIm(HF)$_{2.3}$Fs, the electrochemical windows are significantly extended mainly by the shifts of anodic limits toward positive potentials. A wide electrochemical window of about 5 V in addition to a high ionic conductivity would be advantageous for electrochemical application of the RTILs as electrolytes.

TABLE 18.2 Some Physical Properties of (HF)$_{2.3}$F and RMPip(HF)$_{2.3}$F at 298 K

	Molecular Weight	T_m (K)	T_g (K)	Density (g cm^{-3})	Viscosity (cP)	Conductivity (mS cm^{-1})
EMPyr(HF)$_{2.3}$F	179	<145	<145	1.07	9.9	74.6
PMPyr(HF)$_{2.3}$F	193	< 145	<145	1.05	11.2	58.1
BMPyr(HF)$_{2.3}$F	207	<145	<145	1.04	14.5	35.9
EMPip(HF)$_{2.3}$F	193	217,237	—	1.07	24.2	37.2
PMPip(HF)$_{2.3}$F	207	—	164	1.06	33.0	23.9
BMPip(HF)$_{2.3}$F	221	—	162	1.04	37.1	12.3

Note: EMPyr: *N*-ethyl-*N*-methylpyrrolidinium; PMPyr: *N*-propyl-*N*-methylpyrrolidinium; BMPyr: *N*-butyl-*N*-methylpyrrolidinium; EMPip: *N*-ethyl-*N*-methylpiperidinium; PMPip: *N*-propyl-*N*-methylpiperidinium; BMPip: *N*-butyl-*N*-methylpiperidinium.

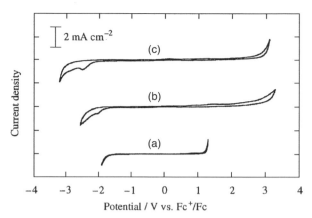

Figure 18.7 *Comparison of the electrochemical windows of (a) EMIm(HF)$_{2.3}$F; (b) EMPip(HF)$_{2.3}$F; (c) EMPyr(HF)$_{2.3}$F. W.E.: GC disk; C.E.: Pt plate or GC rod; scanning rate: 10 mV s^{-1}. The potential is referred to the potential of ferrocene / ferrocenium redox couple in each salt.*

Metathesis of a starting halide and a salt containing the anion to be combined is the most popular and simple method to prepare RTILs, but the complete elimination of the reaction by-product halide often becomes difficult due to the involatile nature of RTILs, which cannot be purified by distillation. RTIL fluorohydrogenates in the present study are synthesized with high purities by the reaction described in equation (18.1), and they also act as good fluorobases against some binary fluorides to give RTIL fluorometallates with high purities [12, 13]. Volatile HF is again easily eliminated from the reaction product.

$$\text{R}^{\curvearrowleft}\text{N}\overset{+}{\bigcirc}\text{N}^{\curvearrowright}_{\text{Me}} \ (\text{HF})_{2.3}\text{F}^- \ + \ \text{MF}_n \ \xrightarrow[-\text{HF}]{} \ \text{R}^{\curvearrowleft}\text{N}\overset{+}{\bigcirc}\text{N}^{\curvearrowright}_{\text{Me}} \ \text{MF}_{n+1}^- \qquad (18.2)$$

Large excess binary fluorides are reacted with RTIL fluorohydrogenates when they are volatile. However, stoichiometric reaction is necessary in the case of involatile fluorides. This Lewis acid–base reaction is usually highly exothermic and precautions are necessary. Table 18.3 shows some physical properties of

TABLE 18.3 Selected Physical Properties of EMIm of Fluorometallate Anion [13 and references therein]

	T_m (K)	Density (g cm^{-3})	Conductivity (mS cm^{-1})	Viscosity (cP)
EMImBF$_4$	288	1.28	13.6	32
EMImPF$_6$	333	1.56	—	—
EMImAsF$_6$	326	1.78	—	—
EMImSbF$_6$	283	1.85	6.2	67
RMImNbF$_6$	272	1.67	8.5	49
EMImTaF$_6$	275	2.17	7.1	51
EMImWF$_7$	258	2.27	3.2	171

1-ethyl-3-methylimidazolium salts prepared with this method. Physical data that have been reported for popular RTILs such as $EMImBF_4$ are somewhat scattered. For example, the highest and the lowest value for the viscosity reported so far for $EMImBF_4$ are 47 and 31.8, respectively [14]. It has been pointed out that contamination of chloride tends to increase the viscosity of the RTIL [15]. The higher values of the viscosity reported for $EMImBF_4$ might be caused by the contamination of chloride from the residual reactant and the reaction by-product in the synthesis by metathesis. It has been found that 1-ethyl-3-methylimidazolium salts of hexa-fluorometallates of group 15 elements (P, As, Sb) are isostructural in the solid-state, $EMImPF_6$-type structure (Figure 18.8). Among them only the SbF_6 salt possesses the melting point of lower than room temperature. The EMIm cation is stable against SbF_6^-, but SbF_5 oxidizes the cation. Moreover, if the excess SbF_5 is supplied, contamination of the salt of dimeric anion, $Sb_2F_{11}^-$, occurs. Therefore conventional metathesis might be exceptionally superior to synthesize $EMImSbF_6$ from the viewpoint of purity. The hexafluorometallate salts of group 5 transition metals (Nb, Ta) have been first synthesized by the reaction of $EMIm(HF)_{2.3}F$ and solid pentafluorides. Almost the same atomic sizes of these elements give similar physical properties of these RTILs in Table 18.3 except their densities, which are of course due to the difference of mass. $EMImWF_7$ is the first example in which hepta-coordinated tungsten species are present in the liquid state. Vibrational spectra resemble those found for solid $CsWF_7$ in which the geometry of the anion has been crystallographically determined to be C_{3v} (monocapped trigonal antiprism) [16].

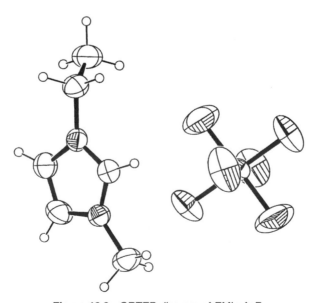

Figure 18.8   *ORTEP diagram of EMImAsF$_6$.*

REFERENCES

1. R. Hagiwara, T. Hirashige, T. Tsuda, Y. Ito, *J. Fluor. Chem.*, **99**, 1 (1999).

2. R. Hagiwara, T. Hirashige, T. Tsuda, Y. Ito, *J. Electrochem. Soc.*, **149**, D1 (2002).

3. R. Hagiwara, K. Matsumoto, Y. Nakamori, T. Tsuda, Y. Ito, H. Matsumoto, K. Momota, *J. Electrochem. Soc.*, **150**, D195 (2003).

4. K. Matsumoto, R. Hagiwara, T. Tsuda, Y. Ito, O. Tamada, *Solid State Sci.*, **4**, 23 (2002).

5. R. Hagiwara, K. Matsumoto, T. Tsuda, Y. Ito, S. Kohara, K. Suzuya, H. Matsumoto, Y. Miyazaki, *J. Noncryst. Solids*, **312–314**, 414 (2002).

6. R. Hagiwara, T. Tsuda, Y. Ito, S. Kohara, K. Suzuya, *Nucl. Inst. Meth. Phys. Res. B*, **199**, 29 (2003).

7. H. Matsumoto, T. Matsuda, T. Tsuda, R. Hagiwara, Y. Ito, Y. Miyazaki, *Chem. Lett.*, 26 (2001).

8. T. Tsuda, T. Nohira, Y. Nakamori, K. Matsumoto, R. Hagiwara, Y. Ito, *Solid State Ionics*, **149**, 295 (2002).

9. M. Ue, M. Takeda, A. Toriumi, A. Kominato, R. Hagiwara, Y. Ito, *J. Electrochem. Soc.*, **150**, A499 (2003).

10. K. Matsumoto, *Ph.D. Thesis*, Kyoto University, 2004.

11. K. Matsumoto, R. Hagiwara, Y. Ito, *Electrochem. Solid State Lett.*, **7**, E41 (2004).

12. K. Matsumoto, R. Hagiwara, Y. Ito, *J. Fluor. Chem.*, **115**, 133 (2002).

13. K. Matsumoto, R. Hagiwara, R. Yoshida, Y. Ito, Z. Mazej, P. Benkič, B. Žemva, O. Tamada, H. Yoshino, S. Matsubara, *Dalton Trans.*, 144 (2004).

14. R. Hagiwara, *Electrochemistry*, **70**, 130 (2002), and references therein.

15. K. R. Seddon, A. Stark, M.-J. Torres, *Pure Appl. Chem.* **72**, 2275 (2000).

16. S. Giese, K. Seppelt, *Angew. Chem., Int. Ed. Engl.*, **33**, 461 (1994).

Chapter *19*

Neutralized Amines

Hiroyuki Ohno

19.1 REQUIREMENT OF EASY PREPARATION

It is important to select for salts suitable ion species with excellent properties. However, the ions must be counter ions that are oppositely charged in ordinary conditions, which makes it impossible to couple a suitable cation and anion. Therefore the salt mix must contain the suitable cation and anion. Then the target salts are isolated from the mixture. Because the solubility (or miscibility) of different salts in molten salts must be taken into account, it is usually difficult to isolate the target salts. Of course, there are some salts that can easily be separated. For example, hydrophobic salts such as some *bis*(trifluoromethanesulfonyl)imide (TFSI) salts can be separated from hydrophilic salts by washing the salt with water. Some other systems are known to separate with fairly good yield. It should be noted here that there is no systematic method to use in preparing any kind of salts from a mixture. This is still a serious problem of ionic liquid preparation, especially for the scientists who wish to create (or find) new ionic liquids.

Salt synthesis is a helpful development for solving this problem. Acid esters are foremost among the salts used for this purpose. Tertiary amines are reacted with acid esters to introduce an alkyl group onto the nitrogen atom and prepare the onium cation. The reacted acid then remains as the counter anion. This method is well covered in several other books on ionic liquids [1]. Several acid esters are known to be effective for direct preparation. Recently even TFSI methyl ester was reported as an effective reagent for the preparation of ionic liquid containing TFSI

Electrochemical Aspects of Ionic Liquids Edited by Hiroyuki Ohno
ISBN 0-471-64851-5 Copyright © 2005 John Wiley & Sons, Inc.

anion [2]. Although there are still lots of anions not suitable for this method, there is strong interest voiced for an easy preparation method of the various ionic liquids.

19.2   NEUTRALIZATION METHOD

It is well known that amines can be neutralized with acids to generate salts. This neutralization process is useful in preparing organic salts with very low melting points. In 1914 Walden reported [3] neutralized ethylamine with nitric acid to prepare a salt that had a melting point of 12°C. This appears to have been the first ionic liquid prepared by a neutralization method [4]. As seen in equation (19.1), there is a slight difference between onium cations prepared by quaternization (a) and neutralization (b) when tertiary amines are used as a starting material:

$$\tag{19.1}$$

This neutralization reaction involves one step and generates nothing. The tertiary amine is dissolved in a suitable solvent like ethyl alcohol. Acid is slowly mixed into the ethyl alcohol solution. At the initial stage it is better not to heat the solution to avoid esterification between ethyl alcohol and the acid. The reaction mixture is evaporated after suitable mixing at room temperature or on an ice bath. The concentrated solution is further dried using an evaporator. Then dehydrated diethyl ether is added to this to wash the salts. The phase separated liquid or solid is further dried in vacuo for two days. The solvent and reaction conditions may vary, depending on the components. The objective is to obtain equimolar mixing, so any excess component is eliminated by evaporation. Because this method can easily be performed under mild conditions, it is the prefered way to prepare polymerizable ionic liquids, meaning ions containing vinyl groups. These polymerizable units are sensitive to heating, and undesired polymerization can occur during preparation of the monomer. Some examples can be found in Chapters 29 to 31.

Table 19.1 shows the characteristics of several amines neutralized with HBF_4. The BF_4 salts cannot be prepared by the acid ester method and are water soluble. Also very pure BF_4 salts are extremely difficult to prepare. However, it is quite easy to get salts by neutralizing the tertiary amines with HBF_4.

The amines neutralized with HBF_4 are stable. They show no weight loss up to 250°C, and the decomposition temperature depends on the starting tertiary amines. This is a remarkable improvement in stability. The melting point (T_m) of neutralized amines is not a simple function of the molar mass of the starting amines, however. The T_m of the starting amines can affect the T_m of the salts. Figure 19.1 shows

TABLE 19.1 Thermal Behavior of Amines Neutralized with HBF$_4$

			Neutralized Amines	
Number	Starting Amines	T_m (°C)	σ_i (S cm^{-1}) at 50°C	σ_i (S cm^{-1}) at 25°C
1	2-Methyl-1-pyrroline	17.1	2.7×10^{-2}	1.6×10^{-2}
2	1-Ethyl-2-phenylindole	29.8	1.6×10^{-2}	8.9×10^{-3}
3	1,2-Dimethylindole	24.5	1.1×10^{-2}	4.3×10^{-3}
4	1-Ethylcarbazole	—	5.1×10^{-3}	2.2×10^{-3}
5	2,4-Lutidine	34.1	5.9×10^{-4}	2.3×10^{-4}
6	2,3-Lutidine	59.4	8.0×10^{-5}	5.9×10^{-6}
7	3,4-Lutidine	45.9	5.3×10^{-5}	3.6×10^{-6}
8	2,6-Lutidine	104.6	1.8×10^{-7}	1.6×10^{-8}
9	N,N'-Dimethylcyclohexylamine	89.0	9.8×10^{-8}	7.3×10^{-9}
10	N,N'-Dimethylcyclohexanmethylamine	143.8	1.3×10^{-8}	—
11	1-Methylindole	—	2.0×10^{-9}	—
12	1-Methylpyrrole	—	1.7×10^{-9}	1.3×10^{-2}
13	1-Methylpyrazole	−5.9	3.5×10^{-2}	1.9×10^{-2}
14	1-Methylbenzimidazole	99.9	1.3×10^{-8}	—
15	2,3-Dimethylindole	—	2.0×10^{-9}	—
16	2-Methylindole	131.0	1.6×10^{-9}	—
17	Pyrrole	—	3.4×10^{-9}	1.4×10^{-9}
18	Carbazole	—	1.3×10^{-9}	—
19	1-Methylimidazole	36.9	2.8×10^{-3}	3.1×10^{-4}
20	1-Ethylpiperidine	164	5.9×10^{-4}	9.4×10^{-5}
21	1-Methylpyrrolidine	−31.9	2.5×10^{-2}	1.6×10^{-2}

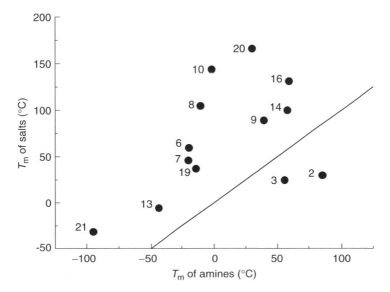

Figure 19.1 Comparison of melting points for starting tertiary amines and ionic liquids obtained by neutralization.

the relation between T_m of the starting amines and corresponding salts. If the T_m of the starting amine dominates the T_m of the salt, a straight line, or parallel lines, are obtained as shown in Figure 19.1. There seems to be some relation between these two T_m's. Therefore, for the preparation of neutralized amines with a low T_m, it is better to use starting amines with a low T_m. It should be mentioned that there are nevertheless some exceptions.

Among these salts, there are two interesting cases, cases 2 and 3 in Figure 19.1. These two salts showed T_m's lower than that of the starting amines. Generally, the T_m of an amine rises considerably after neutralization. The extraordinary drop in T_m evident in cases 2 and 3 cannot be attributed to the starting amine, indole. Further, although these two salts are formed by indole derivatives, other indole derivatives, such as 11, 15, and 16, do not show such an unusual characteristic. A comparison of cases 3 and 16 shows the only difference to be in the N-substituted methyl group. The introduction of a methyl group lowered the T_m by more than 100°C. In contrast, the introduction of an alkyl group is not always effective in lowering the T_m. The structure-property relationships of organic ionic liquids are more complicated. A study on these relationship after the amines are neutralized should soon provide some helpful information for their functional design.

19.3   EFFECT OF ION SPECIES ON THE PROPERTIES OF NEUTRALIZED AMINES

An ion species has a considerable effect on the characteristics of salts as well as on ordinary ionic liquids. The series of amines shown in Table 19.1 was neutralized with HBF_4. The salts, of course, showed considerable T_m differences upon being treated by the acids. For example, the T_m of N-methylimidazole neutralized with HBF_4 (case 19) is 37°C, whereas that of N-methylimidazole neutralized with hydrochloric acid or nitric acid is around 70°C. That of triflate salt showed a T_g at 83.5°C. Clearly, it is important to select a suitable acid in preparing neutralized amines with relatively low T_m's. Although the ability of acids to give excellent salts is roughly understood, there are many exceptions. For example, if instead of N-methylimidazole, N-ethylimidazole is neutralized with HBF_4, there is no change in T_m, whereas when it is neutralized with nitric acid, the T_m is 31°C. Triflate salts show a T_m at 8°C, whereas hydrochloric acid salts show a T_m at 57°C. These and lots of other exceptions prevent us from establishing a reliable rule on the structure-property relationship of neutralized amines. Since the neutralization is a simple process and not difficult to assess, many case studies have been carried out to attempt at summarizing the basic findings that might lead to a generalization about the amine–salt relationship.

19.4   IONIC CONDUCTIVITY

Some neutralized amines are obtained as liquid. Analyses of their ionic conductivity results have provided some interesting. Figure 19.2 shows the Arrhenius plots of

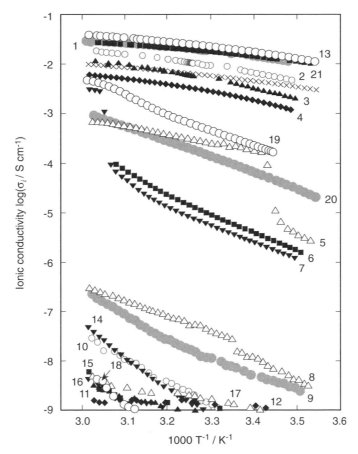

Figure 19.2 *Arrhenius plots of the ionic conductivity for neutralized amines shown in Table 19.1.*

the ionic conductivity of neutralized salts shown in Table 19.1. Generally, ionic conductivity is governed by the mobility of dissociated carrier ions. T_g is used to evaluate the mobility of the matrix. A matrix with low T_g is usually found to have high segmental motion at ambient temperature.

Note that T_g's of neutralized amines in Table 19.2 are quite low—even below $-80°C$. Accordingly, these salts generally show high ionic conductivity. Some of the obtained salts show excellent ionic conductivity around $10^{-2}\mathrm{Scm}^{-1}$ at 25°C. Figure 19.2 provides the Arrhenius plots of ionic conductivity for these neutralized salts. The ionic conductivity of the salts of cases 1, 2, 3, 4, 13, and 21 is very high, which can be expected from their T_m (and T_g, data not shown here). Among these salts, case 21 shows high ionic conductivity over $10^{-2}\mathrm{Scm}^{-1}$ at room temperature because of its very low melting point. These salts keep their ionic conductivity high at low temperatures, as can be seen in Figure 19.1. Some salts such as case 5, show

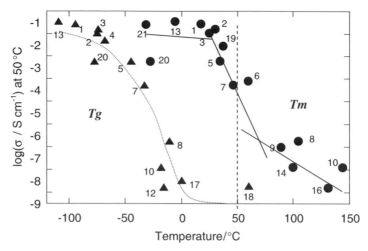

Figure 19.3 *Relation between thermal behavior and the ionic conductivity for neutralized amines.*

considerable conductivity drop at T_m because of a phase change. Despite the moderate T_m at room temperature, a drop of ionic conductivity due to a phase change was not observed for the other salts. This can be explained by a supercooling phenomenon that takes place even below the melting points of salts.

Figure 19.3 shows the relationships between ionic conductivity and the T_g or T_m of a series of neutralized amines. The salts having lower T_g's and T_m's show higher ionic conductivity. The T_m is actually just one factor that governs the motion of ions, so care must be taken in describing this kind of relationship. Since the ionic conductivity plotted in this figure was measured at 50°C, it dropped at around 50°C. Otherwise, there is no particular meaning to the 50°C. The deviations in the plots can be understood to be due to the effect of supercooling. In the relationship between T_g and ionic conductivity, a very clear decrease of ionic conductivity was observed in these neutralized salts with the increasing T_g. This is a clear indication that the ionic conductivity of ionic liquids is mainly governed by the glass transition temperature. A detailed discussion on the effect of T_g on the ionic conductivity can be found in a published paper [5].

19.5 MODEL SYSTEMS FOR ORDINARY QUATERNIZED ONIUM SALTS

Although neutralized amines are useful as models of onium type salts, they are not classified among typical ionic liquids. We have already seen that there is a similarity between the thermal properties and ionic conductivities of neutralized amines and quaternized onium salts [5]. Here we investigate the relationship of ionic conductivity between neutralized and alkylated imidazolium ionic liquids having BF_4^-.

Recall that the ionic conductivity of case 19 in Figure 19.1 was high, 2.8×10^{-3} Scm^{-1} at 50°C. Likewise alkylated imidazolium salts have high ionic conductivities at temperatures above their melting points. The melting point and ionic conductivity of 1-ethyl-3-methylimidazolium tetrafluoroborate (emiBF$_4$), a typical room-temperature ionic liquid, is 15°C and 1.4×10^{-2} Scm^{-1} at 25°C [6]. To test this relationship, we prepared a series of dialkylated imidazolium salts and compared their ionic conductivity with that of neutralized amines having the same structure except for the lack of one alkyl chain. We found a certain relation in their ionic conductivity, which indicated that the neutralized amines are good models for quaternized onium salts of the same structure. Therefore alkylated amine salts, such as those of cases 1, 2, 3, 4, 13, and 21, should form ionic liquids and show high ionic conductivity at room temperature. Already alkylated pyrrolidinium salts having (not BF$_4^-$ but) TFSI$^-$ as the counter anion have been reported as room-temperature molten salts with high ionic conductivity [7]. Our method proved to be convenient for preparing a series of neutralized amines that allowed us to select suitable amines. If these amines are quaternized with alkylating agents, they will form ionic liquids that have excellent properties.

19.6 CONCLUSION

Stable salts can be synthesized that have very low melting points for electrochemical and other applications. The neutralization method introduced in this chapter is an easy way to prepare pure salts and to use them in learning about the structure-property relationships of salts. It has been shown that excellent onium salts can be obtained from the results of neutralized amines. Besides serving as sources of quaternized onium salts, neutralized amines have excellent properties. Some of these are unique and not retained by ordinary onium salts. More vigorous studies on neutralized amines can be expected to be published in the near future.

REFERENCES

1. R. D. Rogers, K. R. Seddon, eds., *Ionic Liquids—Industrial applications for green chemistry*, ACS Symposium Ser., 818, 2002; P. Wasserscheid, T. Welton eds., *Ionic Liquids in Synthesis*, Wiley-VCH, 2003.
2. J. Zhang, G. Robert, D. D. DesMarteau, *Chem. Commun.*, 2334 (2003).
3. P. Walden, *Bull. Acad. Imper. Sci.*, (St. Petersburg) 1800 (1914).
4. T. Welton, *Chem. Rev.*, **99**, 2071 (1999).
5. M. Hirao, H. Sugimoto, H. Ohno, *J. Electrochem. Soc.*, **147**, 4168 (2000).
6. J. Fuller, R. T. Carlin, R. A. Osteryoung, *J. Electrochem. Soc.*, **144**, 3881 (1997).
7. D. R. MacFarlane, P. Meakin, J. Sun, N. Amini, M. Forsyth, *J. Phys. Chem. B*, **103**, 4146 (1999).

Chapter *20*

Zwitterionic Liquids

Masahiro Yoshizawa, Asako Narita, and Hiroyuki Ohno

Molten salts at room temperature, so-called ionic liquids [1, 2], attracting the attention of many researchers because of their excellent properties, such as high ion content, liquid-state over a wide temperature range, low viscosity, nonvolatility, nonflammability, and high ionic conductivity. The current literature on these unique salts can be divided into two areas of research: neoteric solvents as environmentally benign reaction media [3–7], and electrolyte solutions for electrochemical applications, for example, in the lithium-ion battery [8–12], fuel cell [13–15], solar cell [16–18], and capacitor [19–21].

Most electrochemical systems require target carrier ions such as the lithium cation, the proton, or the iodide anion for the construction of corresponding cells. In other words, a matrix that predominantly transports these target ions is required for electrochemical applications. Although ionic liquids show excellent ionic conductivity of over 10^{-2} S cm^{-1} at room temperature, this derives from the migration of component ions [22–24]. These ions are mostly useless as target ions. Even when target ions were added to the ionic liquid, the ionic liquid component ions still migrated along the potential gradient. Further addition of other salts induced increase in both T_g and viscosity, so ionic conductivity was considerably reduced [9, 11].

In recent years some designs of ionic liquids in which the component ions cannot migrate along the potential gradient have been tried [25–27]. One such design is the zwitterion structure in which both cation and anion are tethered [25]. Since a zwitterion has both a cation and an anion, it is expected not to migrate even under the potential gradient. There are a few reports on zwitterions as ion conductive matrices [25, 28–30]. Although the zwitterions serve as an insulator, they show ionic

Electrochemical Aspects of Ionic Liquids   Edited by Hiroyuki Ohno
ISBN 0-471-64851-5   Copyright © 2005 John Wiley & Sons, Inc.

conductivity of 10^{-5}–10^{-7} S cm^{-1} at room temperature when alkali metal salts are added. Of course, their ionic conductivity depends on the structure and properties of both cation and anion in the zwitterion and also those of the added salts. Although there are major efforts to investigate ion conductive poly(zwitterion)s, the analysis of the properties of monomeric zwitterions is far less developed. Details on the polymerized zwitterions can be found in Chapter 30. This chapter covers the relation between structure of zwitterions and both thermal properties and their ionic conductivity after alkali metal salts are added.

Schemes 20.1–20.3 show the synthetic routes of zwitterions with different anion structure [31–34]. As shown in Scheme 20.1, zwitterions are obtained

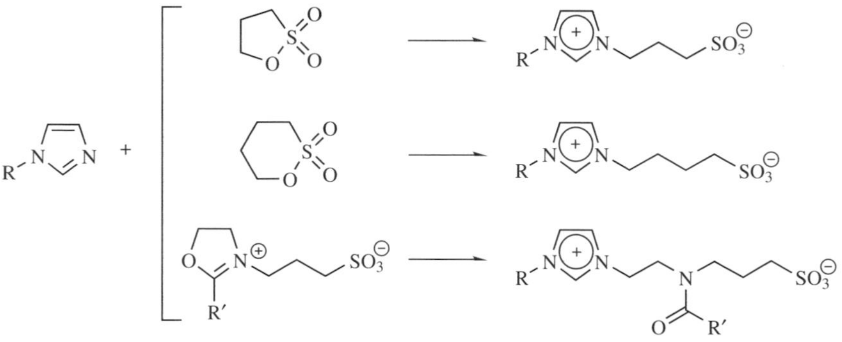

Scheme 20.1 *Synthesis of zwitterions having sulfonate anion.*

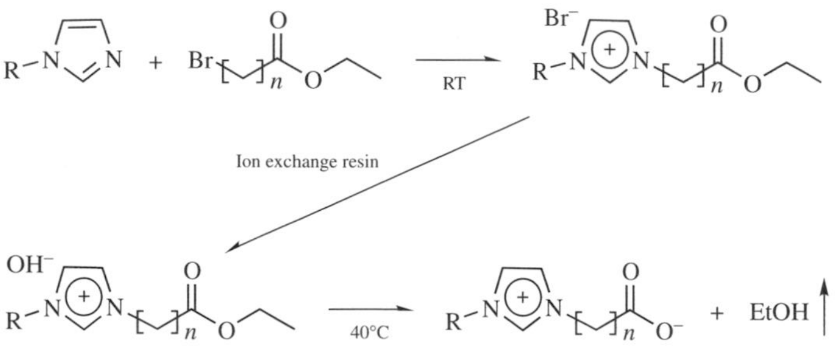

Scheme 20.2 *Synthesis of zwitterions having carboxylate anion.*

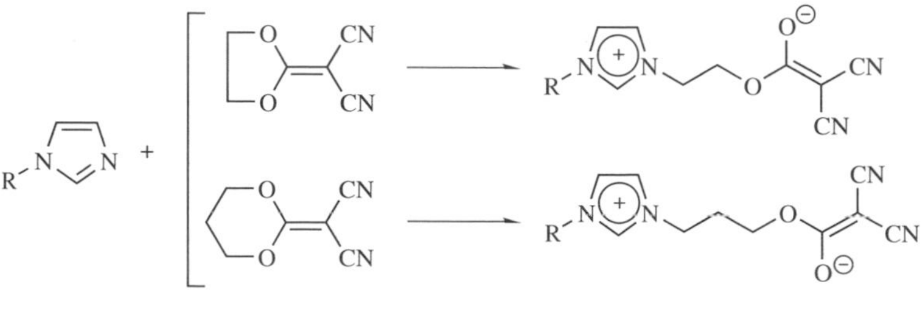

Scheme 20.3 *Synthesis of zwitterions having dicyanoethenolate anion.*

by one-step reaction of a tertiary amine and alkylsultone. This reaction gives no by-products, so the purification is very simple and there is no contamination of the zwitterions by small ions. This one-pot procedure is useful for the preparation of pure zwitterions as well as the ionic liquids prepared by neutralization of tertiary amines and organic acids [23]. In addition the zwitterions having various counter anions, such as carboxylate and dicyanoethenolate, have been prepared, as shown in Schemes 20.2 and 20.3. It is important to change the anion structure of zwitterions in studying the relationship between the zwitterion structure and its properties, and the optimization of their properties.

Table 20.1 summarizes melting point (T_m) of zwitterions having a sulfonate group as the counter anion.

Most zwitterions melt above 100°C, indicating an increase in the T_m after the tethering of cation with anion. Some decomposed before melting, namely zwitterions **1** and **2**. A considerable increase in T_m would be due to an decrease in the motion of each ion by tethering. Moreover the intermolecular electrostatic interaction between oppositely charged groups was observed in a single crystal of trimethylammoniopropane sulfonate, as shown in Figure 20.1 [35].

All these factors helped elevate the T_m. In the cases having sulfonate group as the counter anion, zwitterions **4** and **11** show relatively low T_m compared with other zwitterions. Since zwitterion **4** has a branched spacer structure, it can obstruct the intermolecular ionic association. In the case of zwitterion **11**, the butylimidazolium cation also has a long side-chain, which can prevent the zwitterion from dense packing. However, zwitterions **13** and **15**, which have methyl group at the 2-position on the imidazolium cation ring, show higher T_m than those zwitterions **12** and **14**, respectively. Such tendency agrees with that of simple ionic liquids, namely the steric effect of 2-methyl group.

Figures 20.2 and 20.3 show the DSC charts of zwitterions **9** and **10**. In the first heating scan, only T_m was observed in both systems. In the second heating for zwitterion **9**, the glass transition temperature (T_g) crystallization point (T_c) and T_m were found at 22°C, 91°C, and 177°C. However, zwitterions **10** also shows T_g, T_c, and T_m at 14°C, 134°C, and 175°C in the second heating process. According to these data, zwitterions are strongly indicated to easily become amorphous. Where the T_g of a zwitterion is around 20°C, it is still solid at room temperature. It is thus necessary to analyze the effective and controllable factors that lead to a lowering of the T_m of zwitterions.

The effect of anion structure on the thermal properties of the zwitterions is a good place to begin. Table 20.2 summarizes the T_m of zwitterions having carboxylate anion. To analyze the effect of the anion structure on the T_m of zwitterions, we compared those having sulfonate and carboxylate anions. Note that the T_m of zwitterions **9** and **18** is almost the same. However, zwitterion **19** shows a lower T_m than that of **10**. However, the carboxylate group cannot be concluded to be more suitable for lowering the T_m of zwitterions with only these data. Table 20.2 shows that zwitterions having a longer spacer length between the cation and the anion have lower a T_m (within the range of spacer length studied). This is likely the result of compensation for the increase in the flexibility of ions and decrease in the ion density. Zwitterion **22** having a branched spacer structure shows a lower T_m than does

TABLE 20.1 Melting Points of Zwitterions Having Sulfonate

Number	Structure	T_m (°C)	Reference
1		271.1[a]	31
2		284.7[a]	31
3		293.5[a]	31
4		146.9	31
5		69–170	36
6		274.3	31
7		279.1[a]	31
8		216.5	31
9		178	37
10		175	37
11		158	37
12		190	38

TABLE 20.1 (*Continued*)

Number	Structure	T_m (°C)	Reference
13		240	38
14		227	38
15		248	38
16		224	38

[a]Decomposition temperature.

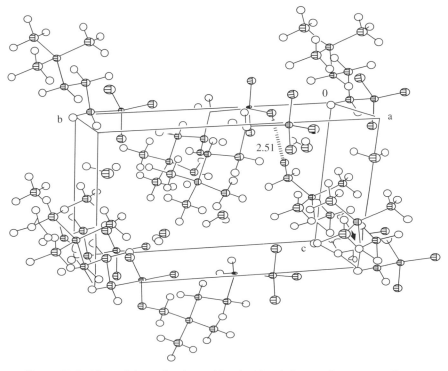

Figure 20.1 View of the molecular packing for trimethylammoniopropane sulfonate.

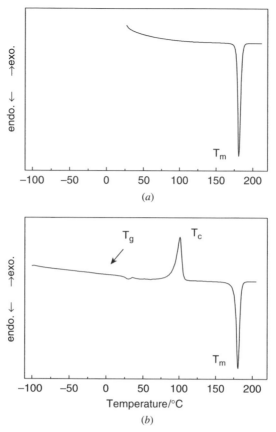

Figure 20.2 *DSC charts of zwitterion* **9**. *(a) First heating; (b) second heating. (Reproduced by permission of CSIRO Publishing.)*

zwitterion **17** and also zwitterions having a sulfonate group as a counter anion. It can be concluded that due to the steric hindrance the branched spacer is effective in lowering the T_m of zwitterions.

Table 20.3 and 20.4 summarize the T_m of reported zwitterions composed of the onium cation and the dicyanoethenolate anion. Zwitterions having a dicyanoethenolate anion show a lower T_m than zwitterions having a sulfonate anion. For instance, **1** and **2** decomposed before melting at 271° and 285°C, whereas zwitterions **35** and **36** showed T_m's at 187° and 148°C. In addition the T_m of **6** was 274°C, whereas that of **29** was 174°C. In particular, zwitterion **37** showed the lowest T_m of 70°C among the systems listed in all four tables. Since a negative charge is delocalized by the electron withdrawing effect of the cyano group, the interaction force between the cation and the dicyanoethenolate anion is weaker than that in the case of the sulfonate and carboxylate anion. Therefore the dicyanoethenolate structure is effective in lowering the T_m of the zwitterions. However, it should be mentioned here that the starting material, dicyanoketene cyclic acetal, is obtained from highly toxic reagents like potassium cyanide.

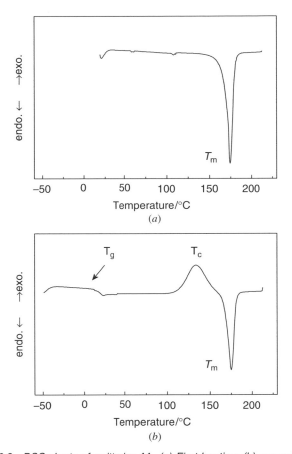

Figure 20.3 *DSC charts of zwitterion* **11**. *(a) First heating; (b) second heating.*

Zwitterions containing the polymerizable group (**5, 12–16, 24, 25, 30–34, 38–45**) are important monomers for the preparation of zwitterionic polymers. Characteristics of zwitterionic polymers will be discussed later. These zwitterions are nonconductive materials that turns conductive when suitable salts are added. So the thermal behaviors of zwitterions become important after the salt addition. Table 20.5 shows the thermal behavior of zwitterion **10** containing an equimolar amount of lithium salts. Five kinds of lithium salts (LiTFSI, lithium *bis*(perfluoroethylsulfonyl)imide (LiBETI), LiCF₃SO₃, LiBF₄, and LiClO₄) were compared to study the effect of added salt species on the thermal behavior of zwitterions. Zwitterion **10** showed a T_g at 18°C in the second heating without a salt addition. When lithium salts were added to the zwitterion, the T_g was the same despite the added salt species. The T_g of zwitterion **10** after mixing with an equimolar amount of LiTFSI was −37°C (see Figure 20.4). This was the lowest T_g among the mixtures of zwitterion **10** with five lithium salts. The T_g of the mixture of **10** and lithium salts showed the following sequence of anions: $TFSI^- < BETI^- < BF_4^- < CF_3SO_3^- < ClO_4^-$. LiTFSI

TABLE 20.2 Melting Points of Zwitterions Having Carboxylate

Number	Structure	T_m (°C)
17		250
18		171
19		165
20		130
21		134
22		168
23		212[a,b]

[a]Decomposition temperature.
[b]Holbrey et al. [39].
(*Reproduced by permission of CSIRO publishing.*)

also gave the lowest T_g for an analogous zwitterion **9** [25]. It is known that the TFSI anion has a plasticizing effect on the polyether matrix [40]. This also lowers the T_g in zwitterions. In Figure 20.4 the ionic conductivity of zwitterions with and without lithium salts was measured at a wide temperature range. It has already been confirmed that zwitterion including an equimolar amount of LiTFSI decompose at around 300°C [41]. Therefore these zwitterionic liquids are all good nonvolatile solvents that are capable of dissolving salts. This feature is important for the electrochemical applications.

Figure 20.5 shows the temperature dependence of the ionic conductivity for zwitterion **10** with and without an equimolar amount of lithium salts. Neat zwitterion **10** shows low ionic conductivity of about 10^{-6} S cm^{-1} at even 200°C from the ac impedance measurement. This is because there are no mobile ions in the system. However, zwitterion **10**, which is mixed with an equimolar amount of lithium salt,

TABLE 20.3 Melting Points of Zwitterions Having Dicyanoethenolate

Number	Structure	T_m (°C)	Reference
24		186–187	32
25		113	32
26		201–202	33
27		174	33
28		168–170	33
29		173–174	33
30		242–246	32
31		213–217	32
32		239–240	32
33		>300	32
34		183–185	32

TABLE 20.4 Melting Points of Zwitterions Having Dicyanoethenolate

Number	Structure	T_m (°C)	Reference
35		186.9	31
36		148.4	31
37		70	33
38		94–95	32
39		143–144	32
40		115	32
41		136–137	32
42		144–145	32
43		129.9	32
44		118–119	32
45		111.4	32

TABLE 20.5 Thermal Behavior of Zwitterion 10 with an Equimolar Amount of LiX*

	T_m (°C)	T_g (°C)
Neat zwitterion **10** (2nd heating)	175	18
+LiTFSI	—	−37
+LiBETI	—	−5
+LiCF$_3$SO$_3$	—	19
+LiBF$_4$	—	5
+LiClO$_4$	—	24

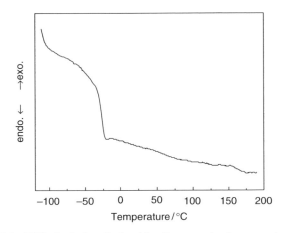

Figure 20.4 DSC chart of zwitterion **10** with an equimolar amount of LiTFSI*.

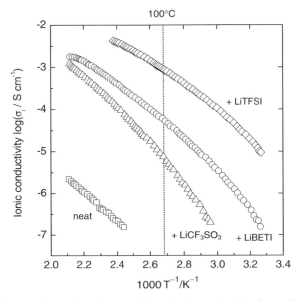

Figure 20.5 Temperature dependence of the ionic conductivity for zwitterion **10** with an equimolar amount of LiX*.

*(Reproduced by permission of CSIRO publishing)

shows much higher ionic conductivity than that of the neat zwitterion. The higher conductivity is due to the increase of carrier ions from the addition of lithium salts. The ionic conductivities of BETI$^-$ and CF$_3$SO$_3^-$ are about 10^{-4} S cm^{-1} and 10^{-5} S cm^{-1} at 100°C. Among these lithium salts, the zwitterion mixture including LiTFSI shows the highest ionic conductivity of 8.9×10^{-4} S cm^{-1} at 100°C, reflecting the lowest T_g of -37°C. This result agrees with that of zwitterion **9**, which contains various lithium salts [25]. For all the lithium salts the ionic conductivity of zwitterion with and without lithium salts show traces out an upper convex curve. This is explained by the ion transport in zwitterions being led by the motion of zwitterion itself. Moreover there is a good relationship between the ionic conductivity and T_g for various zwitterions, including lithium salts (see Figure 20.6). The ionic conductivity of zwitterions containing an equimolar of lithium salt thus increases with the decreasing T_g. The T_g was found to govern the ionic conductivity of zwitterions containing LiTFSI.

The lithium transference number (t_{Li}^+) in zwitterion **10** containing an equimolar amount of LiTFSI is 0.56 [34b], which is higher than the numbers of other organic solvents. This number suggests that zwitterions can transport lithium ions the best. This result seems to based on the imidazolium cation and the TFSI anion interacting in the ionic liquid domain, which then provides the ion conductive path for the lithium cation. However the t_{Li}^+ in zwitterion **18** containing carboxylate is 0.16 [34b], which indicates that it is necessary to introduce a counter anion structure having a high degree of dissociation in order to obtain the high t_{Li}^+ in the zwitterion.

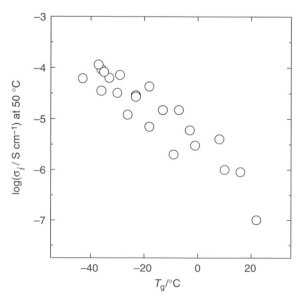

Figure 20.6 *Relation between T_g and the ionic conductivity for zwitterions containing an equimolar LiTFSI.*

Although zwitterions are mainly considered for their novel ion conductive matrix in this chapter, they are being used as not only as solvents and catalysts for organic reactions [42] but also as organogelators [43]. Zwitterions have been screened as solvent/catalysts for several classical acid-promoted organic reactions such as the Fischer esterification, alcohol dehydrodimerization, and the pinacol/benzopinacole rearrangement. The zwitterion containing an equimolar trifluoromethane sulfonic acid is liquid at room temperature. Because they can work as solvent/catalysts, as shown in the reactions discussed in this chapter, zwitterionic liquids should open the door to a whole new area of applications.

REFERENCES

1. J. S. Wilkes, M. J. Zaworotko, *J. Chem. Soc., Chem. Commun.*, 965 (1992).

2. E. I. Cooper, J. M. O'Sullivan, in *Proc. 8th Int. Symp. Ionic Liquids*, R. J. Gale, G. Blomgren, H. Kojima, eds., PV 92-16, Electrochemical Society, 1992, NJ. p. 386.

3. T. Welton, *Chem. Rev.*, **99**, 2071 (1999).

4. P. Wasserscheid, W. Keim, *Angew. Chem. Int. Ed.*, **39**, 3772 (2000).

5. J. Dupont, R. F. de Souza, P. A. Z. Suarez, *Chem. Rev.*, **102**, 3667 (2002).

6. J. G. Huddleston, H. D. Willauer, R. P. Swatloski, A. E. Visser, R. D. Rogers, *Chem Commun.*, 1765 (1998).

7. (a) J. D. Holbrey, K. R. Seddon, *Clean Prod. Proc.*, **1**, 223 (1999).; (b) K. R. Seddon, *J. Chem. Tech. Biotechnol.*, **68**, 351 (1997).

8. V. R. Koch, C. Nanjundiah, G. B. Appetecchi, B. Scrosati, *J. Electrochem. Soc.*, **142**, L116 (1995).

9. H. Nakagawa, S. Izuchi, K. Kuwana, T. Nukuda, Y. Aihara, *J. Electrochem. Soc.*, **150**, A695 (2003).

10. D. R. MacFarlane, J. Huang, M. Forsyth, *Nature*, **402**, 792 (1999).

11. A. Hayashi, M. Yoshizawa, C. A. Angell, F. Mizuno, T. Minami, M. Tatsumisago, *Electrochem. Solid-State Lett.*, **6**, E19 (2003).

12. H. Sakaebe, H. Matsumoto, *Electrochem. Commun.*, **5**, 594 (2003).

13. M. Doyle, S. K. Choi, G. Proulx, *J. Electrochem. Soc.*, **147**, 34 (2000).

14. (a) J. Sun, D. R. MacFarlane, M. Forsyth, *Electrochim. Acta*, **46**, 1673 (2001). (b) J. Sun, L. R. Jordan, M. Forsyth, D. R. MacFarlane, *Electrochim. Acta*, **46**, 1703 (2001).

15. (a) A. Noda, M. A. B. H. Susan, K. Kudo, S. Mitsushima, K. Hayamizu, M. Watanabe, *J. Phys. Chem. B*, **107**, 4024 (2003). (b) M. A. B. H. Susan, A. Noda, S. Mitsushima, M. Watanabe, *Chem. Commun.*, 938 (2003).

16. N. Papageogiou, Y. Athanassov, M. Armand, P. Bonhôte, H. Pettersson, A. Azam, M. Grätzel, *J. Electrochem. Soc.*, **143**, 3099 (1996).

17. R. Kawano, M. Watanabe, *Chem. Commun.*, 330 (2002).

18. W. Kubo, T. Kitamura, K. Hanabusa, Y. Wada, S. Yanagida, *Chem. Commun.*, 374 (2002).

19. A. B. McEwen, H. L. Ngo, K. LeCompte, J. L. Goldman, *J. Electrochem. Soc.*, **146**, 1687 (1999).

20. A. B. McEwen, S. F. McDevitt, V. R. Koch, *J. Electrochem. Soc.*, **144**, L84 (1997).

21. C. Nanjundiah, S. F. McDevitt, V. R. Koch, *J. Electrochem. Soc.*, **144**, 3392 (1997).

22. P. Bonhôte, A.-P. Dias, M. Armand, N. Papageorgiou, K. Kalyanasundaram, M. Grätzel, *Inorg. Chem.*, **35**, 1168 (1996).

23. (a) M. Hirao, H. Sugimoto, H. Ohno, *J. Electrochem. Soc.*, **147**, 4168 (2000). (b) M. Yoshizawa, W. Ogihara, H. Ohno, *Electrochem. Solid-State Lett.*, **4**, E25 (2001). (c) H. Ohno, M. Yoshizawa, *Solid State Ionics*, **154–155**, 303 (2002).

24. R. Hagiwara, Y. Ito, *J. Fluorine Chem.*, **105**, 221 (2000).

25. M. Yoshizawa, M. Hirao, K. Ito-Akita, H. Ohno, *J. Mater. Chem.*, **11**, 1057 (2001).

26. (a) M. Yoshizawa, H. Ohno, *Ionics*, **8**, 267 (2002). (b) W. Ogihara, M. Yoshizawa, H. Ohno, *Chem. Lett.*, 880 (2002).

27. (a) H. Ohno, K. Ito, *Chem. Lett.*, 751 (1998). (b) M. Yoshizawa, H. Ohno, *Chem. Lett.*, 889 (1999). (c) M. Hirao, K. Ito-Akita, H. Ohno, *Polym. Adv. Technol.*, **11**, 534 (2000). (d) H. Ohno, *Electrochim. Acta*, **46**, 1407 (2001). (e) M. Yoshizawa, H. Ohno, *Electrochim. Acta*, **46**, 1723 (2001). (f) M. Yoshizawa, W. Ogihara, H. Ohno, *Polym. Adv. Technol.*, **13**, 589 (2002).

28. J. Cardoso, A. Huanosta, O. Manero, *Macromolecules*, **24**, 2890 (1991).

29. M. Galin, E. Marchal, A. Mathis, J.-C. Galin, *Polym. Adv. Technol.*, **8**, 75 (1997).

30. M. Galin, A. Mathis, J.-C. Galin, *Polym. Adv. Technol.*, **12**, 574 (2001).

31. M. Galin, A. Chapoton, J.-C. Galin, *J. Chem. Soc. Perkin Trans.*, **2**, 545 (1993).

32. M.-L. Pujol-Fortin, J.-C. Galin, *Macromolecules*, **24**, 4523 (1991).

33. W. J. Middleton, V. A. Engelhardt, *J. Am. Chem. Soc.*, **80**, 2788 (1958).

34. (a) Y. Chevalier, P. L. Perchec, *J. Phys. Chem.*, **94**, 1768 (1990). (b) A. Narita, M. Yoshizawa, H. Ohno, submitted.

35. T. Yokoyama, G. Murakami, H. Akashi, M. Zenki, *Anal. Sic.*, **19**, 805 (2003).

36. R. S. Armentrout, C. L. McCormick, *Macromolecules*, **33**, 419 (2000).

37. M. Yoshizawa, A. Narita, H. Ohno, *Aust. J. Chem.*, **57**, 139 (2004).

38. J. C. Salamone, W. Volksen, S. C. Israel, A. P. Olson, D. C. Raia, *Polymer*, **18**, 1058 (1997).

39. J. D. Holbrey, W. M. Reichert, I. Tkatchenko, E. Bouajila, O. Walter, I. Tommasi, R. D. Rogers, *Chem. Commun.*, 28 (2003).

40. S. Besner, A. Vallée, G. Bouchard, J. Prud'homme, *Macromolecules*, **25**, 6480 (1992).

41. H. Ohno, M. Yoshizawa, W. Ogihara, *Electrochim. Acta*, **48**, 2079 (2003).

42. A. C. Cole, J. L. Jensen, I. Ntai, K. L. Tran, K. J. Weaver, D. C. Forbes, J. H. Davis Jr., *J. Am. Chem. Soc.*, **124**, 5962 (2002).

43. C. Wang, A. Robertson, R. G. Weiss, *Langmuir*, **19**, 1036 (2003).

Chapter 21

Alkali Metal Ionic Liquid

Wataru Ogihara, Masahiro Yoshizawa, and Hiroyuki Ohno

Ionic liquids (ILs) inherently have high ionic conductivity. However, their serious drawback is that they provide migration of only component ions. Their structure and combination of ions can form low-temperature molten salts, but it is extremely difficult to prepare ILs composed of the desired (target) carrier ions such as Li^+. Thus target carrier ions such as Li^+ or H^+ must be added into general ILs. There are further upper limits that must be observed for the solubilization of most additives in common ILs. The addition of the target carrier ions raises the viscosity of ILs, and thus decreases their ionic conductivity. Inorganic salts such as LiCl containing useful carrier ions at high temperatures show extremely high ionic conductivity. In this chapter we report on our experiments with ILs that contain useful carrier ions such as Li^+ and H^+.

21.1 ALKALI METAL IONIC LIQUID [1]

Most ILs generally consist of monovalent cations (imidazolium cation or pyridinium cation) and monovalent anions (tetrafluoroborate anion or bis(trifluoromethanesulfonyl)imide anion). It is difficult to form IL with multivalent ions owing to their very strong electrostatic interaction, which causes an elevation of melting point (T_m) or an increase of viscosity. However, we tried to lower the T_m of salts containing divalent anions with aid of organic cations. Many salts are composed of a divalent anion, an organic cation, and an alkali metal cation. Alkali metal ionic

Electrochemical Aspects of Ionic Liquids   Edited by Hiroyuki Ohno
ISBN 0-471-64851-5   Copyright © 2005 John Wiley & Sons, Inc.

liquids (AMILs) consist of an organic cation and various kinds of target carrier cations that interact with the sulfate anion (Scheme 21.1). It has been confirmed

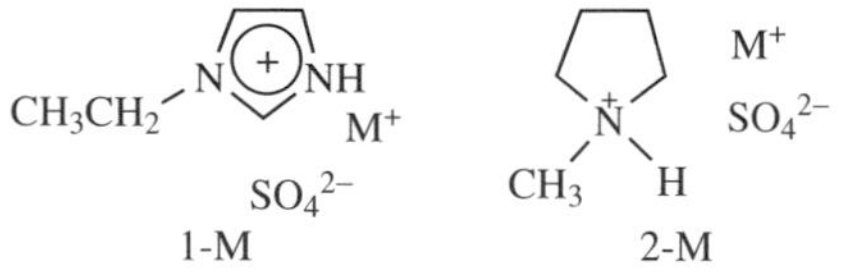

$$CH_3CH_2-N(+)NH \quad M^+ \quad SO_4^{2-}$$

1-M   2-M

$M^+ = H^+, Li^+, Na^+, K^+, Rb^+, Cs^+,$ etc.

Scheme 21.1   *Structure of AMILs.*

that the series of *N*-ethylimidazolium salts (1-M) and 1-methylpyrrolidinium salts (2-M) become liquid at room temperature. It has been recognized that the *N*-ethylimidazolium cation can provide ionic liquid at room temperature with a lot of monovalent anions [2].

Table 21.1 shows the glass transition temperature (T_g) of these AMILs. The obtained AMILs had no T_m and showed only T_g. These liquids showed good thermal stability like other ILs. However, AMILs were highly viscous compared with general ILs such as EMImTFSI or EMImBF$_4$. This high viscosity is thought to be the result of a strong interaction among the divalent anions. Even the lowest viscosity of 754 cP for 2-K at 40°C is far larger than that of general ILs. It is well known that T_g and viscosity are the factors that govern ionic conductivity [3]. A structural re-design is there fore necessary to improve the thermal properties and viscosity.

The ionic conductivity of AMILs is nearly equal regardless of the alkali metal ion species. The value of 1-M is around 10^{-4} S cm^{-1} and that of 2-M was around 10^{-3} S cm^{-1}. Figure 21.1 shows the ionic conductivity of two AMILs containing Li$^+$ as the carrier ion. As the figure shows, the ionic conductivity of AMILs is mainly affected by the structure of the organic cation. This difference is reflected in the T_g. Figure 21.2 shows the relationship between the ionic conductivity and the T_g of all AMILs and the difference in properties due to the organic cation structure.

The Li$^+$ migration in AMILs is confirmed by pfg-NMR measurement of the two salts containing Li$^+$. However, its migration is less diffusive than that of the organic

TABLE 21.1 T_g of AMILs

M=	1-M T_g(°C)	2-M T_g(°C)
H	−75	−96
Li	−60	−78
Na	−68	−82
K	−67	−85
Rb	−55	−85
Cs	−59	−84

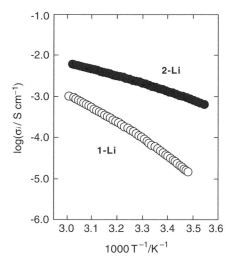

Figure 21.1 *Temperature dependence of the ionic conductivity of 1-Li and 2-Li.*

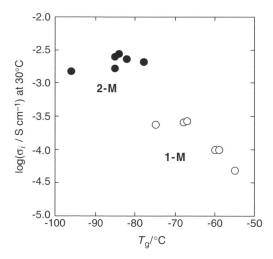

Figure 21.2 *Relationship between the ionic conductivity and the T_g of 1-M and 2-M.*

cation for each salt. The 2-Li ions have a ^7Li diffusion coefficient, which is ten times larger than that of 1-Li, reflecting the difference of ionic conductivity. This result suggests the 2-Li to be superior to 1-Li as a lithium-ion conductor. To study the use of this system as an ion conductive material, it is necessary to consider some other characteristics, such as the identification of the carrier ion species and its dissociative behavior in a matrix.

21.2 GELATION OF ALKALI METAL IONIC LIQUID [4]

In some fields solid (or film) electrolytes are preferred to the liquid electrolytes. This is especially the case in industrial applications. Accordingly gel-type ion conductive polymers are prepared by mixing AMILs with several polymers. AMILs having Li^+ can be mixed with a host of polymers, as shown in Scheme 21.2, to

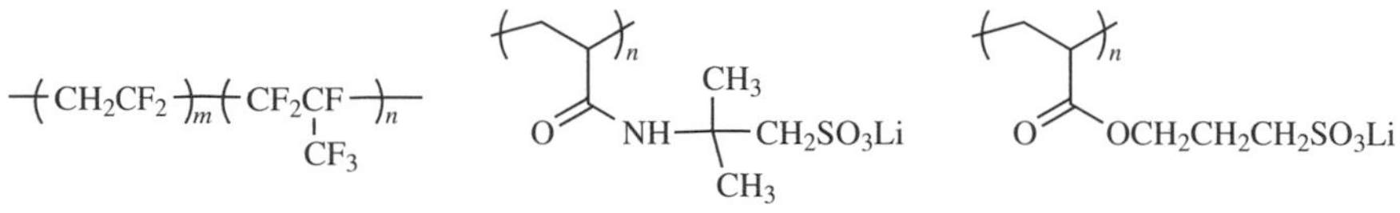

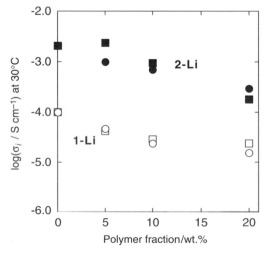

$m : n = 85 : 15$

P(VdF-HFP)   PAMPSLi   PAC3SLi

Scheme 21.2 *Structure of host polymers.*

study their gelation behavior. Although the hydrophilic polymers, PAC3SLi and PAMPSLi, are soluble in AMILs, P(VdF-HFP) is insoluble in these liquids. The compatibility observed for the AMILs with particular polymers is understood to be due to their hydrophobicity.

The gels based on AMILs are prepared by changing the amount of polymers from 5 to 20 wt%. Figure 21.3 shows the ionic conductivity of the obtained gel in each host polymer fraction. These gels show similar ionic conductivity to that of bulk AMILs. Even with 20 wt% the gel based on 2-Li had ionic conductivity greater than $10^{-4}\,S\,cm^{-1}$ at room temperature. The ionic conductivity for each gel is also different by one order of magnitude, depending on the structure of the

Figure 21.3 *Ionic conductivities at 30°C plotted as a function of polymer fraction. Square plot: PAMPSLi-based gel; circle plot: PAC3SLi-based gel.*

organic cation. This result implies that the component ions of AMILs migrate in the gel matrix as carrier ions.

To study the mobility of Li$^+$ on the gel matrix, we measured the ^{7}Li diffusion coefficient before and after gelation of the 2-Li (with 5 wt% PAMPSLi) using pfg-NMR. The ^{7}Li diffusion coefficient in the gel was almost unchanged from the value before gelation, despite the decrease in the ionic conductivity with gelation. Because, as this result suggests, the mobility of Li$^+$ is not suppressed by gelation, this gel system is favorable for Li$^+$ conductive materials.

21.3   TRIPLE ION-TYPE IMIDAZOLIUM SALT [5]

The idea of developing a zwitterionic liquid (ZIL) for alkali metal transport has led to the consideration of a new cation conductive material. As shown in Chapter 20, an increase in cation conduction occurs when LiTFSI is added to ZILs. An equimolar mixture of ZIL and LiTFSI may give us a new model, namely on imidazolium cation containing two tethered anions. This novel system, called triple ion-type imidazolium salt, consists of three charges. Scheme 21.3 shows the structure of such triple ion-type imidazolium salts. These salts are prepared by the reaction of an imidazole analogue with an alkane sultone (see Chapter 20). Besides the imidazolium cation having two tethered sulfonate anions, this salt has a target carrier cation

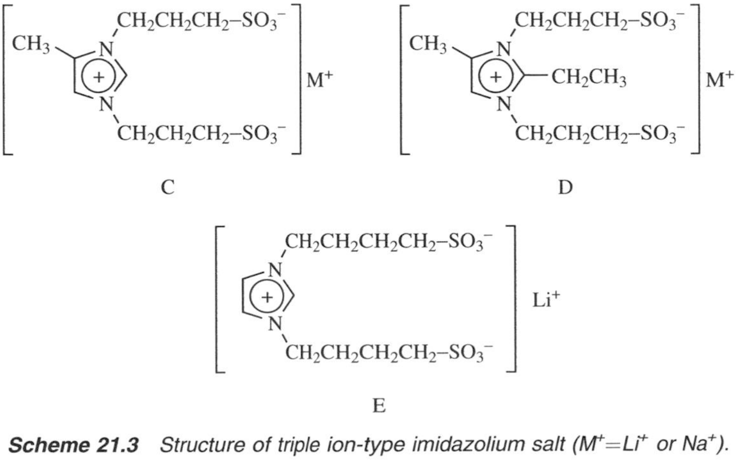

Scheme 21.3 *Structure of triple ion-type imidazolium salt (M$^+$=Li$^+$ or Na$^+$).*

TABLE 21.2   Thermal Properties of Triple Ion-Type Imidazolium Salts

	$M^+ = Li^+$		$M^+ = Na^+$	
	T_m (°C)	T_g (°C)	T_m (°C)	T_g (°C)
A	287	—	254	—
B	—	42	—	20
C	—	40	—	23
D	—	60	—	
E	—	30	—	

Note:—Not detected.

(an alkali metal cation as the counter ion). This system is expected to become a novel single ion conductive material because the alkali metal cation is more mobile than a triple ion.

Table 21.2 shows thermal properties of eight kinds of triple ion-type imidazolium salts. The T_m of the lithium salt of compound A is 287°C, and it is almost 100°C higher than that of corresponding ZIL. This can be regarded as decrease of freedom for the component ion because of the increasing number of immobilized anions. To lower the T_m of lithium, several approaches were attempted. When the alkali metal ion of compound A was Na^+, the T_m of the corresponding salt was 254°C, which is about 30°C lower than that of the lithium salt. This difference is probably caused by the weaker electrostatic interaction between Na^+ and the sulfonate group because the surface charge density of Na^+ is smaller than that of Li^+. On the other hand, the introduction of the alkyl group into some position on the imidazolium ring is another factor that can improve thermal properties. The salts forming the methyl group on the imidazolium ring (compounds B, C, and D) show only a T_g at around ambient temperature. This result suggests that the presence of the alkyl group suppresses the crystallization of the system. The same result is confirmed for compound E, which has a longer spacer between the imidazolium cation and the sulfonate anion. However, both the T_g and T_m of these triple ion-type salts are still high. The thermal properties of these ions should be improved by molecular design.

Figure 21.4 shows the ionic conductivity of triple ion-type imidazolium salt. Among these salts, B and C containing Na^+ show the highest ionic conductivity of around 10^{-3} S cm^{-1} at 200°C. However, the ionic conductivity at room temperature of each salt was quite low value because of their high T_m and T_g. As in the case of the ZIL, the ionic conductivity can be greatly improved by certain additive salt species because the ZIL has no carrier ions. Salt addition into this system can also improve the ionic conductivity, but this approach provides bi-ionic conductive properties, which conflicts with the design concept of this chapter. The details of our studies on the addition of salts will be published separately.

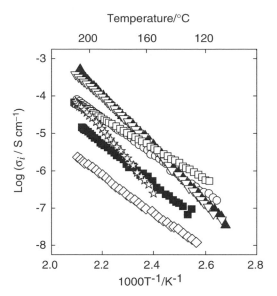

Figure 21.4 *Temperature dependence of the ionic conductivity for triple ion-type imidazolium salts.* ▲: *BNa;* ▽: *CNa;* □: *BLi;* ○: *ANa;* ☆: *ELi;* ■: *ALi;* ◇: *DLi.*

Besides the compounds mentioned in this chapter, some studies have been vigorously changing anion structure, spacer structure, and spacer length to investigate improvements to the characteristics of Imidazolium salts. Room-temperature triple ion-type molten salt is expected to be obtained in the near future with the accumulation of these studies.

21.4 CONCLUSION

In this chapter two methods and their possible applications were discussed and a few examples showed their relevance for the molecular design of ILs. Although the direction ILs will take are difficult to predict, at the present time the ILs designed could be highly functional because many IL are being studied using organic ions that have unlimited ion species and combinations.

REFERENCES

1. W. Ogihara, M. Yoshizawa, H. Ohno, *Chem. Lett.*, 880 (2002).
2. H. Ohno, M. Yoshizawa, *Solid State Ionics*, **154–155**, 303 (2002).
3. (a) P. Bonhôte, A. P. Dias, N. Papageorgiou, K. Kalyanasundaram, M. Grätzel, *Inorg. Chem.* **35**, 1168 (1996). (b) M. Hirao, H. Sugimoto, H. Ohno, *J. Electrochem. Soc.*, **147**, 4168 (2000).
4. W. Ogihara, J. Sun, M. Forsyth, D. R. MacFarlane, M. Yoshizawa, H. Ohno, *Electrochim. Acta*, **49**, 1797 (2004).
5. M. Yoshizawa, H. Ohno, *Ionics*, **8**, 267 (2002).

Chapter 22

Polyether/Salt Hybrids

Tomonobu Mizumo and Hiroyuki Ohno

22.1 ANOTHER IONIC LIQUID

Ambient temperature molten salt can be obtained by several methods. One effective way to obtain a room-temperature molten salt is by the introduction of polyether chains to ions. The term "polyether/salt hybrid" is used in this chapter as a common name for polyether oligomers having anionic or cationic charge(s) on the chain (Figure 22.1). Polyethers, such as poly-(ethylene oxide) (PEO), are known as representative ion conductive polymers [1]. Polyether/salt hybrids have been studied as a kind of room-temperature molten salt apart from the development of onium-type ionic liquids [2]. The preparation of ionic liquids consisting of metal ions has been one of the important goals in this research field. Polyether/salt hybrid derivatives give one such solution for this task.

The ion conductive property of a semi-crystalline PEO/alkali metal salt complex was first reported by Wright in 1975 [3]. Since then, numerous PEO-based polymer electrolytes has been reported for application in solid-state batteries [4]. Polyether derivatives can induce the dissociation of a variety of salts because of their large dipole moment. The dissociated ions can migrate in the polyether matrices assisted by the segmental motion. Generally, The PEO/alkali metal salt complex shows ionic conductivity over 10^{-5} S cm^{-1} above its melting point (T_m) at around 40° to 60°C. However, since the ionic conductivity at the crystalline state is lower than 10^{-8} S cm^{-1}, many researchers have examined amorphous and rubbery systems. In these studies some researchers have observed that PEO oligomers having salt groups became amorphous fluid showing no T_m [5,6]. These salts which have flexible tails,

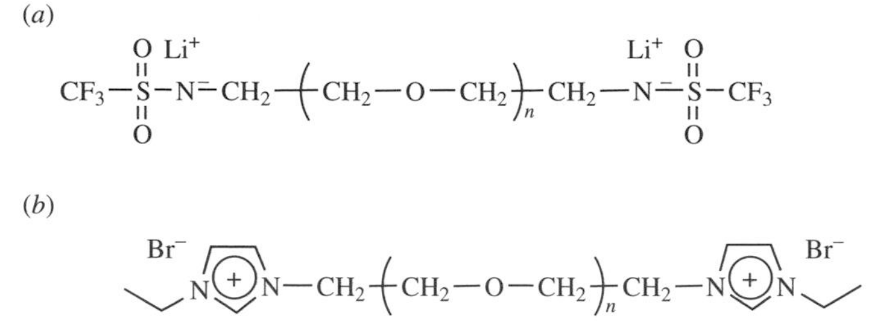

(a)

$$CF_3-\overset{\overset{O}{\|}}{\underset{\|}{S}}-N\overset{Li^+}{=}CH_2-\left(CH_2-O-CH_2\right)_n CH_2-N\overset{Li^+}{=}\overset{\overset{O}{\|}}{\underset{\|}{S}}-CF_3$$

(b)

$$\overset{Br^-}{\underset{}{\diagup}}N\overset{+}{\underset{}{\bigcirc}}N-CH_2-\left(CH_2-O-CH_2\right)_n CH_2-N\overset{+}{\underset{}{\bigcirc}}N\overset{Br^-}{\underset{}{\diagdown}}$$

Figure 22.1   *Representative chemical structures of polyether/salt hybrids. (a) Cation conductor, and (b) anion conductor.*

can be categorized as an ionic liquid. In this chapter we introduce this unique material.

22.2   DESIGN OF POLYETHER/SALT HYBRIDS

Although a number of salts having organic chains have been synthesized to date, only a few salts were obtained as liquid or glass. In the case of the polyether/salt hybrid, the crystalline phase of both the salt and polyether disappear under appropriate conditions (Figure 22.2). The Polyether moiety in the polyether/salt hybrid is assumed to inhibit the crystallization of the salts by its flexible segmental motion. Since polyethers help dissociate the bound salt through their ion–dipole interactions, the polyether/salt hybrid inherently shows ionic conductivity.

Polyether/salt hybrids show some important features:

1. *Can be employed with a large variety of organic salts.* As mentioned in other chapters, most ionic liquids are composed of only a few kinds of ions. It has been experimentally shown that larger and asymmetrically substituted ions

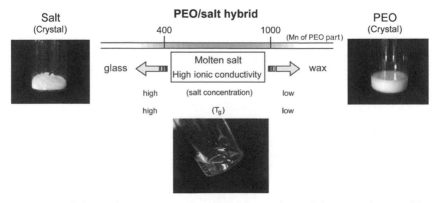

Figure 22.2   *Polyether/salt hybrid can be obtained as molten salt in appropriate conditions.*

with delocalized charge are favored to form ionic liquids with low T_m's. However, there remain a lot of ions that are recognized as unsuitable for ionic liquid formation, despite their useful features [7]. In contrast, the methodology of the polyether/salt hybrid can be applied to a variety of salts including metal salts, such as lithium carboxylate and lithium sulfonate. Generally, the introduction of the polyether enables any salt to melt at ambient temperatures.

2. *Selective charge transfer.* Ionic liquids show excellent ion conductive properties, but the ionic species that constitute the matrix will migrate along the potential gradient. Therefore many ionic liquids are not useful. Ionic liquids that can selectively transport specific ions, such as the lithium ion, are nevertheless needed for energy devices [8]. Against these, in the bulk polyether/salt hybrids, only one ionic species can readily migrate: this is because the counter ion is anchored to the bulky polyether chain, which has a small diffusion coefficient.

These features of polyether/salt hybrids depend on the structure of both the polyether and the salt moiety. In particular, it should be noted that the salt concentration of this system increases with the shortening of the polyether chain length. In the low molecular weight region, the polyether/salt hybrids tend to show a higher glass transition temperature (T_g) and become stiff glass because the salt characteristics become dominant. In contrast, in the higher molecular weight region, the polyether feature is dominant. To obtain a liquid-type polyether/salt hybrid, the polyether chain length must be set in the appropriate region. The appropriate PEO chain length has been empirically shown to be roughly Mn $= 400 - 1000$ when the linear PEO/salt hybrids have salt moieties on both chain ends [2, 5, 6, 9–11]. When only one charge is fixed on the PEO chain, the crystalline phase tends to appear in the region lower than 750. Furthermore we examined several types of polyether/salt hybrid, as shown in Figure 22.3. We found that there are suitable polyether chain lengths depending on the type (Figure 22.3) and salt structure.

22.3 IMPROVEMENT OF IONIC CONDUCTIVITY

PEO oligomers having carboxylic acids at their chain ends are commercially available. The PEO/carboxylate hybrids are obtained simply by the neutralization of PEO-carboxylic acids with their corresponding bases [5]. Since polyether has high affinity with salts, the polyether/salt hybrids should be carefully prepared to avoid contamination or generation of low molecular weight salts. After removing water or any other solvents, the PEO/carboxylate hybrids can be obtained as a viscous liquid with a low T_g below $-50°C$. However, the bulk ionic conductivity of the PEO/carboxylate hybrids is lower than 10^{-7} S cm^{-1} at room temperature because of the poor dissociation of the carboxylate. To improve the ionic conductivity, we investigated the structural design of both polyether and the salt moiety.

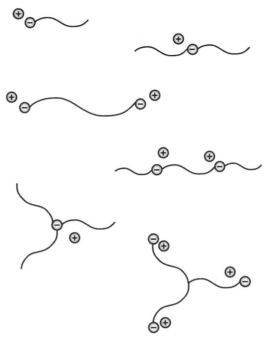

Figure 22.3   Schematic representation for several types of polyether/salt hybrids.

Since ionic conductivity is the function of the number of carrier ions and their mobility (equation 22.1), the dissociation constant of the salt has a great influence on the ionic conductivity:

$$\sigma_i = \sum ne\mu, \qquad (22.1)$$

where n, e, and μ are the number, charge, and mobility of carrier ions, respectively. We examined the ion conductive properties of PEO/salt hybrids having a variety of salts such as sulfonate ($-SO_3^-M^+$; M^+ represents alkali metal ion) [2, 6], benzene sulfonate ($-C_6H_4-SO_3^-M^+$) [9], thiolate ($-S^-M^+$) [10], sulfonamide salt ($-N^--SO_2-R\ M^+$) [11, 12], and sulfonimide salt ($-SO_2-N^--SO_2-R\ M^+$) [13] groups. The ionic conductivity was improved with the increase of the salt dissociation constant [14], while the maximum ionic conductivity was observed to depend on the PEO chain length in every case. Among these salts, the trifluoromethylsulfony-limide salt ($-SO_2-N^--SO_2CF_3\ M^+$) gave a high dissociation constant because of the delocalized anionic charge [15]. The ionic conductivity of α-methoxyethyl PEO ($Mn = 550$) containing this salt structure was $7.0 \times 10^{-5}\ S\ cm^{-1}$ at 30°C (see Figure 22.4).

Although the introduction of the highly dissociative salt group proved to be effective in increasing the number of carrier ions, the increase of the salt fraction significantly elevated the T_g. The characteristics of salt turned dominant by increasing the salt fraction, as mentioned above. Since the carrier ions can migrate faster in

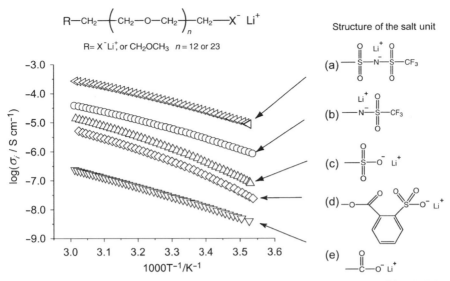

Figure 22.4 *Temperature dependence of the ionic conductivity for a series of PEO/salt hybrids. The number of ethylene oxide repeating unit is 23 for (b) - (d), and 12 for (a). (b) - (d) has salt units on both α, and ω chain ends, while the α chain end of (a) is capped by methoxy group.*

less viscous media, it is also important to optimize the polyether chain length and their structures by considering several factors such as T_m, T_g, and the number of carrier ions. Generally, PEO/lithium salt hybrids show a maximum ionic conductivity in the salt concentration ($[Li^+]$/[ethylene oxide unit]) of around 10 mol%. The introduction of a noncharged group can also enhance their ionic conductivity. In linear-type PEO/salt hybrids, the ionic conductivity is improved by capping one of the chain ends by a methoxy group. The methoxy-capped system shows lower T_g and higher ionic conductivity than those of the PEO/salt hybrids with salt groups on both chain ends, even though both systems contain nearly equal salt concentrations.

Poly(propylene oxide) (PPO) has also been investigated as an interesting alternative to PEO because of its amorphous property [12, 13]. PPO/salt hybrids show no T_m regardless of the chain length and are readily obtained as liquid or glass. The PPO-based system is further favorable for fast ion migration. However, the ionic conductivity of PPO/salt hybrid is generally lower than that of the corresponding PEO system. The lower ionic conductivity of the PPO/salt hybrid is ascribed to the lower polarity of the propylene oxide unit [16]. To improve the ionic conductivity of PPO systems, highly dissociative salt groups, such as sulfonimide salt groups have been employed. The α-n-butoxy PPO/lithium trifluoromethylsulfonylimide hybrid has shown an ionic conductivity of 3.3×10^{-6} S cm^{-1} at 30°C [13].

It is also noteworthy that the ionic conductivity of a simple mixture of PEO and alkali metal salt was improved 10 to 100 times by employing larger cations such as sodium or potassium ions [17], whereas numerous lithium cation conductors are known to date. The improved ionic conductivity is attributed to the weaker ion-dipole interaction, since larger ions have lower charge densities. Higher carrier ion mobility could be realized by the control of such interaction forces.

22.4 POLYMERIZED PEO/SALT HYBRIDS

So far a number of high molecular weight polymer electrolytes consisting of PEO (side-chain and/or main chain) and salt units have been evaluated for application in rechargeable batteries [18]. In particular, polyanionic electrolytes show ion transference numbers of almost unity, whereas the lithium ion transference number of conventional PEO/salt mixtures is no more than 0.3. However, the ionic conductivity of these lithium-ion conductors is generally lower than 10^{-8} S cm^{-1} at ambient temperatures due to the poor dissociation constant of salt moieties such as carboxylate, sulfonate, and phosphate. To improve the ionic conductivity, polyanions having highly dissociative salt structure such as $-R_f-SO_3^-M^+$ [19], $-N^-$-SO_2CF_3 M^+ [20], and $-SO_2-N^--SO_2CF_3$ M^+ [21] have been recently prepared. Sisca et al. prepared comb-type copolymers consising of polysiloxane backbone with alkyl-trifluoromethylsulfonamide and oligo(ethylene oxide) side-chains (Figure 22.5). The flexible polysiloxane backbone was effective in avoiding the elevation of T_g of the system. The obtained polymers showed a low T_g of $-67°C$ ([Li$^+$]/[ethylene oxide unit] was 4.5 mol%) and relatively high ionic conductivity of 1.20×10^{-6} S cm^{-1} at 25°C.

On the other hand, we have reported the preparation of gel-type polymer electrolytes using polyether/salt hybrid oligomers [12]. A network-type PPO film was swollen with the PPO/trifluoromethylsulfonamide lithium salt hybrid. The obtained gel containing 50 wt% of PPO/salt hybrid showed a low T_g of $-62°C$. The PPO/salt hybrid behaved as not only an added salt but also a plasticizer. The ionic conductivity was over 10^{-6} S cm^{-1} at room temperature. Furthermore this gel electrolyte showed a favorable lithium-ion transference number of 0.7–0.8.

22.5 POLYETHER/ALMINATE, OR BORATE SALT HYBRIDS

Recently polyether derivatives having inorganic salt moiety such as alminate or borate salt units have been vigorously investigated. Early works were developed by Shriver's and Fujinami's groups [22]. Doan et al. first examined the network-type

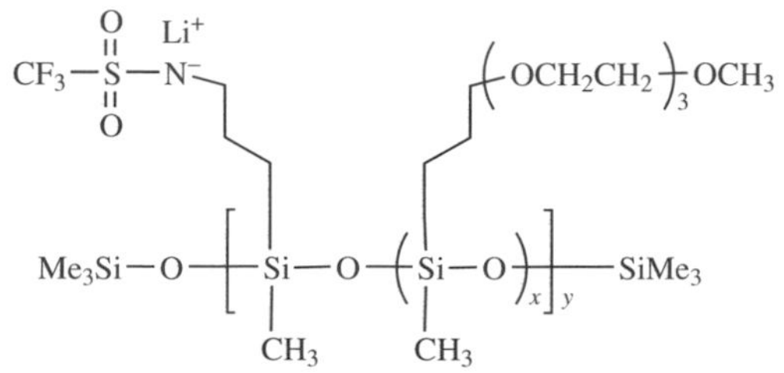

Figure 22.5 *Polyanionic electrolyte.*

(a)   (b)

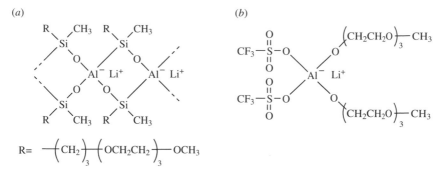

R= —(CH₂)₃—(OCH₂CH₂)₃—OCH₃

Figure 22.6 *Polyethers containing lithium aluminate groups.*

PEO derivatives containing sodium tetraalkoxyalminate groups. The ionic conductivity of this system remains at around 10^{-8} S cm^{-1} at room temperature. However, the ionic conductivity is considerably improved to 2.3×10^{-5} S cm^{-1} at 25°C by employing siloxyalminate in place of alkoxyalminate (Figure 22.6a). Although the lithium-ion transference number of 0.71 suggests that oligomeric derivatives remained in this system, the improved ionic conductivity can be ascribed to the effective charge delocalization of siloxyaluminate via Si–O $p\pi$-$d\pi$ system. Recently Fujinami et al. have prepared alminate salts containing two PEO tails and two electron withdrawing unit such as $-CO_2CF_3$, $-SO_3CF_3$, and $-OC_6F_5$ [23], These compounds were obtained as molten salts. Among them, remarkably high ionic conductivity of 10^{-4} S cm^{-1} at ambient temperature was realized when the $-CF_3SO_3$ group was introduced (Figure 22.6(b)).

In another approach, Lewis acidic boron containing PEO derivatives was investigated as an "anion-trapping" polymer electrolytes [24]. Since Lewis acidic boron compounds have a vacant p-electron orbital, the mobile anions in the polymer electrolytes should effectively be trapped [25]. Mehta and Fujinami et al. reported that a boroxine ring based network polymer/LiSO₃CF₃ mixture showed a high lithium-ion transference number of over 0.7 (Figure 22.7a). Angell et al. prepared anion-trapping polymers by the dehydrative condensation reaction between phenylboric acid and PEO (Figure 22.7b). They pointed out that polyelectrolytes containing borate groups can be simply prepared by adding a variety of salts or bases to these acidic boron containing polyethers. Since borate anions tend to coordinate weakly to cations, organoboron-based anionic species are favorable to develop high ionic conductivities [26]. Recently Xu and Angell prepared poly(lithium mono-oxalato borate), a new type of polyanionic electrolytes (Figure 22.7c) [27]. The obtained polymer showed the ionic conductivity of over 10^{-5} S cm^{-1} at ambient temperatures.

Finally, well-defined boron-containing polymers by dehydrocoupling reactions have been prepared for a hydroboration reaction (Figure 22.7d) [28, 29]. These polymers can be readily obtained in mild conditions without generating water or other by-products. This synthetic route is preferred because it avoids the hydrolysis and disproportionation of boron-containing polymers. Recently we examined a

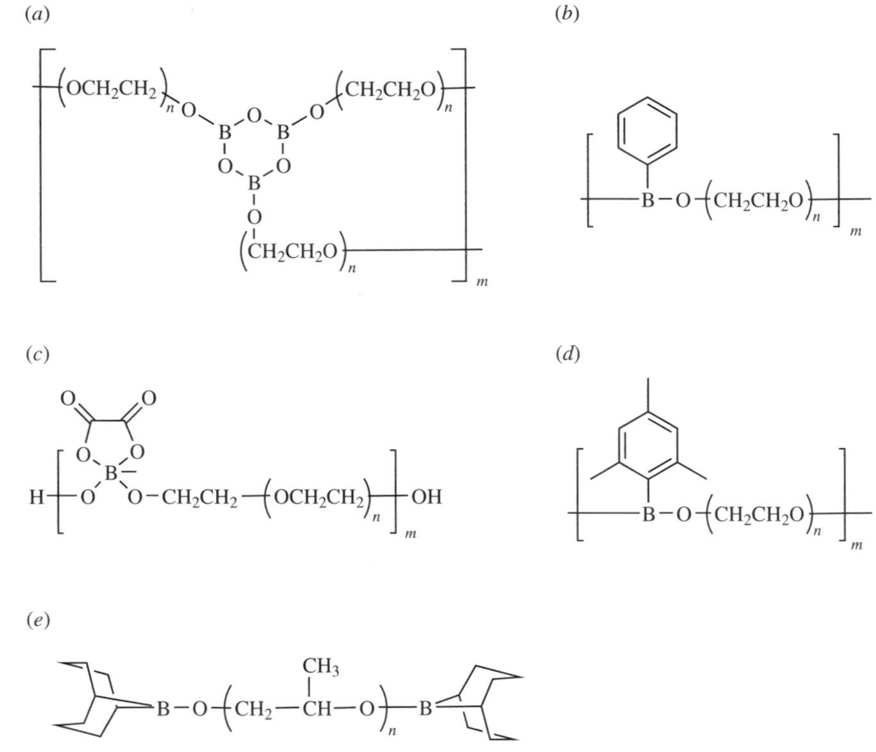

(a)

(b)

(c)

(d)

(e)

Figure 22.7 Polyethers containing boron atoms.

facile preparation of anion-trapping oligomers by the one-pot reaction between 9-borabicyclo[3.3.1]nonane (9-BBN) and PPO [30]. The 9-BBN is a common, commercially available hydroboration reagents [31]. The obtained oligomer/LiSO$_3$CF$_3$ mixture showed ionic conductivity of over 10^{-6} S cm^{-1} at room temperature and high lithium-ion transference numbers of over 0.7; both values are comparable to the conventional PPO/salt hybrids (Figure 22.7e).

22.6 ANION CONDUCTIVE POLYETHER/SALT HYBRID

As mentioned above, cation-conductive polymer electrolytes have been earnestly researched for practical use. However, polyethers are originally suitable for anion conduction, since the ion–dipole interaction is firmer for cations rather than anions. We have investigated the preparation and ion conductive properties of PEOs having onium salt moieties such as imidazolium, pyridinium, and triethylammonium salts on chain ends (Figure 22.5b) [2, 32]. These oligomers were readily obtained as viscous liquid with a low T_g, even when the PEO chain length was shorter than 400. The ionic conductivity of each system was higher than that of corresponding cation conductors. Both low T_g and high ionic conductivity were attributed to weak anion–dipole interaction. In the case of halogenic ion conductor, the ionic conductivity was improved to 10^{-5} S cm^{-1} by employing smaller anions such as Cl$^-$. The ionic conductivity was further improved by employing bulky and charge delocalized anions such as SCN$^-$ and CF$_3$SO$_3^-$. At the same time significantly low T_g at about $-70°$C was observed. These results suggest that smaller anions can migrate faster than bulky ions in PEO matrices, whereas the bulky ions are effective both to increase the number of carriers, and to avoid the ion–dipole interactions.

22.7 CONCLUSION

Polyether/salt hybrids have considerable possibility as new molten salts. The most advantageous point is that any salt can be prepared as a liquid by introducing a polyether chain to the ions. Some drawbacks such as relatively low conductivity, high viscosity, and so on, remain to be improved.

REFERENCES

1. (a) F. M. Gray, ed., *Solid Polymer Electrolytes*, VCH, 1991. (b) D. Baril, C. Michot, M. Armand, *Solid State Ionics*, **94**, 35 (1997).

2. (a) K. Ito, N. Nishina, Y. Tominaga, H. Ohno, *Solid State Ionics*, **86–88**, 325 (1996). (b) K. Ito, H. Ohno, *Electrochim. Acta*, **43**, 1247 (1998).

3. (a) D.E. Fenton, J. M. Parker, P. V. Wright, *Polymer*, **14**, 589 (1973).(b) P. V. Wright, *Br. Polym. J.*, **7**, 319 (1975). (c) P. V. Wright, *Electrochim. Acta*, **43**, 1137 (1998).

4. J. M. Tarascon, M. Armand, *Nature*, **414**, 359 (2001).

5. H. Ohno, K. Ito, *Polymer*, **36**, 891 (1995).

6. K. Xu, C. A. Angell, *Electrochim. Acta*, **40**, 2401 (1995).

7. W. Ogihara, M. Yoshizawa, H. Ohno, *Chem. Lett.*, 880, **2002**.

8. (a) M. Yoshizawa, M. Hirao, K. I. Akita, H. Ohno, *J. Mater. Chem.*, **11**, 1057 (2001). (b) M. Yoshizawa, A. Narita, and H. Ohno, *Aust. J. Chem.*, **57**, 139 (2004).

9. (a) K. Ito, H. Ohno, *Solid State Ionics*, **79**, 300 (1995). (b) K. Ito, Y. Tominaga, H. Ohno, *Electrochim. Acta*, **42**, 1561 (1997).

10. K. Kato, K. I. Akita, H. Ohno, *J. Solid State Electrochem*, **4**, 141 (2000).

11. (a) Y. Tominaga, K. Ito, H. Ohno, *Polymer*, **38**, 1949 (1997). (b) Y. Tominaga, H. Ohno, *Solid State Ionics*, **124**, 323 (1999). (c) Y. Tominaga, H. Ohno, *Chem. Lett.*, 955 (1998). (d) Y. Tominaga, H. Ohno, *Electrochim. Acta*, **45**, 3081 (2000).

12. Y. Tominaga, T. Mizumo, H. Ohno, *Polym. Adv. Technol.*, **11**, 524 (2000).

13. T. Mizumo, H. Ohno, *Polymer*, **45**, 861 (2004).

14. H. Ohno, H. Kawanabe, *Polym. Adv. Technol.*, **7**, 754 (1996).

15. (a) J. Foropoulos, D. D. Desmarteau, *Inorg. Chem.*, **23**, 3720 (1984). (b) A. Webber, *J. Electrochem. Soc.*, **138**, 2586 (1991). (c) P. Johansson, P. Jaccobsson, *Electrochim. Acta*, **46**, 1545 (2001).

16. (a) D. Teeters, S. L. Stewart, L. Svoboda, *Solid State Ionics*, **28–30**, 1054 (1988). (b) M. Begin, C. Vachon, C. Labreche, B. Goulet, J. Prud'homme, *Macromolecules*, **31**, 96 (1998). (c) A. Ferry, M. Tian, *Macromolecules*, **30**, 1214 (1997).

17. (a) G. Zouu, I. M. Khan, J. Smid, *Macromolecules*, **26**, 2202 (1993). (b) H. Ohno, K. Ito, *Polym. Adv. Technol.*, **2**, 97 (1991).

18. (a) E. Tsuchida, N. Kobayashi, H. Ohno, *Macromolecules*, **21**, 96 (1986). (b) M. Watanabe, S. Nagano, K. Sanui, N. Ogata, *Solid State Ionics*, **28–30**, 911 (1988). (c) H. Derand, B. Wesslen, B. E. Mellander, *Electrochim. Acta*, **43**, 1525 (1998).

19. (a) D. Benrabah, S. Sylla, F. Alloin, J. Y. Sanchez, M. Armand, *Electrochim. Acta*, **40**, 2259 (1995). (b) J. F. Snyder, J. C. Hutchison, M. A. Ratner, D. F. Shriver, *Chem. Mater.*, **15**, 4223 (2003).

20. D. P. Siska, D. F. Shriver, *Chem. Mater.*, **13**, 4698 (2001).

21. (a) A. E. Feiring, S. K. Choi, M. Doyle, E. R. Wonchoba, *Macromolecules*, **33**, 9262 (2000).(b) O. E. Geiculescu, J. Yang, H. Blau, R. Bauley-Walsh, S. E. Creager, W. T. Pennington, D. D. Desmarteau, *Solid State Ionics*, **148**, 173 (2002). (c) M. A. Hofmann, C. M. Ambler, A. E. Maher, E. Chalkova, X. Y. Zhou, S. N. Lvov, H. R. Allcock, *Macromolecules*, **35**, 6490 (2002).

22. (a) K. E. Doan, M. A. Ratner, D. F. Shriver, *Chem. Mater.*, **3**, 418 (1991). (b) T. Fujinami, A. Tokimune, M. A. Mehta, D. F. Shriver, G. C. Rawsky, *Chem. Mater.*, **9**, 2236 (1997). (c) T. Fujinami, M. A. Mehta, K. Sugie, M. Mori, *Electrochim. Acta*, **45**, 1181 (2000).

23. T. Fujinami, Y. Buzoujima, *J. Power Sources*, **119–121**, 438 (2003).

24. (a) S. S. Zhang, C. A. Angell, *J. Electrochem. Soc.*, **143**, 4047 (1996). (b) H. S. Lee, X. Q. Yang, C. L. Xiang, J. M. McBreen, L. S. Choi, *J. Electrochem. Soc.*, **145**, 2813 (1998). (c) J. McBreen, H. S. Lee, X. Q. Yang, X. Sun, *J. Power Sources*, **89**, 163 (2000). (d) H. S. Lee, X. Q. Yang, X. Sun, J. McBreen, *J. Power Sources*, **97–98**, 566 (2001).

25. (a) M. A. Mehta, T. Fujinami, *Chem. Lett.*, 915, **1997**. (b) M. A. Mehta, T. Fujinami, *Solid State Ionics*, **113–115**, 187 (1998). (c) M. A. Mehta, T. Fukonami, S. Inoue, K. Matsushita, T. Miwa, T. Inoue, *Electrochim. Acta*, **45**, 1175 (2000). (d) X. Sun, C. A. Angell, *Electrochim. Acta*, **46**, 1467 (2001). (e) W. Xu, X. Sun, C. A. Angell, *Electrochim. Acta*, **48**, 2255 (2003). (f) T. Hirakimoto, M. Nishiura, M. Watanabe, *Electrochim. Acta*, **46**, 1609 (2001). (g) S. Tabata, T. Hirakimoto, M. Nishiura, M. Watanabe, *Electrochim. Acta*, **48**, 2105 (2003).

26. (a) W. Xu, C. A. Angell, *Electrochem. Solid State Lett.*, **4**, E1 (2001). (b) H. Yamaguchi, H. Takahashi, M. Kato, J. Arai, *J. Electrochem. Soc.*, **150**, A312 (2003).

27. (a) W. Xu, C. A. Angell, *Solid State Ionics*, **147**, 295 (2002). (b) W. Xu, D. Williams, C. A. Angell, *Chem. Mater.*, **14**, 401 (2002). (c) W. Xu, L. Wang, C. A. Angell, *Electrochim. Acta*, **48**, 2037 (2003).

28. (a) N. Matsumi, K. Sugai, H. Ohno, *Macromolecules*, **35**, 5731 (2002). (b) N. Matsumi, K. Sugai, H. Ohno, *Macromolecules*, **36**, 2321 (2003).

29. (a) Y. Chujo, I. Tomita, Y. Hashiguchi, H. Tanigawa, E. Ihara, T. Saegusa, *Macromolecules*, **24**, 345 (1991). (b) Y. Chujo, I. Tomita, N. Murata, H. Mauermann, T. Saegusa, *Macromolecules*, **25**, 27 (1992).

30. T. Mizumo, K. Sakamoto, N. Matsumi, H. Ohno, *Chem. Lett.*, **33**, 396 (2004).

31. (a) E. F. Knights, H. C. Brown, *J. Am. Chem. Soc.*, **90**, 5280 (1968). (b) H. C. Brown, R. Liotta, G. W. Kramer, *J. Am. Chem. Soc.*, **101**, 2966 (1979). (c) H. C. Brown, J. C. Chen, *J. Org. Chem.*, **46**, 3978 (1981).

32. M. Yoshizawa, K. I. Akita, H. Ohno, *Electrochim. Acta*, **45**, 1617 (2000).

Chapter 23

Electric Conductivity and Magnetic Ionic Liquids

Gunzi Saito

In the 1950s an increase of electric conductivity by about 10^2 to 10^4 times was observed in the melting of insulating aromatic hydrocarbons such as naphthalene [1]. The increase of electric conductivity was ascribed not to the increase of carrier concentration and the contribution of ionic conduction of the molecular ions but to the increase of carrier mobility [1]. The first highly conductive organic semiconductor, developed by Akamatu, Inokuchi, and Matsunaga in the 1950s, was perylene • bromide, in which the cation radical perylene molecules contributed to the electronic conduction. This compound was classified as a cation radical salt [2]. The first organic metal tetrathiafulvalene (TTF, the main molecules are depicted in Figure 23.1) • tetra-cyanoquinodimethane (TCNQ) was developed in 1970s [3]. The TTF molecule is an electron donor (D) while the TCNQ molecule is an electron acceptor (A). The complex is classified as an ionic DA charge transfer complex having a partial charge transfer state $TTF^{0.59+} \cdot TCNQ^{0.59-}$ and both components contribute to the electronic conduction. The first organic superconductor $(TMTSF)_2PF_6$ (TMTSF: tetramethyl-tetraselenafulvalene) and the 10K class organic superconductor $\kappa\text{-}(BEDT\text{-}TTF)_2\text{-}Cu(NCS)_2$ (BEDT-TTF: bis(ethylenedithio)-TTF) developed in 1980s are cation radical salts [4, 5]. The component molecules or ions in these conducting crystals are associated by the ionic and van der Waals interactions in addition to the metallic bond, and in general, the crystals are mechanically fragile and thermally not durable. During heating they either decompose or provide an unstable molten complex

Electrochemical Aspects of Ionic Liquids   Edited by Hiroyuki Ohno
ISBN 0-471-64851-5   Copyright © 2005 John Wiley & Sons, Inc.

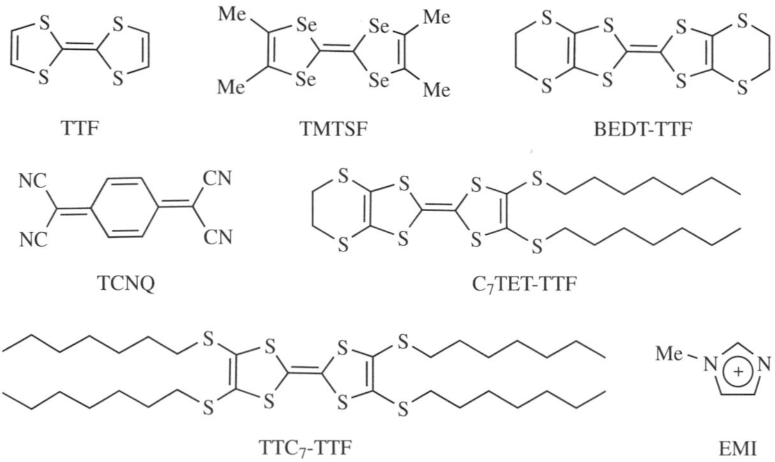

TTF   TMTSF   BEDT-TTF

TCNQ   C₇TET-TTF

TTC₇-TTF   EMI

Figure 23.1 *Chemicals in this chapter.*

that decomposes or dissociates into its component molecules, resulting in the rapid decrease of electric conductivity. However, although the neutral DA charge transfer complexes have rather poor electric conductivity, some of them provide stable melts, and a rapid increase of conductivity has been observed during the melting [6].

So far no liquid organic metals of the charge transfer type have been developed. The organic semiconductive condenser (OSCON) is based on the melting phenomenon of the solid complex, N-n-butylisoquinolinium (TCNQ)$_2$. Generally, the use of molecules having long alkyl chains makes it difficult to grow single crystals, but the obtained solid complex has both low melting point and thermally stable melting state. The solidified complex N-n-butylisoquinolinium (TCNQ)$_2$ (anion radical salt in which TCNQ$^{0.5-}$ contributes the electric conduction) after melting (m.p. = 215°C) keeps high conductivity (room temperature conductivity σ_{RT} = 0.3 S cm^{-1}). The frequency dependence of the impedance of the alumina condenser including the above-mentioned solidified complex as an electrolyte is considerably better than that of the tantalum condenser in the frequency region of 10 to about 1000 kHz [7].

In this chapter the neutral DA charge transfer complex that exhibits a rapid increase of conductivity during the melt, the cation radical salts of TTF derivatives with low melting point, the anion radical salts of TCNQ with imidazolium cations with low melting point, and 1-ethyl-3-methylimidazolium (EMI) salts containing complexes with paramagnetic metals will be described.

23.1 NEUTRAL DA CHARGE TRANSFER COMPLEX

Neutral DA charge transfer complexes $D^{\delta+} \cdot A^{\delta-}$ ($\delta \approx 0$) have a lower melting point than those of ionic DA complexes ($\delta \geq 0.5$), since the Madelung energy in the neutral DA makes a much smaller contribution to the lattice energy. For the neutral DA complex, (1) the concentration of conduction carriers is negligible at low temperatures since $\delta \approx 0$ and (2) there is no conduction path with a small activation energy since the lattice is made of alternating stacks composed of D and A molecules having different potential energies. Consequently the neutral DA complex has a small n (carrier concentration) and μ (carrier mobility), resulting in poor conductivity, $\sigma = ne\mu$, in the solid state. Many studies on the transport properties of the charge transfer solids have been reported, whereas for the liquid state there have been no such studies. Only a few studies on the optical characterization of the molten charge transfer complexes have been reported [8].

We observed a rapid increase of dc conductivity up to 10^{-4} S cm^{-1} above about 60°C in a 1:1 neutral charge transfer DA complex ($\sigma_{RT} < 10^{-9}$ S cm^{-1}) composed of donor C_7TET-TTF (m.p. = 12°C) and acceptor TCNQ molecules. At about 60°C the complex became a fluid that coexisted with the microcrystals of TCNQ. Concomitantly the spin concentration exhibited an increase by about 100 times, which corresponded to about $10^{-2}\%$ of the total complexes in terms of EPR measurement. The fluid also presented a charge transfer band similar to that observed in the solution's spectrum. The contribution of ionic conductivity was not large in the fluid because the conductivity was kept nearly constant for a long period of time under constant dc voltage. Based on these observations, the mechanism and design of liquid organic semiconductors (liquid electronic conductors) can be described as follows.

The simultaneous occurrence of the rapid increase of conductivity and spin concentration upon the melting of solid complex ($D^{\delta+} \cdot A^{\delta-}$) indicates that the carriers ($D^{\bullet+}$ and $A^{\bullet-}$) are generated in the following sequence: The molten complex ($D^{\delta+} \cdot A^{\delta-}$) dissociates into liquid neutral donors and neutral TCNQ molecules. These have an equilibrium with the radical species ($D^{\bullet+}$ and $A^{\bullet-}$), and the radicals migrate in the liquid. However, the migration of the ionic species may be different from conventional ionic conduction since the ionic conductivity was a minor contribution.

Since the C_7TET-TTF molecules have long alkyl chains, it is possible that they assemble in the melt to form a self-assembled matrix in which the π-moieties and alkyl chains of C_7TET-TTF are arranged as depicted in Figure 23.2. The electron

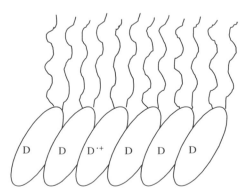

Figure 23.2 *Schematic representation of a self-assembled array which consists of $D^{\bullet+}$ and D (C_7TET-TTF) molecules. Each molecule is composed of a π-electron part (ellipse) and two alkyl chains (wavy lines).*

transfer is surmised to be as shown by equation (23.1). The donor molecules in this donor array are close to that observed in a segregated column in a highly conductive charge transfer solid:

$$D^{\bullet+}DDD \rightarrow DD^{\bullet+}DD \rightarrow DDD^{\bullet+}D \rightarrow DDDD^{\bullet+}. \qquad (23.1)$$

To obtain a conductive melt based on charge transfer complex, it is advantageous to use the neutral DA charge transfer solid because the Madelung energy is not dominant in the lattice energy. Although a high concentration of carrier density is not present in a neutral complex, it is known that for the extrinsic semiconductor some carrier doping remarkably enhances the conductivity [9].

The high conductivity in liquid crystal ($1.2 \times 10^{-3}\,\mathrm{S\,cm^{-1}}$ at 110°C) due to the viologen derivative has been interpreted in terms of the interaction of ionic conduction of the viologen dication, the cation radical, and iodine species [10].

23.2   CATION RADICAL SALTS WITH LOW MELTING POINT

Recently cation radical salts with a low melting point have been prepared using the TTC_7–TTF molecule (m.p. $= 44$–46°C), in which the *n*-heptylthio groups attached to both sides of TTF skeleton [11].

Mixing neutral TTC_7–TTF with radical salt $(TTC_7$–$TTF^{2+})(Br_3^-)_2$ in suitable ratio gave mixed valence salts $(TTC_7$–$TTF^{n+})(Br_3^-)_n$, and their characteristics varied depending on the dopant concentration. In the case of $1 < n < 2$, the resulting compounds were purple crystallites, while dark brown viscous liquids were obtained in the case of $0 < n < 1$. The conductivity at RT was in the range of 10^{-5} to $10^{-7}\,\mathrm{S\,cm^{-1}}$ for the liquid form (maximum value is $8 \times 10^{-6}\,\mathrm{S\,cm^{-1}}$), and the ion aggregations led to a slight increase of conductivity (maximum value is $2 \times 10^{-5}\,\mathrm{S\,cm^{-1}}$). It is thus apparent that the conduction is entirely between the TTC_7–TTF^{n+} molecules instead of the Br_3^- anions and is therefore by way of the electrons (holes).

When neutral TTC_7–TTF and nitrosonium $NOBF_4$ were reacted in hexane, viscous liquids $(TTC_7$–$TTF^{n+})(BF_4^-)_n$ were obtained. As the doped concentration increased, the color changed from initially dark brown to deep green and then blue. Although the conductivity at RT increased as the mixing ratio of the starting materials approached unity ($n = 1$; the maximum value is $4 \times 10^{-7}\,\mathrm{S\,cm^{-1}}$), the values were much lower than those of the tribromide salts.

23.3   TCNQ ANION RADICAL SALTS WITH LOW MELTING POINT

The series of TCNQ complexes is well known to be conductive organic anion radical salts. Among them, only those with a partial charge transfer state, where the average charge of TCNQ is a noninteger, exhibit high conductivity. In general, TCNQ molecules form a one-dimensional column, which exhibits the physical phenomena characteristic of low-dimensional electron systems (e.g., Peierls and spin-Peierls transitions). Since most of the melting (or decomposition) points of TCNQ complexes are over 200°C, the studies of TCNQ complexes have mainly been carried out in the solid state. For example, of the TCNQ complexes with a low melting point, a mixed crystal of propyl-isoquinolinium, methyl-isoquinolinium, TCNQ, and Me-TCNQ (m.p. = 80°C) was reported [12]. The average charge of the TCNQ derivative molecules in the mixed crystal was −0.5, and its conductivity at RT was 5×10^{-5} S cm^{-1}, which was much lower than that of other 1:2 TCNQ salts (e.g., (N-n-butyl-isoquinolinium)(TCNQ)$_2$: $\sigma_{RT} = 3 \times 10^{-1}$ S cm^{-1}, m.p. = 215°C [7]). This mixed crystal shows semiconductive behavior, and the activation energies (E_a) in the solid and liquid states were 0.14 eV and 0.11 eV, respectively, and the conductivity increased twice by melting.

In the ionic liquids based on the low-symmetric imidazolium cations, inorganic closed-shell anions have often been used as counter anions. The replacement of the inorganic anion with TCNQ anion radical should lead to the development of liquid conductors in which ionic and electronic conductions occur simultaneously. TCNQ salts of 1-alkyl-3-methylimidazolium (alkyl = methyl, ethyl, n-propyl, n-butyl, and n-hexyl) have been reported [12], and 1:1 and 2:3 stoichiometry salts, where the average charge of TCNQ molecules were −1 and −0.67, respectively, were obtained for each of cations. The conductivities at RT were 10^{-6} to 10^{-9} S cm^{-1} and 10^{-1} to 10^{-5} S cm^{-1} for 1:1 and 2:3 salts, respectively (see Table 23.1), and both liquids exhibited semiconductive behavior. For the 1:1 salt of 1-ethyl-3-methylimidazolium (EMI), dimers of the TCNQ molecules formed a one-dimensional column [14]. In Table 23.2 the melting points of the 1:1 salts are presented; those of the 2:3 salts were not reported [13]. We obtained 1:2 salts of 1-alkyl-3-methylimidazolium (alkyl = ethyl and n-butyl) [15] and their decomposition points are also presented in the table. All salts are solid at RT and have melting points higher than 100°C. The 1:2 salts have higher melting (or decomposition) points than those of the corresponding 1:1 salts, owing to their higher cohesive energy. A decrease of melting point was observed when TCNQ derivatives (R^2–TCNQ) were used instead of the pristine TCNQ. In the 1:1 salts of EMI, the melting points decreased in the order $R^2 = CF_3$, H, CH_3, and n-$C_{10}H_{21}$ (see Table 23.2, right).

TABLE 23.1 List of Conductivities at Room Temperature (σ_{RT}/S cm^{-1}) of TCNQ Anion Radical Salts of 1-alkyl(R^1)-3-methylimidazolium and TCNQ

R^1	1:1[a]	2:3[a]	1:2
CH_3	8×10^{-8}	7×10^{-5}	
C_2H_5	4×10^{-8}	2×10^{-3}	5×10^{-4}
n-C_3H_7	3×10^{-9}	2×10^{-2}	
n-C_4H_9	3×10^{-7}	2×10^{-1}	2×10^{-1}
n-C_6H_{13}	7×10^{-6}	4×10^{-2}	

[a]From Šorm et al. [13].

TABLE 23.2 **List of Melting Points (°C) of TCNQ Anion Radical Salts with Imidazolium Cations. Left Panel: Salts of 1-alkyl(R¹)-3-methylimidazolium and TCNQ; Right Panel: Salts 1-ethyl-3-methylimidazolium and Substituted TCNQs (R²-TCNQ)**

R^1	$1:1^a$	$1:2^b$	R^2	$1:1$
CH_3	174–177		H	139.9–140.2
C_2H_5	143–146	225–226	CH_3	128.8–129.4
$n\text{-}C_3H_7$	159–161		CF_3	144.3–144.7
$n\text{-}C_4H_9$	154–156	190–194	$n\text{-}C_{10}H_{21}$	~40
$n\text{-}C_6H_{13}$	124–126			

^{a}From Šorm et al. [13].
^{b}Decomposition point.

The conducting behavior in the liquid state has not been reported for these TCNQ salts. We found that the conductivity of the 1:1 salt of EMI–TCNQ increased by four orders of magnitude by melting and decreased by three orders of magnitude by freezing (Figure 23.3) [15]. The arrows in the figure show the hysteresis of the temperature dependence of conductivity. The conducting behavior was semiconductive with $E_a = 0.41$ and 0.38 eV for the solid and liquid states, respectively. The value of conductivity in the liquid state ($\sim 10^{-2}$ S cm^{-1}) was quite high compared to those of ordinary monovalent TCNQ salts, namely $< 10^{-4}$ S cm^{-1}. It seems that the high conductivity in the liquid state was realized by the increase of conducting carriers by the dissociation of dimers $((TCNQ^{\bullet -})_2 \rightleftarrows 2TCNQ^{\bullet -})$ and the contribution of ionic conductivity in addition to electronic conductivity

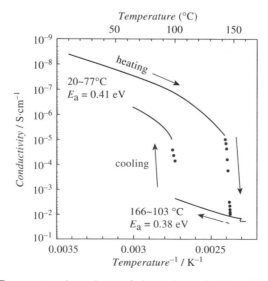

Figure 23.3 *Temperature dependence of electronic conductivity of EMI·TCNQ (1:1).*

of TCNQ$^{\cdot -}$. This salt kept the supercooled state down to around 100°C, which was lower than its melting point by 40°C.

23.4 EMI SALTS CONTAINING COMPLEXES WITH PARAMAGNETIC METALS

Although numerous reports have appeared describing catalytic, spectrochemical, and electrochemical investigations of halometalates in N,N'-dialkylimidazolium (e.g., EMI) chloride/AlCl$_3$ ionic liquid, there are only a few RT ionic liquids composed of N,N'-dialkylimidazolium cation and complexes with paramagnetic metals, for which their physical properties were investigated.

Table 23.3 shows the EMI-based RT ionic liquid containing paramagnetic metals [EMI][FeCl$_4$] [16–18], and Table 23.4 the structurally characterized EMI-based salts containing paramagnetic metals [18–24]. All the metal complexes listed in the tables are based on halometallates, although some EMI-based salts with diamagnetic Ag^I(CN)$_2$ [25], Cd$^{II}_2$(SCN)$_6$ [26] and Pb^{II}(O$_2$CMe)$_4$ [27] anions, and BMI-based (BMI: 1-butyl-3-methylimidazolium) salts with paramagnetic M^{II}(d-mit)$_2$ (M = Ni and Pd) [28] anions have been structurally reported. This implies that EMI-based salts with paramagnetic metals are at a primitive level of development. One of the most pronounced features is that the salts with monoanionic complexes have a lower melting point, and this marked difference can be explained by the interionic coulomb interactions. Physical properties of [EMI] [FeCl$_4$] have been recently investigated by our group [17,18], and it appears that its ionic conductivity

TABLE 23.3 EMI-based RT Ionic Liquids Containing Paramagnetic Metals

Compound	Central Metal	Melting Point	References
[EMI][FeCl$_4$]	Fe(III)	18°C	16–18

TABLE 23.4 Structurally Characterized EMI-based Salts Containing Paramagnetic Metals

Anion	Central Metal	Melting Point (°C)	References
[VOCl$_4$]$^{2-}$	V(IV)	99	19
[Zr$_6$MnCl$_{18}$]$^{5-}$	Mn(II)^a		20,21
[FeCl$_4$]$^{2-}$	Fe(II)	86	18
[Zr$_6$FeCl$_{18}$]$^{4-}$	Fe(III)^a		20
[CoCl$_4$]$^{2-}$	Co(II)	100–102	22
[NiCl$_4$]$^{2-}$	Ni(II)	92–93	22
[Ru$_2$Br$_9$]$^{3-}$	Ru(III)		23
[IrCl$_6$]$^{2-}$	Ir(IV)	220^b	24

^{a}Valence state for starting material.
^{b}Decomposition temperature.

(1.8×10^{-2} S cm^{-1} at 20°C) is comparable to that of the well-known [EMI][AlCl$_4$] (2.3×10^{-2} S cm^{-1} at 25°C [29]). From the ac and dc susceptibility measurements we learned that the salt has an uncorrelated high-spin system ($S = 5/2$) at around RT regardless of its form, indicating the first conductive-paramagnetic bifunctional ionic liquid. Also it passes through an antiferromagnetic transition at 4.2 K (spin-flop field is ca. 8 kOe). The transition temperature is considerably lower than the value of 9.75 K observed for neat FeCl$_3$ [30], possibly owing to the presence of EMI cations. On the other hand, the Fe^{III}Cl$_4$ salt magnetically diluted by diamagnetic Ga^{III}Cl$_4$ anions exhibits no magnetic transition down 1.9 K, and follows the Curie-Weiss law.

An EMI-based salt with dicationic irons, [EMI]$_2$[Fe^{II}Cl$_4$], was also prepared and characterized by our group [18]. The powder X-ray diffraction data were collected using a synchrotron radiation on the beam line BL02B2 at SPring-8, and the crystal structure was refined by Rietveld method. It appears that the salt is isomorphous to the reported [EMI]$_2$[M^{II}Cl$_4$] (M = Co and Ni) salts [22], where each EMI cation has hydrogen bonds with three tetrahedral MCl$_4$ anions. The unit cell volume (3929.03 Å^3) is apparently larger than those of the two isomorphous salts (3890 and 3871 Å^3), which may be related to its lower melting point than those of the Co^{II}Cl$_4$ and Ni^{II}Cl$_4$ salts. An estimated magnetic moment at RT (5.22μ$_B$) indicates the uncorrelated high-spin system ($S = 2$), as observed for the Co^{II}Cl$_4$ and Ni^{II}Cl$_4$ salts, and the spin configurations are largely caused by the tetrahedral geometry of anions.

The intercalation of ionic liquids containing paramagnetic anions into materials with layer (e.g., graphite) or porous (e.g., zeolite) structures and research concerning transport properties under the magnetic field are both promising areas for future exploration.

Notes: After submittion of this work and Ref. 17, a work on the paramagnetic behavior of [BMI] [FeCl$_4$] was submitted and recently published [31].

REFERENCES

1. J. Kommandeur, G. J. Korinek, W. G. Schneider, *Can. J. Chem.*, **36**, 513 (1958).

2. H. Akamatu, H. Inokuchi, Y. Matsunaga, *Nature*, **173**, 168 (1954).

3. J. P. Ferraris, D. O. Cowan, V. Walatka Jr., J. H. Perlstein, *J. Am. Chem. Soc.*, **95**, 948 (1973).

4. D. Jerome, K. Bechgaard, *Sci. Am.*, **247**, 50 (1982).

5. T. Ishiguro, K. Yamaji, G. Saito, *Organic Superconductors*, 2nd ed., Springer, 1998.

6. A. Otsuka, G. Saito, T. Nakamura, M. Matsumoto, Y. Kawabata, K. Honda, M. Goto, M. Kurahashi, *Synth. Met.*, **27**, B575 (1988).

7. (a) S. Niwa, *Synth. Met.*, **18**, 665 (1987). (b) I. Isa, *Nikkei New Materials*, Issue Jan. 30, 48 (1989). (c) S. Niwa, Y. Taketani, *J. Power Sources*, **60**, 165 (1996).

8. (a) J. Aihara, A. Sasaki, Y. Matsunaga, *Bull. Chem. Soc. Jpn.*, **43**, 3323 (1970). (b) J. Aihara, *ibid.*, **48**, 1031 (1975).

9. C. Kittel, *Introduction to Solid State Physics*, 7th ed., Chapter 8, John Wiley and Sons, 1996.

10. I. Tabushi, K. Yamamura, K. Kominami, *J. Am. Chem. Soc.*, **108**, 6409 (1986).

11. Y. Yoshida, A. Otsuka, G. Saito, in preparation.

12. V. A. Starodub, E. M. Gluzman, K. I. Pokhodnya, M. Ya. Valakh, *Theor. Exp. Chem.*, **29**, 240 (1993).

13. M. Šorm, S. Nešpůrek, M. Procházka, I. Koropecký, *Coll. Czech. Chem. Commun.*, **48**, 103 (1983).

14. M. C. Grossel, P. B. Hitchcock, K. R. Seddon, T. Welton, S. C. Weston, *Chem. Mater.*, **6**, 1106 (1994).

15. K. Nishimura, G. Saito, *Synth. Met.*, in press.

16. Y. Katayama, I. Konishiike, T. Miura, T. Kishi, *J. Power Sources*, **109**, 327 (2002).

17. Y. Yoshida, J. Fujii, K. Muroi, A. Otsuka, G. Saito, M. Takahashi, T. Yoko, *Synth. Met.*, in press.

18. Y. Yoshida, A. Otsuka, G. Saito, S. Natsume, E. Nishibori, M. Takata, M. Sakata, M. Takahashi, T. Yoko, submitted.

19. P. B. Hitchcock, R. J. Lewis, T. Welton, *Polyhedron*, **12**, 2039 (1993).

20. C. E. Runyan, Jr., T. Hughbanks, *J. Am. Chem. Soc.*, **116**, 7909 (1994).

21. D. Sun, T. Hughbanks, *Inorg. Chem.*, **39**, 1964 (2000).

22. P. B. Hitchcock, K. R. Seddon, T. Welton, *J. Chem. Soc., Dalton Trans.*, 2639 (1993).

23. D. Appleby, R. I. Crisp, P. B. Hitchcock, C. L. Hussey, T. A. Ryan, J. R. Sanders, K. R. Seddon, J. E. Turp, J. A. Zora, *J. Chem. Soc., Chem. Commun.*, 483 (1986).

24. M. Hasan, I. V. Kozhevnikov, M. R. H. Siddiqui, C. Femoni, A. Steiner, N. Winterton, *Inorg. Chem.*, **40**, 795 (2001).

25. Y. Yoshida, K. Muroi, A. Otsuka, G. Saito, M. Takahashi, T. Yoko, *Inorg. Chem.*, **43**, 1458 (2004).

26. F. Liu, W. Chen, X. You, *J. Solid State Chem.*, **169**, 199 (2002).

27. J. T. Hamill, C. Hardacre, M. Nieuwenhuyzen, K. R. Seddon, S. A. Thompson, B. Ellis, *Chem. Commun.*, 1929 (2000).

28. (a) W. Xu, D. Zhang, C. Yang, X. Jin, Y. Li, D. Zhu, *Synth. Met.*, **122**, 409 (2001) (b) C. Jia, D. Zhang, W. Xu, X. Shao, D. Zhu, *Synth. Met.*, **137**, 1345 (2003).

29. A. A. Fannin, D. A. Floreani, L. A. King, J. S. Landers, B. J. Piersma, D. J. Stech, R. L. Vaughn, J. S. Wilkes, J. L. Williams, *J. Phys. Chem.*, **88**, 2614 (1984).

30. E. R. Jones, O. B. Morton, L. Cathey, T. Auel, E. L. Amma, *J. Chem. Phys.*, **50**, 4755 (1969).

31. S. Hayashi, H. Hamaguchi, *Chem. Lett.*, 1590 (2004).

Part V

Ionic Liquids in Ordered Structures

Chapter 24

Ion Conduction in Plastic Crystals

Maria Forsyth, Jennifer M. Pringle, and Douglas R. MacFarlane

A plastic crystal can be defined as a material that consists of a regular three-dimensional crystalline lattice such that the long-range position of species is well defined, but orientational and/or rotational disorder exists with respect to the molecular species or molecular ions. These materials thus possess a larger number of degrees of freedom and hence display a higher entropy than a fully ordered solid [1–4]. As a result of this disorder a significant degree of motion is available that leads to a high degree of plasticity (i.e., ready deformation under an applied stress) and high diffusivity, which, in the case of ionic substances, leads to ionically conducting phases [5–11]. Fast ion conduction in plastic crystal phases of inorganic materials including Li_2SO_4 [12–15] and Na_3PO_4 [16, 17] have been of interest for some time; these materials conduct lithium ions and sodium ions, respectively, via a paddle wheel motion of the SO_4^{2-} and PO_4^{3-} anions.

Plastic crystal phases in organic materials have been known since the time of Timmermans [1], and these phases are often reached via a solid–solid transition below the final melting point of the crystal. These transitions often represent the onset of rotational motions of the molecules within the crystalline lattice and the resultant phases are sometimes referred to as rotator phases [18–23]. Timmermans proposed a general rule that plastic phases have a low final entropy of fusion ($\Delta S_f < 20 \, \text{J K}^{-1} \, \text{mol}^{-1}$) because the rotational component of the entropy of fusion of the fully ordered phase is already present in the plastic phase. The bulk of the

Electrochemical Aspects of Ionic Liquids Edited by Hiroyuki Ohno
ISBN 0-471-64851-5 Copyright © 2005 John Wiley & Sons, Inc.

early work on plastic crystals was on a wide range of organic molecular compounds [2] such as cyclohexane, succinonitril [24], and pentaglycerine [25]. Only more recently has it been discovered that ionic compounds can show plastic crystal behavior. One such family of compounds discovered in our laboratories are those based on substituted pyrrolidinium cations (**I**) and anions including *bis*(trifluoromethanesulfonyl) amide (TFSA) (**II**), BF_4^-, PF_6^-, I^-, SCN^-, and dicyanamide (DCA) (**III**) as shown in Scheme 24.1. These compounds usually display a series of sub-melting phase transitions and conduction of one or both ions in the higher temperature solid phases [6, 7, 26–29].

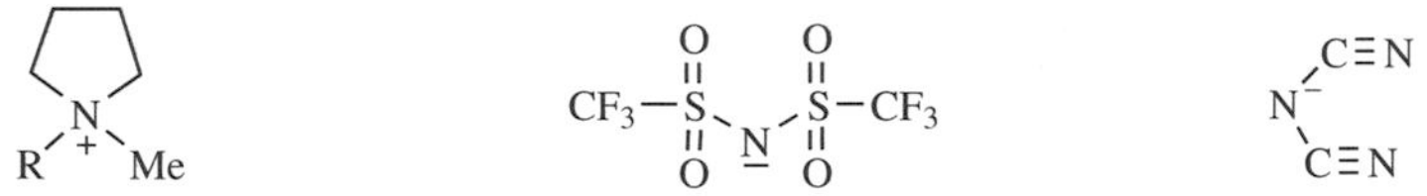

Pyrrolidinium

R = Me(P_{11}), Et(P_{12}), Pr(P_{13}) etc.

Bis(trifluoromethanesulfonyl)amide, TFSA

Dicyanamide (DCA)

(**I**)   (**II**)   (**III**)

Scheme 24.1

A further interesting effect discovered in our laboratories is that the addition of low levels of a second component, or "dopant" ion, can lead to significant increases in the ionic conductivity [6, 30, 31]. Typically these dopant species, for example, Li^+, OH^-, and H^+, are much smaller than the organic ions of the matrix, and since the relaxation times characterizing the motion of these ions are more rapid than those of the bulk matrix itself, these materials may represent a new class of fast ion conductor. The dopant ion effect can be used to design materials for specific applications, for example, Li^+ for lithium batteries and H^+/OH^- for fuel cells or other specific sensor applications. Finally, we have recently discovered that this dopant effect can also be applied to molecular plastic crystals such as succinonitrile [32]. Such materials have the added advantage that the ionic conductivity is purely a result of the dopant ions and not of the solvent matrix itself.

24.1 PLASTIC CRYSTAL PHASES—BACKGROUND

Since the initial work by Timmermans more than 40 years ago [1], plastic crystals have been extensively studied [18–21, 33–77]. The entropy of fusion criterion developed by Timmermans is not always obeyed in ionic plastic crystal materials, where the ion possesses internal degrees of freedom that only become activated at the melting point. However, in most cases where plastic crystal behavior is observed, the final entropy of fusion represents only a fraction of the total entropy change in the material from its lowest energy, low temperature phase to the final melting. Figure 24.1 gives a typical thermal analysis trace for a family of pyrrolidinium compounds. One mechanism whereby the entropy of the solid can increase, either suddenly at a phase transition or gradually over a range of temperatures, is through rotational disorder, meaning the onset of rotator motions around one or

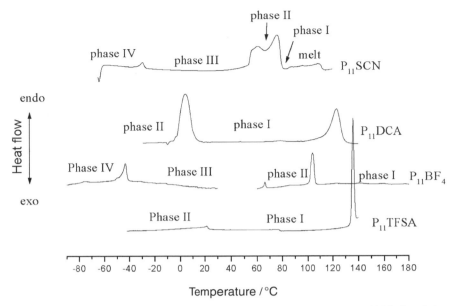

Figure 24.1 *Differential scanning calorimetry (DSC) traces for the different N,N'-dimethylpyr-rolidinium (P_{11}) species. The y-axis has been adjusted to allow comparison between species. Thermal transitions are recorded as peaks (endothermic) or troughs (exothermic) and indicate changes of phase. The structurally similar dicyanamide (DCA) and thiocyanate species both show broad phase transitions in comparison to the TFSA species.*

more molecular axes. Another, related alternative suggested is that molecules can find themselves in different, energetically equivalent positions between which they can interchange, the positions being related by rotational symmetry operators associated with the individual species (orientational disorder) [2]. The onset of rotator motions in the phases below the melting point then raises the entropy of the solid state closer to that of the liquid, resulting in a lower ΔS_f compared to that for a fully ordered, static crystal. It is proposed that with the onset of these rotations comes the formation of vacancy defects allowing high diffusivity and, ultimately, high "plasticity" meaning easy deformation under an applied stress. These materials are often seen to flow under their own weight (creep) [2].

Many salts composed of quaternary alkylammonium cations in combination with anions of high symmetry have been found to display rotator phases and/or plastic crystal phases, as determined by thermal analysis and nuclear magnetic resonance (NMR) measurements [19–21, 33, 53–56, 74–77]. High ionic conductivities have been measured in some of these systems, attributed to vacancy migration. Tansho et al. [10] have investigated plastic phases in NH_4Cl, NH_4Br, and NH_4NO_3, which all display isotropic rotation of the ammonium cation in the plastic phase and show cation diffusion via either a direct or indirect (interstitial intermediate) vacancy migration process. More complex species such as $(CH_3NH_3)_2ZnBr_4$ have also been shown to posses ionic plastic crystal phases at elevated temperatures [78, 79].

24.2   SYNTHESIS AND THERMAL PROPERTIES OF IONIC PLASTIC CRYSTAL ELECTROLYTES

The compounds prepared and characterized in our laboratories can all be described as salts of quaternized nitrogen cations combined with a variety of anions. The pyrrolidinium family of cations, in particular, shows rich phase behavior in the solid state. The anions themselves often possess some degree of rotational symmetry, and the degree of plasticity is certainly dependent on the nature of the anion as well as the cation, as discussed later.

The compounds are prepared by a metathesis reaction from the corresponding iodide or bromide salt. The compounds are typically low melting point salts (in many cases molten at room temperature), which range from being quite hydrophobic (TFSA, PF_6^-) to completely water soluble and hygroscopic (DCA, BF_4^-). The purity of the samples is established via solution NMR, electrospray mass spectroscopy, and chemical analytical analysis.

Figures 24.1 and 24.2 present the thermal analysis traces for a range of typical plastic crystal materials based on the pyrrolidinium cation. The highest temperature phase is denoted as phase I, with subsequent lower temperature phases denoted as phases II, III, and so on. It can be seen that the nature of the anion plays a significant role in the thermal behavior of these compounds. For example, $P_{11}TFSA$ has a barely detectable solid–solid transition at 20°C, followed by a melt at above

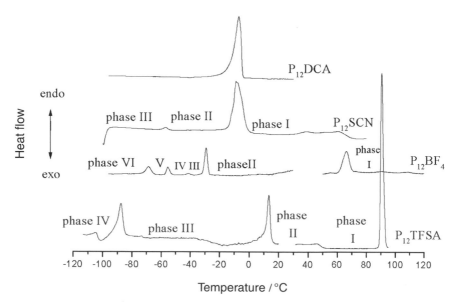

Figure 24.2   *DSC traces for the N-methyl-N-ethylpyrrolidinium (P_{12}) species. The y-axis has been adjusted to allow comparison between species. As observed for the P_{11} series, the DCA and thiocyanate species show broad phase transitions in comparison to the very sharp TFSA salt. The significant differences observed between the species highlights the strong influence of the anion in determining the phase behavior.*

130°C, whereas $P_{11}BF_4$ has a large transition from phase IV to phase III at $-43°C$, a small transition from phase III to phase II at 61°C, followed by phase II to phase I at 112°C, and finally decomposes at 340°C. $P_{11}DCA$ has a similarly rich thermal behavior. Comparison of Figures 24.1 and 24.2 also demonstrates that the substituent on the pyrrolidinium ring has a strong influence on the solid state behavior of the salts.

In many of these compounds, thermal analysis reveals metastable behavior, where rapid quenching can trap the material in higher temperature phases and glass transition temperatures can often be observed. The $P_{14}PF_6$ system (Figure 24.3) is one example where quenching from ambient temperature leads to the formation of a phase II glass, where the rotationally disordered state found in phase II is quenched into an arrested state and subsequently undergoes a transition (glass transition) at $-47°C$, and it is at this temperature that onset of the rotational motions of phase II occurs. Given that phase II is metastable at this temperature, the sample undergoes a cold crystallization to the stable phase III state before subsequently transforming to equilibrium phase II at higher temperatures. This behavior can lead to confusion when interpreting thermal analysis (and conductivity) traces of plastic crystals: one needs to be confident that the behavior is reproducible equilibrium behavior as opposed to metastable behavior.

Another interesting feature of plastic crystal thermal behavior is that the shape of the transitions is often nonsymmetric, and strongly resembles a lambda-like transition. Such transitions arise from the cooperative nature of the phenomenon leading to the phase transition. For example, rotational motion may initially occur in only some fraction of the molecules/ions in the crystal lattice, but with increasing temperature a larger fraction of molecules undergo the same motions, assisted by those already rotating, until at some culminating temperature the entire sample is undergoing identical motions. Figure 24.4 shows an example of this in the thermal

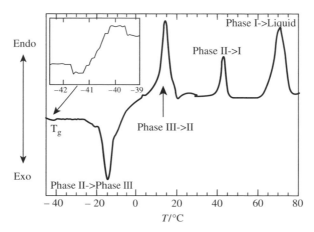

Figure 24.3 DSC trace of N-methyl-N-butylpyrrolidinium hexafluorophosphate ($P_{14}PF_6$) showing the low temperature glass transition, the exothermic crystallisation and the subsequent endothermic transitions up to the melt.

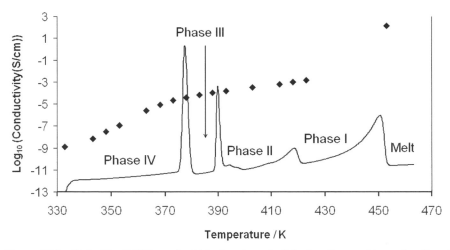

Figure 24.4 *Plot of the DSC trace (solid line) and conductivity (scatter graph) as a function of temperature for tetramethylammonium dicyanamide. The changes in conductivity with temperature mirror the thermal transitions, with an increase up to phase III and a dramatic increase on melting. As the temperature increases, there is a greater internal motion within the material and consequently the conductivity increases.*

analysis behavior of tetramethylammonium DCA. This behavior is consistent with NMR second moment analyses (Figure 24.5) [80] and with heat capacity data of a related compound, tetraethylammonium DCA (Figure 24.6) [81]. The broad melting associated with many organic plastic crystal compounds investigated in our laboratories is suggested to arise from the large fraction of vacancies present in phase I of the most plastic materials, which can lead to an impurity effect where the impurities are essentially the vacancy defects. This is consistent with the variable shape of the final melt that is often observed for the most plastic samples.

Doping of lithium salts into organic ionic plastic crystals usually leaves the solid–solid phases unchanged but decreases and broadens the final melting point of the compound, consistent with liquidus behavior. For higher dopant concentrations, extra thermal transitions often appear, consistent with eutectic melting and the presence of a new lithium-rich phase. Figure 24.7 shows the change in thermal behavior of P_{11}DCA when doped with LiDCA. The appearance of a second, isothermal peak at about 32°C, which is present even at 2 wt% of LiDCA, is suggestive of a relatively low solubility limit of this lithium salt in P_{11}DCA. The final melting peak shifts to lower temperatures, as is expected when a second component (or impurity) is mixed into a pure system.

24.3 CONDUCTIVITY IN ORGANIC IONIC PLASTIC CRYSTALS

The conductivity behavior of a number of different dimethylpyrrolidinium salts is illustrated in Figure 24.8. In all cases high conductivity is observed in phase I, the

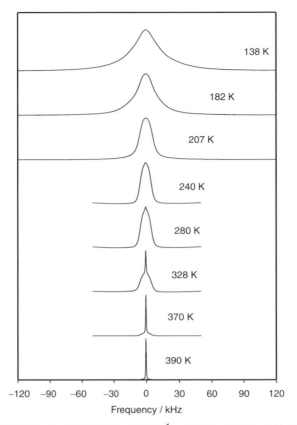

Figure 24.5a *Change in single-pulse static ^1H NMR spectra with temperature for tetramethylammonium dicyanamide. The spectra show line narrowing as the temperature increases, indicating the onset of various rotational modes within the material. The presence of multiple peaks within each spectrum indicates the existence of multiple modes of rotation.*

highest being displayed by the BF_4^- salt, which shows conductivity in excess of 5×10^{-3} S cm^{-1} at 110°C. Some salts show a distinct step in conductivity at one or more of the solid–solid phase transition temperatures. This behavior was also seen in Et$_4$NDCA, presented in Figure 24.6, and may be associated with an increased number of vacancy defects, as discussed below.

The effect of doping these plastic crystals with lithium salts containing the same anion has been investigated. An increase in the conductivity of phases I and II of P$_{12}$TFSA of up to three orders of magnitude has been observed when the material is doped with LiTFSA [6, 31]. Similar enhancements of ion conductivity are observed when LiDCA is doped into the plastic crystal phase of P$_{11}$DCA (Figure 24.9). Since the anion is common to both the additive and the matrix, the lithium can be thought

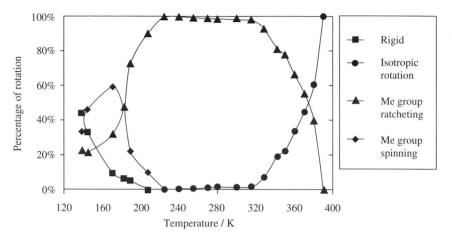

Figure 24.5b *Relative contributions of the predominant modes of cation rotation within the tetramethylammonium dicyanamide salt across a range of temperatures. The static state decreases rapidly with increasing temperature, as does the methyl group spinning. Isotropic tumbling begins to a very small degree at 240 K and increases dramatically above 315 K. These rotational transitions are unusual for the tetramethylammonium cation.*

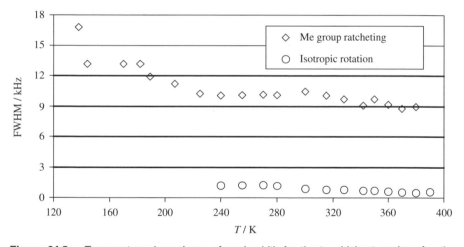

Figure 24.5c *Temperature dependence of peak width for the two highest modes of cation rotation within tetramethylammonium dicyanamide. The data show that the onset of rotations is not sudden but occurs over a broad temperature range. As the temperature increases the number of species undergoing isotropic rotations increases until a phase transition occurs.*

of as the cation dopant. The effect of the dopant on the conductivity is dramatic given the very small amounts of lithium involved.

In order to further improve the mechanical properties of these organic plastic crystals, especially at higher temperatures, mixtures with compatible polymers

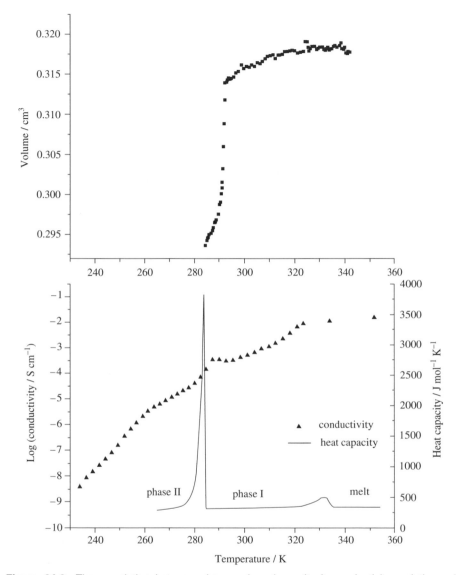

Figure 24.6 *The correlation between changes in volume (top), conductivity and thermal behavior with temperature for tetraethylammonium dicyanamide. There is an extremely large volume expansion on moving from phase II to phase I indicating considerable vacancy formation. This is mirrored by the simultaneous increase in conductivity.*

such as poly(vinyl pyrrolidone) (PVP) have been investigated. It is interesting to observe that just 2 wt% of PVP is capable of suppressing plastic behavior and reducing the ionic conductivity. Optical microscopy comparisons of $P_{13}BF_4$ based systems suggest that the crystal habitat changes substantially with polymer additions, although an explanation of this phenomenon is yet to be established.

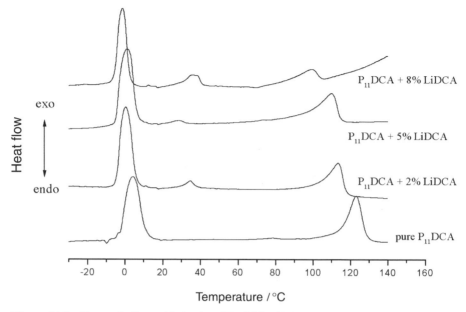

Figure 24.7 *Change in thermal behavior of $P_{11}DCA$ with increasing lithium concentration. As the concentration increases, the melting transition gets broader and weaker and an additional transition appears at around $30\,^\circ C$. This is a result of the eutectic melting and the formation of new phase rich in lithium.*

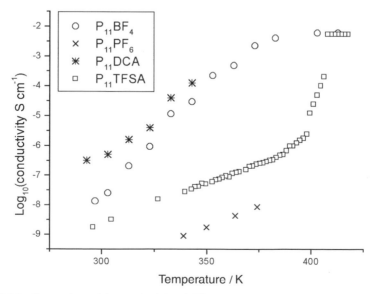

Figure 24.8 *Comparison of the conductivity of different N,N-dimethylpyrrolidinium species. The significant differences in conductivity for the four species again highlight the importance of anion type in determining the physical properties of the material. The conductivity of $P_{11}PF_6$ is considerably worse than the BF_4 species, as is the TFSA species, until the melt at $400\,K$.*

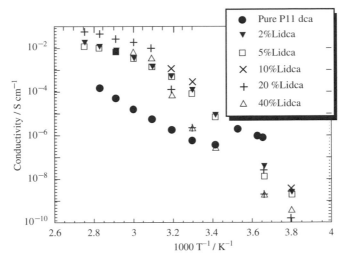

Figure 24.9 *Effect of lithium doping on the conductivity of $P_{11}DCA$. The addition of even small amounts (2%) LiDCA results in a dramatic increase in the conductivity of the material.*

24.4 TRANSPORT MECHANISMS IN PLASTIC CRYSTAL PHASES

The NMR data presented in Figure 24.5 indicates that in the high temperature phases of Me_4NDCA, rotational motions of some or all of the ions take place. Figure 24.10 shows how such rotations are certainly possible in the case of, for

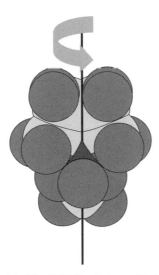

Figure 24.10 *Space-filled model of the N,N-dimethylpyrrolidinium cation. The 3D model shows the spherical nature of the cation and gives an indication as to how easy rotation about the axis should be given the lack of steric hindrance.*

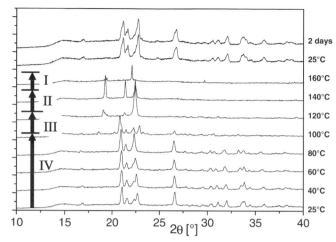

Figure 24.11 *X-ray diffraction pattern of tetramethylammonium dicyanamide shows the change in lattice parameters on moving between phases. As the temperature increases, there is a loss of some peaks and the appearance of new ones, indicating a change in unit cell. Generally, the diffraction patterns become simpler on moving to lower phases, culminating in only one peak for the phase I species. Leaving the material for two days does not appear to have any effect on the diffraction pattern and hence on the structure of the salt.*

example, dimethyl pyrrolidinium salts. In this case the rotation about the twofold axis appears to involve very little steric hindrance. Such rotations cause the crystal lattice symmetry to become of a higher order, since the rotational asymmetry disappears. As a result the X-ray diffraction pattern exhibits fewer reflections in the higher temperature phases, as exemplified by Figure 24.11. In the situation where full isotropic motion of both ions occurs, a simple cubic symmetry can be observed, as in phase I of Me_4NDCA.

The onset of such rotations have previously been linked to phase changes into what are known as "rotator phases" [2, 53], and this in turn has been associated with the formation of vacancies, presumably due to the expansion of the lattice and subsequent decrease in the strength of intermolecular/interionic interactions. Interestingly in the pyrrolidinium-based systems, and in tetraalkylammonium dicyanamide, the onset of these rotations is not instantaneous but rather occurs over a broad temperature range, as shown in Figure 24.5 for Me_4NDCA. In this case NMR second moment analyses have shown that a number of intermediate situations arise, where some species are undergoing restricted rotations and others isotropic rotation. With increasing temperature the number of species undergoing isotropic rotations increases dramatically until finally this is the dominant process, coinciding with the peak of the thermal phase transition.

Interestingly in the case of Et_4NDCA the transition into phase I is accompanied by a substantial increase in volume and a subsequent decrease in the rate of change of volume for phase I compared with phase II. Although we have not yet quantified the vacancy size and number in this DCA system, other work on the pyrrolidinium

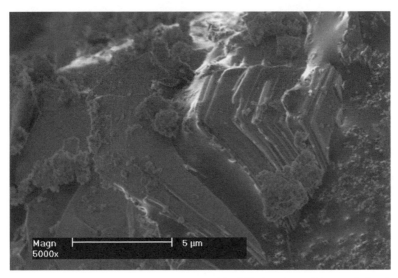

Figure 24.12 *Scanning electron micrograph of $P_{13}BF_4$ showing evidence of slip planes.*

tetrafluoroborate family of salts using positron annihilation lifetime spectroscopy has suggested that a significant step in the vacancy size occurs concomitantly with a drop in apparent vacancy number. This suggests that coalescence of vacancies may occur, leading to larger scale defects such as dislocations. Evidence for this has also been found through SEM analysis [82]. Figure 24.12 presents an SEM micrograph for $P_{13}BF_4$ showing evidence of slip planes separated by about 50 nm. Such defects are known to enhance self-diffusion by a process known as pipe-diffusion whereby the defect acts as a pipe, along which molecules can diffuse [2, 83].

Diffusion coefficients for different species in organic ionic plastic crystals have been determined via pulsed field gradient NMR measurements. These coefficients have shown that despite its apparently large size, the pyrrolidinium cation diffuses quite rapidly in the higher temperature solid-state phases. For example, in the case of $P_{12}TFSA$ diffusion coefficients of the order of 10^{-11} m^2 s^{-1} have been measured [30, 82]. Figure 24.13 shows the temperature dependent diffusion coefficients for cation and anion in the $P_{13}BF_4$ system. It can be seen that while both the cation and anion are quite diffusive in this system, the pyrrolidinium cation is more mobile despite its larger size. The addition of LiBF$_4$ to $P_{13}BF_4$ leads to a slight decrease in the diffusivity of the matrix ions in this case (which is in contrast to the effect of LiTFSA additions to $P_{12}TFSA$ [31]). In this system the lithium and pyrrolidinium cation diffusivities are comparable, and the anion is diffusing at a faster rate. In previous work where LiTFSA was doped into $P_{12}TFSA$, ^7Li NMR line-width measurements showed that even in the lower temperature phases some fraction of the lithium species is mobile, until eventually in phase I all the lithium ions are diffusing, as shown by a single, very narrow resonance [31].

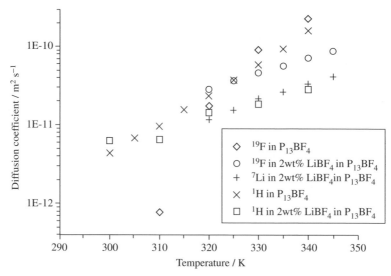

Figure 24.13   *Effect of lithium doping and temperature on the ^7Li, ^{19}F, and ^1H diffusion in $P_{13}DCA$. Doping the material with 2% $LiBF_4$ results in a decrease in cation diffusion. It is believed that an increase in salt concentration results in increased ion interaction and thus in a loss of mobility.*

24.5   CONCLUDING REMARKS

Plastic crystalline materials that are ion-conductors are promising electrolyte materials for a variety of electrochemical device applications. In their soft, plastic form they posses close to the ideal material properties for a solid electrolyte. They are nonflammable and as ionic materials they have negligible vapor pressures, so loss of the electrolyte by evaporation is not a concern. However, the ideal plastic crystalline electrolyte material is yet to be found. For an ambient temperature electrolyte, a material that exists in its plastic phase over a wide temperature range, namely between 0° and 100°C, would be desirable. For Li and Li-ion battery use, a lithium-doped material would be ideal, but it would also be preferable for the lithium ion to be the dominant charge carrier, whereas in the present systems the matrix ions also contribute to the conductivity in the higher temperature phases. One option being investigated is use of a neutral molecular plastic crystal matrix and doping this with an inorganic salt [32]. We have recently shown that lithium salts and some acids can be dissolved in the plastic phase of succinonitrile, which leads to conductivities around 10^{-4} S cm^{-1} at room temperature.

Plastic crystals present many challenges in terms of elucidating the mechanisms of rotational motion, conduction, diffusion, mechanical deformation, and the interrelationship between these mechanisms. Then there is the broader challenge of understanding the effect of doping or mixing these systems with additional components such as acids, inorganic salts, and polymers on these transport mechanisms.

REFERENCES

1. J. Timmermans, *J. Phys. Chem. Solids*, **18**, 1 (1961).

2. J. Sherwood, *The Plastically Crystalline State: Orientationally Disordered Crystals*, Wiley, 1979.

3. G. J. Kabo, A. V. Blolhin, M. B. Charapennikau, A. G. Kabo, U. M. Sevruk, *Thermochim. Acta*, **345**, 125 (2000).

4. G. Guthrie, J. McCullough, *J. Phys. Chem. Solids*, **18**, 53 (1961).

5. D. R. MacFarlane, P. Meakin, J. Sun, M. Forsyth, *J. Phys. Chem., B*, **103**, 4164 (1999).

6. D. R. MacFarlane, J. Huang, M. Forsyth, *Nature*, **402**, 792 (1999).

7. D. R. MacFarlane, J. Sun, J. Golding, P. Meakin, M. Forsyth, *Electrochim. Acta*, **45**, 1271 (2000).

8. M. Tansho, D. Nakamura, R. Ikeda, *Ber. Bunsenges Phys. Chem.*, **95**, 1643 (1991).

9. M. Tansho, D. Nakamura, H. Ikeda, *J. Chem. Soc. Faraday Trans.*, **87**, 3255 (1991).

10. M. Tansho, Y. Furukawa, D. Nakamura, R. Ikeda, *Ber. Bunsenges Phys. Chem.*, **96**, 550 (1992).

11. E. I. Cooper, C. A. Angell, *Solid State Ionics*, **18/19**, 570 (1986).

12. R. Aronsson, B. Jansson, H. Knape, A. Lunden, L. Nilsson, C. Sjoblom, L. Torell, *J. de Phys.*, **C6**, 35 (1980).

13. R. Aronsson, H. Knape, L. Torell, *J. Chem. Phys.*, **77**, 677 (1982).

14. R. Aronsson, L. Borjesson, L. Torell, *Phys. Lett.*, **98A**, 205 (1983).

15. L. Nilsson, J. Thomas, B. Tofield, *J. Phys. C: Solid State Phys.*, **13**, 6441 (1980).

16. M. Witschas, H. Eckert, D. Wilmer, R. Banhatti, K. Funke, J. Fitter, R. Lechner, G. Korus, M. Jansen, *Zeit. Physik. Chem.*, **214**, 643 (2000).

17. M. Witshas, H. Eckert, *J. Phys. Chem. A*, **103**, 10764 (1999).

18. S. Fukada, H. Yamamoto, R. Ikeda, D. Nakamura, *J. Chem. Soc., Faraday Trans. 1*, **83**, 3207 (1987).

19. H. Ishida, R. Ikeda, D. Nakamura, *J. Phys. Chem.*, **86**, 1003 (1982).

20. H. Ishida, T. Iwachido, R. Ikeda, *Ber. Bunsenges Phys. Chem.*, **96**, 1468 (1992).

21. H. Ishida, N. Hayama, R. Ikeda, *Chem. Lett.*, 1331 (1992).

22. H. Suga, M. Sugisaki, S. Seki, *Mol. Cryst.*, **1**, 377 (1966).

23. J. M. Chezeau, J. H. Strange, *Physics Reports (Review Section of Physics Letters)*, **53**, 32 (1979).

24. H. M. Hawthorne, J. N. Sherwood, *J. Chem. Soc. Faraday Trans.*, **66**, 1792 (1970); P. Derollez, J. Lefebve, M. Descamps, W. Press, H. Fontaine, *J. Phys.: Condens. Matter.*, **2**, 6893 (1990); A. Masui, S. Yoshioka, and S. Kinoshita *Chem. Phys. Lett.*, **341**, 299 (2001).

25. J. H. Helms, A. Majumdar, D. Chandra, *J. Electrochem. Soc.*, **141**, 1921 (1994).

26. D. R. MacFarlane, S. A. Forsyth, J. Golding, G. B. Deacon, *Green Chem.*, **4**, 444 (2002).

27. J. Golding, N. Hamid, D. R. MacFarlane, M. Forsyth, C. Forsyth, C. Collins, J. Huang, *J. Chem. Mater.*, **13**, 558 (2001).

28. S. Forsyth, J. Golding, D. R. MacFarlane, M. Forsyth, *Electrochim. Acta*, **46**, 1753 (2001).

29. J. M. Pringle, J. Golding, C. M. Forsyth, G. B. Deacon, M. Forsyth, D. R. MacFarlane, *J. Mater. Chem.*, **12**, 3475 (2002).

30. H. Every, *Ph.D. Thesis*, 2001, Monash University, Australia.

31. M. Forsyth, J. Huang, D. R. MacFarlane, *J. Mater. Chem.*, **10**, 2259 (2000).

32. S. Long, D. R. MacFarlane, M. Forsyth, *Solid State Ionics*, **161**, 105 (2003).

33. C. Angell, L. Busse, E. Cooper, R. Kadiyala, A. Dworkin, M. Ghelfenstein, H. Szwarc, A. Vassal, *J. de Chim. Phys.*, **82**, 267 (1985).

34. T. Asanuma, H. Nakayama, T. Eguchi, N. Nakamura, *Mol. Cryst. Liq. Cryst.*, **326**, 395 (1999).

35. V. Balevicius, A. Marsalka, V. Sablinskas, L. Kimtys, *J. Molec. Struct.*, **219**, 123 (1990).

36. R. Baughman, D. Turnbull, *J. Phys. Chem. Solids*, **33**, 121 (1972).

37. B. Beck, E. Roduner, H. Dilger, P. Czarnecki, D. Fleming, I. Reid, C. Rhodes, *Physica B: Cond. Matter*, **289–290**, 607 (2000).

38. M. Bee, J. Amoureux, *Molec. Phys.*, **48**, 63 (1983).

39. M. Bee, *J. de Chim. Phys.*, **82**, 205 (1985).

40. P. Bladon, N. Lockhart, J. Sherwood, *Molec. Phys.*, **20**, 577 (1971).

41. K. Adamic, S. Greenbaum, K. Abraham, M. Alamgir, M. Wintersgill, J. Fontanella, *Chem. Mater.*, **3**, 534 (1991).

42. N. Boden, J. Cohen, P. Davis, *Molec. Phys.*, **23**, 819 (1972).

43. S. Brauninger, S. Dou, H. Fuess, W. Schmahl, R. Staub, A. Weiss, *Ber. Bunsenges Phys. Chem.*, **98**, 1096 (1994).

44. A. Brichter, J. Strange, *Molec. Phys.*, **37**, 181 (1979).

45. J. Chezeau, J. Strange, *Phys. Rep.*, **53**, 1 (1979).

46. E. Cooper, C. Angell, *Solid State Ionics*, **18–19**, 570 (1986).

47. H. Hawthorne, J. Sherwood, *Trans. Faraday Soc.*, **66**, 1792 (1970)

48. H. Hawthorne, J. Sherwood, *Trans. Faraday Soc.*, **66**, 1783 (1970).

49. H. Hawthorne, *Trans. Faraday Soc.*, **66**, 1799 (1970).

50. H. Honda, S. Ishimaru, N. Onoda, R. Ikeda, *Naturforsch.*, **50a**, 871 (1995).

51. H. Honda, M. Kenmotsu, H. Ohki, R. Ikeda, Y. Furukawa, *Ber. Bunsenges Phys. Chem.*, **99**, 1009 (1995).

52. R. Hooper, J. Sherwood, *Surf. Defect Prop. Solids*, **6**, 308 (1977).

53. K. Horiuchi, H. Takayama, S. Ishimaru, R. Ikeda, *Bull. Chem. Soc. Jpn.*, **73**, 307 (2000).

54. H. Ishida, R. Ikeda, D. Nakamura, *Bull. Chem. Soc. Jpn.*, **59**, 915 (1986).

55. H. Ishida, R. Ikeda, D. Nakamura, *Bull. Chem. Soc. Jpn.*, **60**, 467 (1987).

56. H. Ishida, N. Matsuhashi, R. Ikeda, D. Nakamura, *J. Chem. Soc., Faraday Trans.*, **85**, 111 (1989).

57. H. Ishida, K. Takagi, R. Ikeda, *Chem. Lett.*, 605 (1992).

58. H. Ishida, Y. Furukawa, S. Kashino, S. Sato, R. Ikeda, *Ber. Bunsenges Phys. Chem.*, **100**, 433 (1996).

59. S. Iwai, R. Ikeda, D. Nakamura, *Can. J. Chem.*, **66**, 1961 (1988).

60. S. Iwai, M. Hattori, D. Nakamura, R. Ikeda, *J. Chem. Soc., Faraday Trans.*, **89**, 827 (1993).

61. D. Larsen, B. Soltz, F. Stary, R. West, *J. C. S. Chem. Comm.*, 1093 (1978).

62. K. McGrath, R. Weiss, *Langmuir*, **13**, 4474 (1997).

63. P. McKay, J. Sherwood, *J. Chem. Soc., Faraday Trans., Pt. 1*, **71**, 2331 (1975).

64. H. Ono, R. Seki, R. Ikeda, H. Ishida, *J. Mol. Struct.*, **345**, 235 (1995).

65. H. Ono, R. Ikeda, H. Ishida, *Ber. Bunsenges Phys. Chem.*, **100**, 1833 (1996).

66. H. Ono, S. Ishimaru, R. Ikeda, H. Ishida, *Bull. Chem. Soc. Jpn.*, **70**, 2963 (1997).

67. H. Ono, S. Ishimaru, R. Ikeda, H. Ishida, *Ber. Bunsenges. Phys. Chem.*, **102**, 650 (1998).

68. H. Ono, S. Ishimaru, R. Ikeda, H. Ishida, *Bull. Chem. Soc. Jpn.*, **72**, 2049 (1999).

69. L. Prabhumirashi, R. Ikeda, D. Nakamura, *Ber. Bunsenges Phys. Chem.*, **85**, 1142 (1981).

70. S. Sato, R. Ikeda, D. Nakamura, *Ber. Bunsenges Phys. Chem.*, **86**, 936 (1982).

71. T. Shimizu, S. Tanaka, N. Onoda-Yamamamuro, S. Ishimaru, R. Ikeda, *J. Chem. Soc. Faraday Trans.*, **93**, 321 (1997).

72. T. Tanabe, D. Nakamura, R. Ikeda, *J. Chem. Soc. Faraday Trans.*, **87**, 987 (1991).

73. S. Tanaka, N. Onoda-Yamamamuro, S. Ishimaru, R. Ikeda, *Bull. Chem. Soc. Jpn.*, **70**, 2981 (1997).

74. H. Ishida, K. Takagi, R. Ikeda, *Chem. Lett.*, 605 (1992).

75. H. Ishida, Y. Furukawa, S. Kashino, S. Sato, R. Ikeda, *Ber. Bunsenges Phys. Chem.*, **100**, 433 (1996).

76. S. Iwai, R. Ikeda, D. Nakamura, *Can. J. Chem.*, **66**, 1961 (1988).

77. S. Iwai, M. Hattori, D. Nakamura, R. Ikeda, *J. Chem. Soc., Faraday Trans.*, **89**, 827 (1993).

78. H. Ishida, T. Iwachido, N. Hayama, R. Ikeda, M. Terashima, D. Nakamura, *Z. Naturforsch.*, **44a**, 741 (1989).

79. H. Ishida, K. Tagaki, T. Iwachido, M. Terashima, D. Nakamura, R. Ikeda, *Z. Naturforsch.*, **45a**, 923 (1990).

80. A. J. Seeber, M. Forsyth, C. M. Forsyth, S. A. Forsyth, G. Annat, D. R. MacFarlane, *Physical Chemistry Chemical Physics*, submitted.

81. G. Annat, *Honors thesis*, Monash University 2002.

82. D. R. MacFarlane, M. Forsyth, *Adv. Mater.*, **13**, 957 (2001).

83. O. A. Porter, K. E. Easterling, *Phase Transformations in Metals and Alloys*, 2nd ed., Chapman & Hall, 1992.

Chapter 25

Liquid Crystalline Ionic Liquids

Takashi Kato and Masafumi Yoshio

Liquid crystals are anisotropically ordered and mobile materials [1] and have been successful in the field of electrooptic devices and high-strength fibers [1]. In addition they are expected to serve as anisotropic conductors due to their self-organized structures. Recently transportation of charges [2–5] and ions [6–12] in liquid crystals has attracted much attention. For this purpose the design and control of molecular interactions and microphase-segregated structures in liquid crystals is essential [13–16]. Moreover the macroscopic orientation of self-organized monodomains plays a key role in the enhancement of properties because the boundary in randomly oriented polydomains disturbs high and anisotropic transportation of charges and ions. Percec [17–19], Nolte [20], Lehn [21], and Kimura [22] incorporated oligo(ethylene oxide)s (PEOs), crown ethers, and azacrowns, which function as ion-conducting moieties into mesogenic rodlike and fanlike molecules. These materials form layer and columnar structures. The ionic conductivities of some of these materials were measured. However, the ionic conductivities of these earlier examples were lower than expected because no macroscopic orientation had been achieved. Recently Kato, Ohno, and coworkers have achieved macroscopic orientation of smectic liquid crystals containing PEO moieties with lithium salts and succeeded in anisotropic (two dimensional) measurements of the ionic conductivities [7, 9, 23–25]. For the PEO-based columnar materials, one-dimensional ion

Electrochemical Aspects of Ionic Liquids   Edited by Hiroyuki Ohno
ISBN 0-471-64851-5   Copyright © 2005 John Wiley & Sons, Inc.

conduction in a macroscopic monodomain has not yet been measured. These PEO-based anisotropic materials have potential applications as functional electrolytes in lithium ion batteries.

Recently ionic liquid-based liquid crystalline materials have been developed as a new family of anisotropic (one or two dimensional) electrolytes [8, 26–29]. Ionic liquids are isotropic organic liquids composed entirely of ions [30–34]. They have advantages for applications as nonvolatile liquid electrolytes in a variety of electrochemical devices such as lithium ion batteries, capacitors, dye-sensitized solar cells, organic light-emitting diodes, and actuators.

In this chapter we review the development of liquid crystalline materials on the basis of ionic liquids. Self-organization behavior and ion-conductive properties of these anisotropic materials are described.

25.1 LIQUID CRYSTALLINE IONIC LIQUIDS BY THE CHEMICAL MODIFICATION OF IONIC LIQUIDS

Ionic liquid crystals are liquid crystals containing ionic moieties. Ionic liquid crystals having ammonium and pyridinium salts were reported to show stable thermotropic liquid crystalline behavior [35–40]. Seddon, Bruce, and coworkers showed that some imidazolium and pyridinium salts exhibit low melting points and liquid crystalline behavior [41]. Ionic liquid crystals with lower temperature mesophase ranges are called liquid crystalline ionic liquids. A large variety of thermotropic liquid crystalline ionic liquids exhibiting anisotropic fluid states were prepared. In particular, the ionic liquid crystals based on imidazolium salts **1** [42–46], **2** [47], **3** [48], and **4** [49] and pyridinium salts **5** [42] containing weakly coordinating perfluorinated anions, such as BF_4^- and PF_6^- are representative liquid crystalline ionic liquids due to their thermal and electrochemical stabilities (Figure 25.1). Liquid crystalline ionic liquids having a benzimidazolium unit **6** were also prepared [50]. For these materials, liquid crystalline phases are induced by microphase-segregation of ionic moieties and long alkyl or perfluoroalkyl chains. The types of liquid crystalline phases exhibited are dependent on the molecular shape and the volume ratio of incompatible ionic and nonionic parts. Most of the ionic molecules containing single chains form bilayer smectic structures with the interdigitation of alkyl chains. The influence of the anion type and chain length on the liquid crystalline phases has been systematically studied for 1-alkyl-3-methylimidazolium salts **1**(n/X^-), where n is the alkyl chain length and X^- indicates the anion species [42–45, 51]. The salts with $X^- = Cl^-$, Br^-, BF_4^-, PF_6^-, and $CF_3SO_3^-$ ($n = 12, 14, 16, 18$) form smectic A phases. For example, **1**($12/BF_4^-$) shows a smectic A phase from 26° to 39°C, and **1**($18/BF_4^-$) exhibits a smectic A phase from 67° to 215°C. The melting and clearing points increase with increasing chain length. The triflate salts **1**($12/CF_3SO_3^-$) and **1**($14/CF_3SO_3^-$) and the *bis*(trifluoromethanesulfonyl)imide (TFSI) salts **1**($n/TFSI^-$) with alkyl chains of $n = 12$–18 are

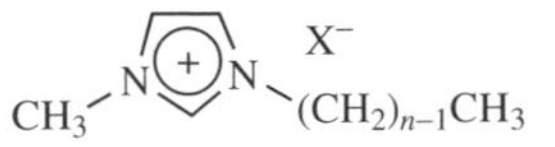

$X^- = Cl^-, Br^-, BF_4^-, PF_6^-, CF_3SO_3^-$

$n = 12, 14, 16, 18$

1

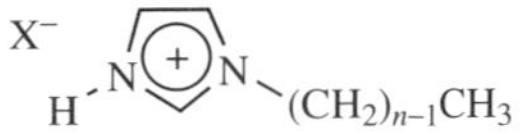

$X^- = Cl^-, NO_3^-, BF_4^-$

$n = 10, 12, 14, 16, 18$

2

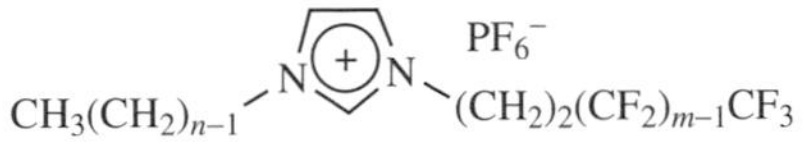

$n = 1,\ m = 6,8$

$n = 4,\ m = 6,8$

3

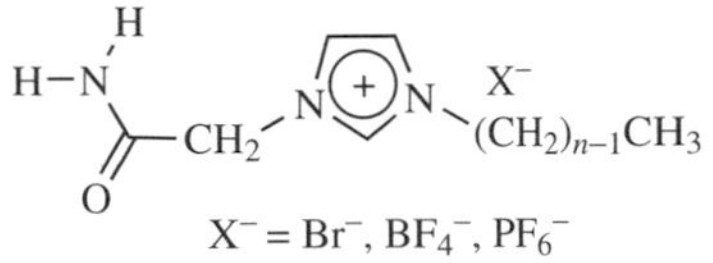

$X^- = Br^-, BF_4^-, PF_6^-$

$n = 10, 12, 14, 16, 18$

4

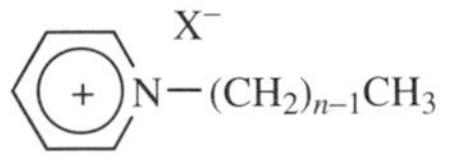

$X^- = Cl^-, PF_6^-$

$n = 12, 14, 16, 18$

5

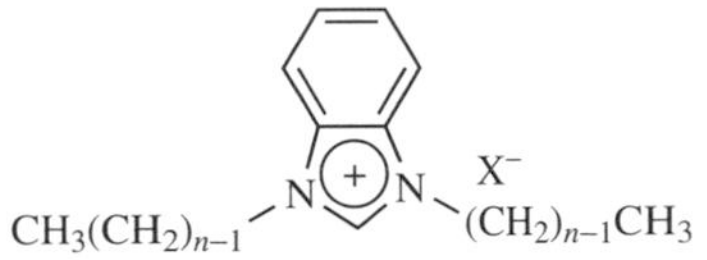

$X^- = Cl^-, Br^-$

$n = 12, 14, 16, 18$

6

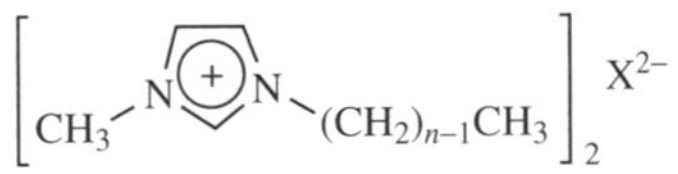

$X^{2-} = [MCl_4]^{2-}$(M: Co, Ni, Cu, Pd)

$n = 12, 14, 16, 18$

7

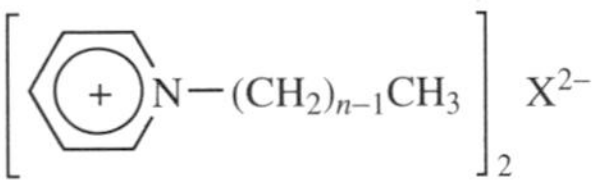

$X^{2-} = [MCl_4]^{2-}, [MBr_4]^2, [MI_4]^{2-}$
(M: Co, Ni, Cu, Pd)

$n = 12, 14, 16, 18$

8

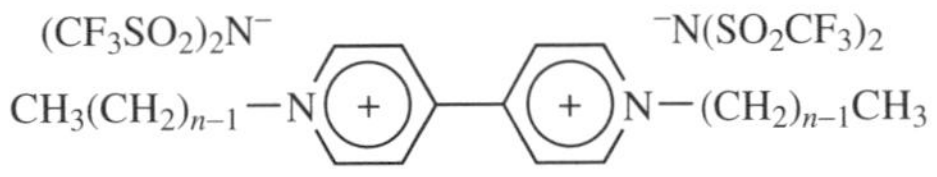

$n = 7, 8$

9

Figure 25.1 *Molecular structures of ionic liquid crystals.*

nonmesomorphic. The order of the stabilization effects on liquid crystalline structures is $Cl^- > Br^- > BF_4^- > PF_6^- > CF_3SO_3^- > (CF_3SO_2)_2N^-$.

The formation of layered assemblies can be induced by the addition of a small amount of water to isotropic ionic liquid **1**(10/Br$^-$) [52, 53]. A lyotropic liquid crystalline gel consisting of **1**(10/Br$^-$) and water of 16 wt% has been prepared. Addition of water to **1**(10/Br$^-$) induces the formation of a lamellar structure with hydrogen bonding between the imidazolium ring (in particular, the proton on C_2) and Br^-, and Br^- and water, while a dried **1**(10/Br^-) including water of less than 1.6 wt% shows no liquid crystalline phases.

Metal-based ionic liquid crystals containing a tetrahedral tetrahalometalate ion (MX_4^{2-}; $X = Cl$, Br, I; $M = Co$, Ni, Cu, Pd, etc.) **7** [41, 54] and **8** [55–57] were studied (Figure 25.1). They provide various coordination geometries. For example, *N*-alkylpyridinium salts **8** often denoted as [C_nPy]$_2$[CuCl$_4$] ($n = 12$–18) show a variety of liquid crystalline phases from smectic to columnar or even cubic phases [57]. These metal-based materials can exhibit properties as metal complexes such as chromism, magnetism, polarizability, redox behavior, and catalysis.

Bipyridinium cation **9** [58], phosphonium [59–62], guanidinium [63], and vinamidinium [64] cations were used to form thermotropic ionic liquid crystals. For example, the mixture of diheptyl and dioctyl viologens **9** (20:80 by wt%) with TFSI anion forms a smectic A phase between 22° and 132°C. It was reported that the viologens show fluorescence in both non-polar and polar organic solvents, which can be useful for the development of biological and chemical sensors [65].

Thermotropic columnar liquid crystalline phases are formed by self-organization of fan-shaped imidazolium molecules **10a,b** (Figure 25.2) [29]. Compound **10a** shows a columnar phase between −29° and 133°C. These ionic liquids exhibit fluid columnar liquid crystalline phases over a wide range of temperatures, including room temperature. In this structure the imidazolium part forms a one-dimensional ionic path inside the column. These columnar materials are macroscopically aligned by shearing on the glass substrate. Figure 25.3 shows polarizing microscope images of compound **10b** before and after shearing. The monodomain of the columnar phase is formed between Au electrodes after the polydomain material is sheared in the sandwiched glasses without rubbing treatment. One-dimensional ionic conductivities were measured by the cell with comb-shaped gold electrodes, which is described in a succeeding section.

Oriented structures of self-organized ionic liquids can be preserved in solid films [28]. Self-standing nanostructured two-dimensional polymer films were prepared by in situ photopolymerization of ionic liquid crystalline monomer **11** that forms homeotropic monodomains of the smectic A phase on a glass plate (Figure 25.4). The film of **12** has a macroscopically oriented layered nanostructure as presented in Figure 25.5.

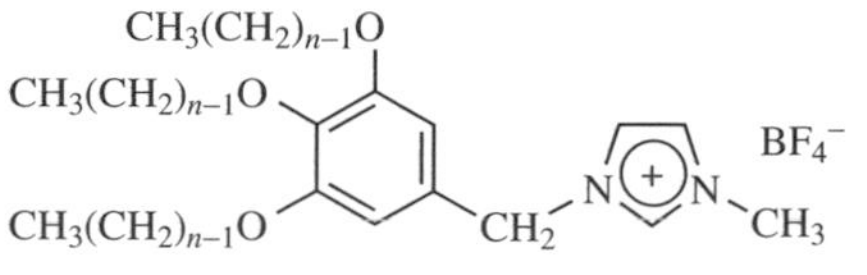

CH$_3$(CH$_2$)$_{n-1}$O

CH$_3$(CH$_2$)$_{n-1}$O

CH$_3$(CH$_2$)$_{n-1}$O

BF$_4^-$

CH$_2$

N⊕N

CH$_3$

10a:$n = 8$
10b:$n = 12$

Figure 25.2 *Molecular structure of fan-shaped ionic liquid crystal* **10**.

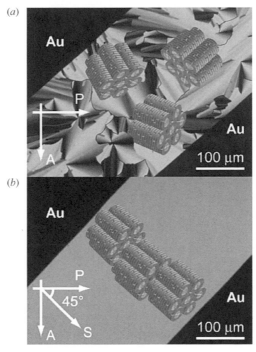

(a)

Au

P

A

Au

100 µm

(b)

Au

P

45°

A   S

Au

100 µm

Figure 25.3 *Polarizing optical microscopic images and schematic illustrations of the oriented and self-assembled structures of* **10b** *in the hexagonal columnar state: (a) Before shearing; (b) after shearing the material along the direction perpendicular to the Au electrodes. Directions of A: analyzer; P: polarizer; S: shearing. (Reproduced with permission from J. Am. Chem. Soc.,* **126***, 994–995 (2004). Copyright 2004 American Chemical Society)*

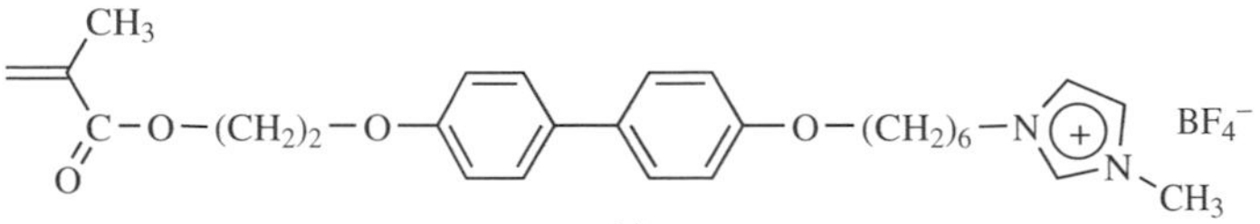

11

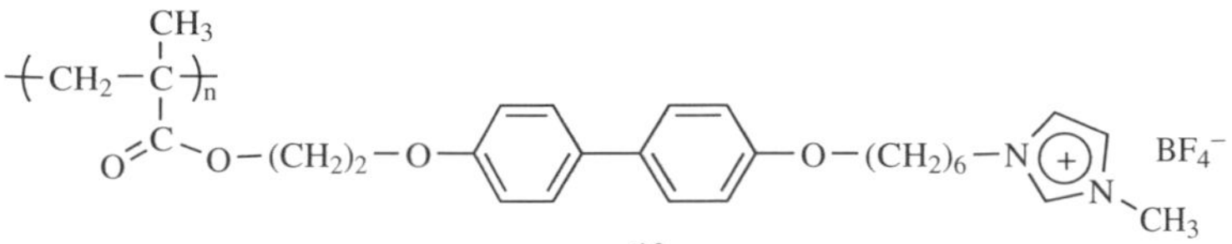

12

Figure 25.4 *Molecular structures of ionic liquid crystalline monomer* **11** *and polymer* **12***.*

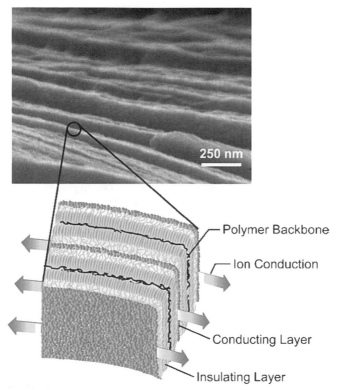

Figure 25.5 *SEM image and schematic illustration of a nanostructured ion conductive flexible film obtained by in situ photopolymerization of ionic liquid crystalline monomer* **11** *in the smectic A phase.*

25.2 SELF-ASSEMBLY OF IONIC LIQUIDS WITH LIQUID CRYSTAL MOLECULES

Simple ionic liquids can be self-organized by mixing with liquid crystalline molecules that have functional groups interacting with ionic liquids [8, 26]. Mesogenic compounds **13**, **14**, and **15** were found to be partially compatible with ionic liquids. The mixtures of 1-ethyl-3-methylimidazolium tetrafluoroborate **16** and compounds **13**, **14**, or **15** form stable layered assemblies (Figures 25.6 and 25.7). In this case the intermolecular interactions between the hydroxyl groups of mesogenic compounds and the ionic liquid enhance the miscibility and lead to the formation of microphase-segregated layered structures at the nanometer scale, as illustrated in Figure 25.7.

The phase transition behavior for the mixture of **13** and **16** is shown as a function of the mole fraction of **16** in Figure 25.8 [8]. Compounds **13** and **16** are miscible up to the mole fraction of 0.7 for **16** in the mixture. These mixtures exhibit smectic phases, while compound **13** alone gives a columnar phase (Col) from 79° to

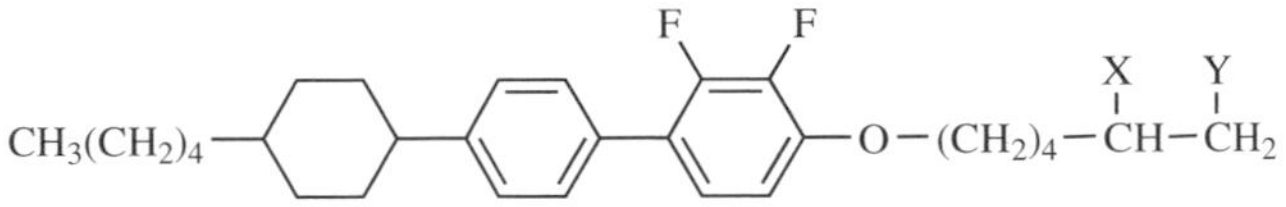

CH₃(CH₂)₄ ⬡—⬡—⬡(F, F) —O—(CH₂)₄—CH—CH₂ (X, Y)

13:X = OH, Y = OH
14:X = H, Y = OH

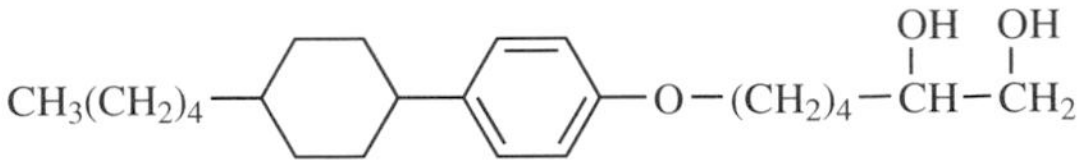

CH₃(CH₂)₄ ⬡—⬡ —O—(CH₂)₄—CH—CH₂ (OH, OH)

15

CH₃—N⁺N—CH₂CH₃   BF₄⁻

16

CH₃(CH₂)₄ ⬡—⬡—⬡(F, F) —O—(CH₂)₅—CH₃

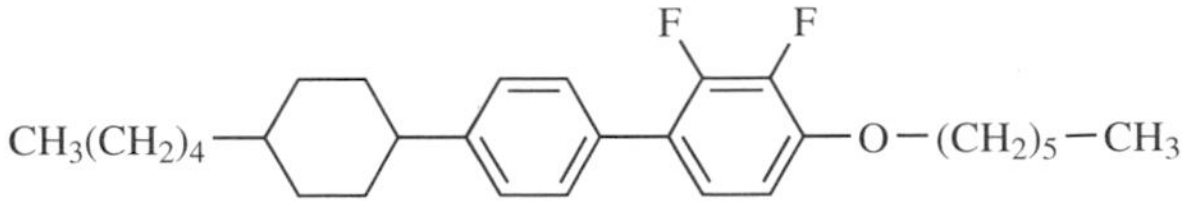

17

Figure 25.6 *Molecular structures of hydroxyl-terminated mesogenic compounds* **13–15**, *ionic liquid* **16**, *and mesogenic compound* **17** *without functional groups.*

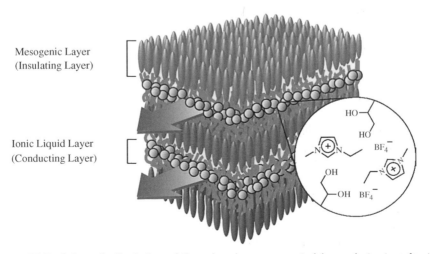

Mesogenic Layer
(Insulating Layer)

Ionic Liquid Layer
(Conducting Layer)

Figure 25.7 *Schematic illustration of the microphase-segregated layered structure for the mixtures of hydroxyl-terminated mesogenic compounds* **13** *and* **15** *with ionic liquid* **16**.

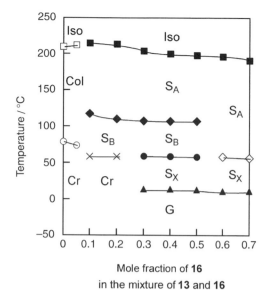

Figure 25.8 *Phase transition behavior for a mixture of* **13** *and* **16** *on the heating runs. Col: columnar; Cr: crystalline; G: glassy; Iso: isotropic; S_A: smectic A; S_B: smectic B; S_X: unidentified smectic.*

210°C. For example, an equimolar mixture of **13** and **16** (**13/16**) shows a glass transition at 14°C and subsequent smectic phases until 198°C on heating. In addition the mixtures of **14** and **16** form thermally stable layered structures. The layer spacing of the smectic A phase for the equimolar mixture of **14** and **16** (**14/16**) at 100°C is 5.0 nm, while the spacing for **14** alone is 4.6 nm at the same temperature. The incorporation of the ionic liquid layers between the mesogenic layers may extend the layer spacing for the mixture. In contrast, phase separation in the macroscopic scale is observed for the mixture of **16** and **17** without hydroxyl groups because of the lack of intermolecular interactions.

25.3 ANISOTROPIC IONIC CONDUCTIVITIES OF IONIC LIQUID CRYSTALS

Liquid crystalline ionic liquids can be used as low dimensional ion conductors [6, 8, 26–29]. For this purpose the formation of oriented monodomains in the macroscopic scale is important because the boundary in the randomly oriented polydomains disturbs high and anisotropic ion conduction.

For the smectic liquid crystalline structures of **13** and **16** (Figure 25.7), alternating ionic liquid layers (conducting parts) and mesogenic layers (insulating parts) microphase-segregate at the nanometer scale. Successful control of the orientation of the layered nanostructure on substrates leads to the preparation of two-dimensionally ion-conductive materials. The layered assemblies spontaneously form a homeotropic alignment on glass and indium tin oxide (ITO) surfaces. The

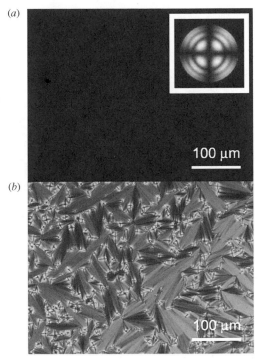

Figure 25.9 *Polarizing optical micrographs of the equimolar mixture of* **14** *and* **16** *in the smectic A liquid crystalline state: (a) homeotropically aligned monodomain is formed on a glass substrate when the sample is cooled slowly from the isotropic state at the rate of 3°C min⁻¹. Inset is a conoscopic image; (b) unaligned polydomains are formed on a glass substrate when the sample is cooled from the isotropic state at the rate of 20°C min⁻¹.*

homeotropic orientation of the materials to the substrates is confirmed by using a polarizing microscope. Figure 25.9a shows orthoscopic and conoscopic images of the sample oriented homeotropically in the smectic A phase. No birefringence is seen under the crossed Nicols condition. On the other hand, focal conic fan texture is observed when the sample is randomly aligned in the smectic A phase (Figure 25.9b).

Anisotropic ionic conductivities were measured by an alternating current impedance method [66, 67] for the samples forming oriented monodomains in two types of cells shown in Figure 25.10. The comb-shaped gold electrodes with thickness of 0.8 micrometer on a glass plate are used for the measurement of ionic conductivity along with the direction parallel ($\sigma_\parallel$) to the layer (Figure 25.10a). A pair of ITO electrodes fixed by a Teflon spacer of 30 micrometer can be useful for the measurement of ionic conductivity along with the direction perpendicular ($\sigma_\perp$) to the layer (Figure 25.10b). Figure 25.11 shows anisotropic ionic conductivities of the equimolar assemblies (**13/16, 14/16**) of mesogenic compounds **13** and **14** with ionic liquid **16** as a function of temperature. The ionic conductivities parallel to the smectic layer ($\sigma_\parallel$) are higher than those perpendicular to the layer ($\sigma_\perp$). For **13/16**, the highest $\sigma_\parallel$ value of 4.1×10^{-3} S cm^{-1} is observed in the smectic A phase at

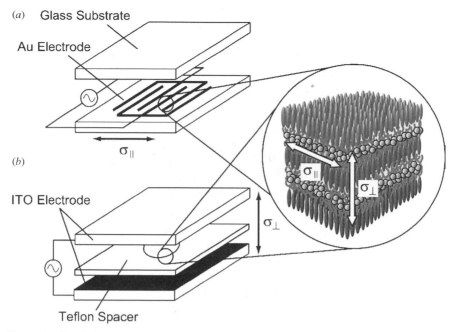

Figure 25.10 *Schematic illustration of the cells for the anisotropic ion conductive measurements. The samples forming homeotropically aligned monodomains in the smectic liquid crystalline states are filled between electrodes.*

192°C. The magnitude of anisotropy ($\sigma_{\parallel}/\sigma_{\perp}$) at 192°C is 2.9 × 10. The magnitude of maximal anisotropy is about 3.1 × 10^3 at 73°C in the smectic B phase. In the smectic B phase, hexagonally ordered packing of the rigid-rod mesogen may effectively disturb the ion conduction perpendicular to the layer. The liquid crystal assembly of **14/16** shows the highest $\sigma_{\parallel}$ value of 2.8 × 10^{-3} S cm^{-1} and the $\sigma_{\parallel}/\sigma_{\perp}$ value of 8.2 × 10 are obtained in the smectic A phase at 132°C. For the anisotropic self-assembled materials containing ionic liquids, the discontinuous changes of ionic conductivities are observed at the phase transition temperatures. In particular, abrupt changes are seen at the isotropic (disorder)–liquid crystalline (order) transitions.

To obtain information on the mobility of ionic liquids in the layered nanostructures, the activation energies in the smectic A phases were estimated from the slope of the ionic conductivities ($\sigma_{\parallel}$) (Figure 25.11). The energy required for **13/16** is about twice greater than that for **14/16**. The ionic conductivities along the direction parallel to the smectic layer are higher than those of **13/16**, which suggests the conductivities depend on the number of hydroxyl groups of the mesogenic molecules in the mixtures. It was found that the increase of the intermolecular interactions between the hydroxyl groups and ionic liquids decreases the mobility. No anisotropy of the ionic conductivity is seen when the assemblies form isotropic states. The formation of the smectic layered structures through the microphase segregation at the nanometer level between ionic liquid and mesogenic compounds results in the spontaneous formation of anisotropic ion-conductive pathways with long-range order.

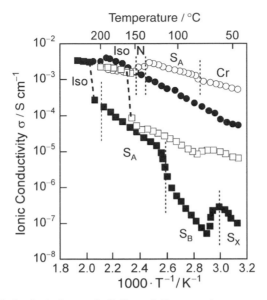

Figure 25.11 *Anisotropic ionic conductivities of the complexes composed of equimolar amounts of mesogenic compounds* **13**, **14** *and ionic liquid* **16** *as a function of temperature: (•) parallel and (■) perpendicular to the smectic layer for the material based on* **13**; *(○) parallel and (□) perpendicular to the smectic layer for the material based on* **14**.

Anisotropic ionic conductivities were also observed for nanostructured 2D films consisting of polymer **12** [28]. The conductivities parallel to the layer are 3.2×10^{-2} S cm^{-1} in the smectic A phase at 209°C and 5.9×10^{-6} S cm^{-1} in the solid state at 50°C.

One-dimensional ion conduction is achieved for columnar liquid crystalline ionic liquids **10a,b** [29]. In the macroscopically ordered states of these columnar materials, ionic conductivities parallel to the columnar axis ($\sigma_{\parallel}$) is higher than those perpendicular to the axis ($\sigma_{\perp}$). For example, compound **10b** shows the conductivities of 3.1×10^{-5} S cm^{-1} ($\sigma_{\parallel}$), 7.5×10^{-7} S cm^{-1} ($\sigma_{\perp}$), and anisotropy ($\sigma_{\parallel}/\sigma_{\perp}$) of 41 at 100°C. These materials function as self-organized electrolytes. They dissolve a variety of ionic species such as lithium salts. Compound **10b** containing LiBF$_4$ (molar ratio of LiBF$_4$ to **10b**: 0.25) exhibits the conductivities of 7.5×10^{-5} S cm^{-1} ($\sigma_{\parallel}$) and 8.0×10^{-7} S cm^{-1} ($\sigma_{\perp}$) at 100°C. The value of $\sigma_{\parallel}/\sigma_{\perp}$ at 100°C is 94. It should be noted that by adding salts only conductivities of $\sigma_{\parallel}$ are enhanced.

25.4 RELATED MATERIALS

Metallomesogens are materials related to liquid crystalline ionic liquids because some of them have ionic structures [68, 69]. Gels based on ionic liquids are also one of related materials. Physical gelation of ionic liquids is accompanied by the formation of bilayer structures of glycolipids and a cationic amphiphile in ionic liquids [70, 71]. Ionic liquids are also physically gelled by the low molecular weight organogelators based on cholesterol and amino acid to tune the fluidity of ionic liquids for application in electrochemical devices [72, 73]. Moreover, gelation of ionic liquids is induced by physical cross-linking of single-walled carbon nanotubes, which is mediated by the local molecular ordering of ionic liquids around the nanotubes [74]. Chemical gels of ionic liquids have been also prepared [75–77].

25.5 SUMMARY

In this chapter we have described the mesomorphic behavior and ionic conductivities of ionic liquid-based liquid crystalline materials. These ion-active anisotropic materials have great potentials for applications not only as electrolytes that anisotropically transport ions at the nanometer scale but also as ordered solvents for reactions. Ionic liquid crystals have also been studied for uses as diverse as nonliner optoelectronic materials [61, 62], photoluminescent materials [78], structure-directing reagents for mesoporous materials [79, 80] and ordered solvents for organic reactions [47, 81]. Approaches to self-organization of ionic liquids may open a new avenue in the field of material science and supramolecular chemistry.

REFERENCES

1. D. Demus, J. W. Goodby, G. W. Gray, H.-W. Spiess, V. Vill, eds., *Handbook of Liquid Crystals*, Wiley-VCH, 1998.

2. D. Adam, P. Schuhmacher, J. Simmerer, L. Häussking, K. Siemensmeyer, K. H. Etzbach, H. Ringsdorf, D. Haarer, *Nature*, **371**, 141 (1994).

3. N. Boden, R. J. Bushby, J. Clements, B. Movaghar, *J. Mater. Chem.*, **9**, 2081 (1999).

4. R. J. Bushby, O. R. Lozman, *Curr. Opin. Solid State Mater. Sci.*, **6**, 569 (2002).

5. M. O'Neill, S. M. Kelly, *Adv. Mater.*, **15**, 1135 (2003).

6. T. Kato, *Science*, **295**, 2414 (2002).

7. T. Ohtake, M. Ogasawara, K. Ito-Akita, N. Nishina, S. Ujiie, H. Ohno, T. Kato, *Chem. Mater.*, **12**, 782 (2000).

8. M. Yoshio, T. Mukai, K. Kanie, M. Yoshizawa, H. Ohno, T. Kato, *Adv. Mater.*, **14**, 351 (2002).

9. K. Kishimoto, M. Yoshio, T. Mukai, M. Yoshizawa, H. Ohno, T. Kato, *J. Am. Chem. Soc.*, **125**, 3196 (2003).

10. F. B. Dias, S. V. Batty, A. Gupta, G. Ungar, J. P. Voss, P. V. Wright, *Electrochim. Acta*, **43**, 1217 (1998).

11. H. V. St. A. Hubbard, S. A. Sills, G. R. Davies, J. E. MacIntyre, I. M. Ward, *Electrochim. Acta*, **43**, 1239 (1998).

12. C. T. Imrie, M. D. Ingram, G. S. McHattie, *Adv. Mater.*, **11**, 832 (1999).

13. T. Kato, *Struct. Bonding*, **96**, 95 (2000).

14. T. Kato, N. Mizoshita, *Curr. Opin. Solid State Mater. Sci.*, **6**, 579 (2002).

15. C. Tschierske, *J. Mater. Chem.*, **8**, 1485 (1998).

16. C. Tschierske, *J. Mater. Chem.*, **11**, 2647 (2001).

17. V. Percec, G. Johansson, J. Heck, G. Ungar, S. V. Batty, *J. Chem. Soc., Perkin Trans. 1*, 1411 (1993).

18. V. Percec, J. A. Heck, D. Tomazos, G. Ungar, *J. Chem. Soc., Perkin Trans. 2*, 2381 (1993).

19. V. Percec, G. Johansson, R. Rodenhouse, *Macromolecules*, **25**, 2563 (1992).

20. C. F. van Nostrum, R. J. M. Nolte, *Chem. Commun.*, 2385 (1996).

21. J.-M. Lehn, J. Malthête, A.-M. Levelut, *J. Chem. Soc., Chem. Commun.*, 1794 (1985).

22. H. Tokuhisa, M. Yokoyama, K. Kimura, *J. Mater. Chem.*, **8**, 889 (1998).

23. T. Ohtake, K. Ito, N. Nishina, H. Kihara, H. Ohno, T. Kato, *Polym. J.*, **31**, 1155 (1999).

24. T. Ohtake, Y. Takamitsu, K. Ito-Akita, K. Kanie, M. Yoshizawa, T. Mukai, H. Ohno, T. Kato, *Macromolecules*, **33**, 8109 (2000).

25. K. Hoshino, K. Kanie, T. Ohtake, T. Mukai, M. Yoshizawa, S. Ujiie, H. Ohno, T. Kato, *Macromol. Chem. Phys.*, **203**, 1547 (2002).

26. M. Yoshio, T. Mukai, M. Yoshizawa, H. Ohno, T. Kato, *Mol. Cryst. Liq. Cryst.*, **413**, 2235 (2004).

27. M. Yoshio, T. Mukai, K. Kanie, M. Yoshizawa, H. Ohno, T. Kato, *Chem. Lett.*, 320 (2002).

28. K. Hoshino, M. Yoshio, T. Mukai, K. Kishimoto, H. Ohno, T. Kato, *J. Polym. Sci., Part A: Polym. Chem.*, **41**, 3486 (2003).

29. M. Yoshio, T. Mukai, H. Ohno, T. Kato, *J. Am. Chem. Soc.*, **126**, 994 (2004).

30. T. Welton, *Chem. Rev.*, **99**, 2071 (1999).

31. P. Wasserscheid, W. Keim, *Angew. Chem. Int. Ed.*, **39**, 3772 (2000).

32. R. D. Rogers, K. R. Seddon, *Science*, **302**, 792 (2003).

33. P. Bonhôte, A.-P. Dias, N. Papageorgiou, K. Kalyanasundaram, M. Grätzel, *Inorg. Chem.*, **35**, 1168 (1996).

34. H. Ohno, M. Yoshizawa, *Solid State Ionics*, **154–155**, 303 (2002).

35. C. G. Bazuin, D. Guillon, A. Skoulios, J.-F. Nicoud, *Liq. Cryst.*, **1**, 181 (1986).

36. S. Ujiie, K. Iimura, *Macromolecules*, **25**, 3174 (1992).

37. V. Hessel, H. Ringsdorf, *Makromol. Chem. Rapid Commun.*, **14**, 707 (1993).

38. Y. Kosaka, T. Kato, T. Uryu, *Liq. Cryst.*, **18**, 693 (1995).

39. C. Chovino, Y. Frere, D. Guillon, P. Gramain, *J. Polym. Sci., A: Polym. Chem.*, **35**, 2569 (1997).

40. Y. Haramoto, Y. Akiyama, R. Segawa, S. Ujiie, M. Nanasawa, *J. Mater. Chem.*, **8**, 275 (1998).

41. C. J. Bowlas, D. W. Bruce, K. R. Seddon, *Chem. Commun.*, 1625 (1996).

42. C. M. Gordon, J. D. Holbrey, A. R. Kennedy, K. R. Seddon, *J. Mater. Chem.*, **8**, 2627 (1998).

43. J. D. Holbrey, K. R. Seddon, *J. Chem. Soc., Dalton Trans.*, 2133 (1999).

44. A. E. Bradley, C. Hardacre, J. D. Holbrey, S. Johnston, S. E. J. McMath, M. Nieuwenhuyzen, *Chem. Mater.*, **14**, 629 (2002).

45. C. Hardacre, J. D. Holbrey, S. E. J. McMath, M. Nieuwenhuyzen, *ACS Symp. Ser.*, **818**, 400 (2002).

46. J. De Roche, C. M. Gordon, C. T. Imrie, M. D. Ingram, A. R. Kennedy, F. Lo Celso, A. Triolo, *Chem. Mater.*, **15**, 3089 (2003).

47. C. K. Lee, H. W. Huang, I. J. B. Lin, *Chem. Commun.*, 1911 (2000).

48. T. L. Merrigan, E. D. Bates, S. C. Dorman, J. H. Davis Jr., *Chem. Commun.*, 2051 (2000).

49. K.-M. Lee, Y.-T. Lee, I. J. B. Lin, *J. Mater. Chem.*, **13**, 1079 (2003).

50. K. M. Lee, C. K. Lee, I. J. B. Lin, *Chem. Commun.*, 899 (1997).

51. A. Downard, M. J. Earle, C. Hardacre, S. E. J. McMath, M. Nieuwenhuyzen, S. J. Teat, *Chem. Mater.*, **16**, 43 (2004).

52. M. A. Firestone, J. A. Dzielawa, P. Zapol, L. A. Curtiss, S. Seifert, M. L. Dietz, *Langmuir*, **18**, 7258 (2002).

53. M. L. Dietz, J. A. Dzielawa, M. P. Jensen, M. A. Firestone, *ACS Symp. Ser.*, **856**, 526 (2003).

54. C. Hardacre, J. D. Holbrey, P. B. McCormac, S. E. J. McMath, M. Nieuwenhuyzen, K. R. Seddon, *J. Mater. Chem.*, **11**, 346 (2001).

55. F. Neve, A. Crispini, S. Armentano, O. Francescangeli, *Chem. Mater.*, **10**, 1904 (1998).

56. F. Neve, A. Crispini, O. Francescangeli, *Inorg. Chem.*, **39**, 1187 (2000).

57. F. Neve, O. Francescangeli, A. Crispini, J. Charmant, *Chem. Mater.*, **13**, 2032 (2001).

58. P. K. Bhowmik, H. Han, J. J. Cebe, R. A. Burchett, *Liq. Cryst.*, **30**, 1433 (2003).

59. D. J. Abdallah, A. Robertson, H.-F. Hsu, R. G. Weiss, *J. Am. Chem. Soc.*, **122**, 3053 (2000).

60. H. Chen, D. C. Kwait, Z. S. Gönen, B. T. Weslowski, D. J. Abdallah, R. G. Weiss, *Chem. Mater.*, **14**, 4063 (2002).

61. A. Kanazawa, T. Ikeda, J. Abe, *Angew. Chem. Int. Ed.*, **39**, 612 (2000).

62. A. Kanazawa, T. Ikeda, J. Abe, *J. Am. Chem. Soc.*, **123**, 1748 (2001).

63. F. Mathevet, P. Masson, J.-F. Nicoud, A. Skoulios, *Chem. Eur. J.*, **8**, 2248 (2002).

64. C. P. Roll, B. Donnio, D. Guillon, W. Weigand, *J. Mater. Chem.*, **13**, 1883 (2003).

65. D. Wang, J. Wang, D. Moses, G. C. Bazan, A. J. Heeger, *Langmuir*, **17**, 1262 (2001).

66. J. R. Macdonald, *Impedance Spectroscopy Emphasizing Solid Materials and Systems*, Wiley, 1987.

67. Y. Tominaga, H. Ohno, *Electrochim. Acta*, **45**, 3081 (2000).

68. D. W. Bruce, *Acc. Chem. Res.*, **33**, 831 (2000).

69. R. Giménez, D. P. Lydon, J. L. Serrano, *Curr. Opin. Solid State Mater. Sci.*, **6**, 527 (2002).

70. N. Kimizuka, T. Nakashima, *Langmuir*, **17**, 6759 (2001).

71. T. Nakashima, N. Kimizuka, *Chem. Lett.*, 1018 (2002).

72. A. Ikeda, K. Sonoda, M. Ayabe, S. Tamaru, T. Nakashima, N. Kimizuka, S. Shinkai, *Chem. Lett.*, 1154 (2001).

73. W. Kubo, T. Kitamura, K. Hanabusa, Y. Wada, S. Yanagida, *Chem. Commun.*, 374 (2002).

74. T. Fukushima, A. Kosaka, Y. Ishimura, T. Yamamoto, T. Takigawa, N. Ishii, T. Aida, *Science*, **300**, 2072 (2003).

75. R. T. Carlin, J. Fuller, *Chem. Commun.*, 1345 (1997).

76. A. Noda, M. Watanabe, *Electrochim. Acta*, **45**, 1265 (2000).

77. H. Ohno, M. Yoshizawa, W. Ogihara, *Electrochim. Acta*, **48**, 2079 (2003).

78. D. Haristoy, D. Tsiourvas, *Chem. Mater.*, **15**, 2079 (2003).

79. Y. Zhou, M. Antonietti, *Adv. Mater.*, **15**, 1452 (2003).

80. Y. Zhou, M. Antonietti, *Chem. Mater.*, **16**, 544 (2004).

81. R. G. Weiss, *Tetrahedron*, **44**, 3413 (1988).

Part VI

Gel-Type Polymer Electrolytes

Chapter *26*

Ionic Liquid Gels

Kenji Hanabusa

Very recently ionic liquids have received much attention for materials in green-sustainable chemistry [1–3]. Ionic liquids composed of ions only have good properties such as extremely low volatility, high thermal stability, wide temperature range for liquid phase, nonflammability, high chemical stability, high ionic conductivity, and wide electrochemical window. Here the preparation of ionic liquid gels and the properties of formed gels are described.

26.1 GELATION OF IONIC LIQUIDS BY LOW MOLECULAR WEIGHT GELATORS

Low molecular weight compounds capable of hardening organic liquids, which are called "gelators," are of special interest for not only academic objects but also potential applications [4, 5]. Gelators have unique characteristics of both good solubility upon heating and inducement of smooth gelation of organic fluids at low concentration. For example, when the hot solution containing gelators is cooled to room temperature, the gel is easily formed in the course of cooling process. The formed gels always exhibit thermally reversible sol-to-gel phase transition. Because the driving forces of physical gelation by gelator are cooperating noncovalent interactions of gelator molecules, such as hydrogen bonding, hydrophobic interaction, π–π interaction, and electrostatic interaction.

With low melting point in mind, nine ionic liquids shown in Figure 26.1 were tested for gelation. The six kinds of imidazolium salts, the two kinds of ammonium

Electrochemical Aspects of Ionic Liquids   Edited by Hiroyuki Ohno
ISBN 0-471-64851-5   Copyright © 2005 John Wiley & Sons, Inc.

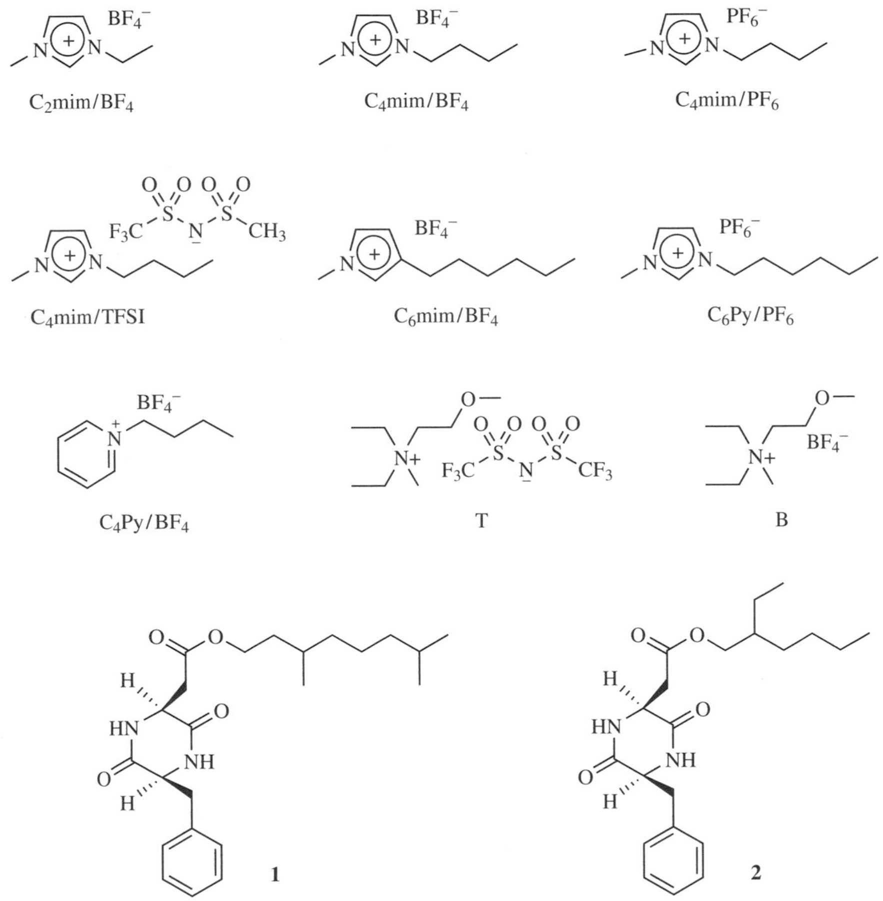

Figure 26.1 *Structures of ionic liquids and gelators.*

salts, and a pyridinium salt are tested. The imidazolium and pyridinium salts were prepared by ordinary methods [6] and two ammonium salts were supplied by a company. From screening test of many gelators, it was found that cyclo(L-β-3,7-dimethyloctylasparaginyl-L-phenylalanyl) (**1**) and cyclo(L-β-2-ethylhexylasparaginyl-L-phenylalanyl) (**2**), prepared from an artificial sweetener "Aspartame$^{®}$," were excellent gelators for the ionic liquids (Figure 26.1) [7]. Other gelators [8–12] could not gel up ionic liquids owing to the low solubility.

The results of gelation test of ionic liquids by gelators **1** and **2** are summarized in Table 26.1, where values mean minimum gel concentrations (MGC) necessary to harden ionic liquids. The gelators gelled up all of ionic liquids and formed translucent or transparent gels by adding less than 1 wt% (gelator/ionic liquid). For example, 3 g of gelator **1** can form 1 L of translucent gel of C_2mim/BF_4.

Gel strength is an important factor in application of gels. Gel strength of C_4mim/BF_4 and C_4Py/BF_4 gels are plotted against the concentration of **1** (Figure 26.2). The

TABLE 26.1   Results of Gelation Test of Ionic Liquids by Gelators 1 and 2 at 25°C

Ionic Liquid	1	2
C_2mim/BF_4	G-Tl (3)	G-Tl (5)
C_4mim/BF_4	G-Tl (5)	G-Tl (7)
C_4mim/PF_6	G-Tl (5)	G-Tl
$C_4mim/TFSI$	G-Tl (8)	G-Tl (12)
C_6mim/BF_4	G-Tp (8)	G-Tp (12)
C_6mim/PF_6	G-Tl (8)	G-Tl
C_4Py/BF_4	G-Tp (3)	G-Tp (5)
T	G-Tl (9)	G-Tl (13)
B	G-Tl (5)	G-Tl (10)

G-Tp: transparent gel, G-Tl: translucent gel.
Values mean minimum gel concentrations, whose unit is g/L (gelator/ionic liquid).

gel strength is evaluated as the power necessary to sink a cylinder bar (10 mm in diameter) 4 mm deep in the gels. It is clear that gel strength increases with increasing concentration of added gelator. The addition of 60 g of gelator **1** to 1 L of C_4Py/BF_4 gave transparent gel, whose gel strength was around 1500 g cm^{-2}.

The formed gels showed thermally reversible sol-to-gel transition due to the noncovalent interactions. The sol-to-gel phase diagram for C_4mim/BF_4 and C_4Py/BF_4

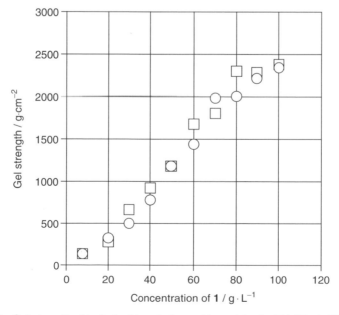

Figure 26.2 *Gel strength of ionic liquids gels formed by gelator **1**: (□) C4mim/BF4 and (○) C4Py/BF4.*

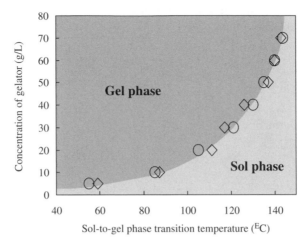

Figure 26.3   *Sol-to-gel phase transition diagram: ($\diamond$) gelator 1 in C$_4$mim/BF$_4$ and ($\bigcirc$) gelator 1 in C$_4$Py/BF$_4$.*

are shown in Figure 26.3, where the MGC are plotted against temperature of ionic liquids. The area above the plots is the gel phase and the area below is the sol phase. The transition temperature from gel to sol increases with increasing concentration of gelator. For instance, when the gel is prepared at the concentration of 70 g L^{-1}, the phase transition temperature of the gel-to-sol transition is more than 140°C. This means the very thermally stable gels can be obtained by using gelators.

26.2   CONDUCTIVITY OF IONIC LIQUID GELS

Solid electrolytes, which are electrically conductive solids with ionic carriers, have received special attention because of their potential use in solid-state batteries, fuel cells, energy storage, and chemical sensors [13, 14]. The gelation of ionic liquids by gelator is a conventional method for making gel electrolytes; therefore it is expected to play a role in the field of solid electrolytes. Figure 26.4 shows ionic conductivity of the various ionic liquids gels formed by **1**. In general, trifluoro-methanesulfonimide and tetrafluoroborate of 1-butyl-3-methylimidazolium (C$_4$mim) show higher ionic conductivity. The plots at left end give the ionic conductivities of the neat ionic liquids without gelator. The ionic conductivities of gels are nearly similar to those of the neat ionic liquids. Namely the ionic conductivity decreased very slightly with increasing concentration of gelator. The gels of ionic liquids were characterized by high ionic conductivity. For instance, the ionic conductivity of the C$_4$mim/BF$_4$ gel formed by **1** (20 g L^{-1}) was 3.20 mS cm^{-1}. This result indicates that the gelator molecules hardly interfere with the mobility of anion and cation.

When in practice ionic liquids are used for electrolytes in a device, the addition of polar solvents is recommended. This is because the addition of polar solvent reduces the viscosity and promotes the dissociation of ionic liquids. The addition

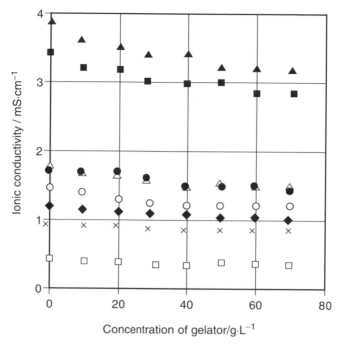

Figure 26.4 *Dependence of ionic liquid gel conductivity on concentration of gelator* **1** *at room temperature:* (▲)*C₄mim/TFSI,* (■) *C₄mim/BF₄,* (●) *C₄mim/PF₆,* (△) *T,* (○) *C₄Py/BF₄,* (◆) *C₆mim/BF₄,* (×) *B, and* (□) *C₆Py/PF₆.*

of a polar solvent consequently improves conductivity. The gelation of a mixture containing C₄mim/BF₄ and propylene carbonate (PC) by gelator **1** was studied. In this case the ionic liquids in PC were thought to be supporting electrolytes. The MGC of gelator **1** were plotted against the PC ratio, as shown in Figure 26.5. In the figure the 0% of the PC ratio means 100% of the ionic liquid, and 100% of PC ratio means pure PC. The gelator was found to gel up mixtures regardless of the PC ratio, and the MGC were almost proportional to the ratio of PC in the mixture.

Ionic conductivity of gels composed of C₄mim/BF₄, PC, and **1** are shown in Figure 26.6. The plots at the left end show the ionic conductivities of the neat mixture of C₄mim/BF₄ and PC without the gelator. It is apparent that the ionic conductivities of the mixture increase with the amount of added PC. It is noteworthy that the ionic conductivities decrease very slightly with the increasing concentration of gelator. The very slight decrease of ionic conductivity suggests that the gelators can be useful in making ionic liquid gels.

26.3 POLYMER GELS OF IONIC LIQUIDS

Zwitterionic molten salts composed of imidazolium cations containing covalently bound anionic sites were synthesized. The radical polymerization of the

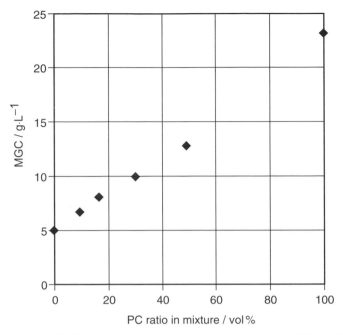

Figure 26.5 *MGC necessary for gelation of mixtures of C$_4$mim/BF$_4$ and PC.*

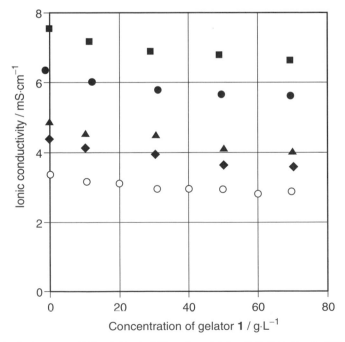

Figure 26.6 *Ionic conductivity of C$_4$mim/BF$_4$ gel containing PC: (■) C$_4$mim/BF$_4$: PC = 1:1; (●) C$_4$mim/BF$_4$: PC = 2:1; (▲) C$_4$mim/BF$_4$: PC = 5:1; (◆) C$_4$mim/BF$_4$: PC = 10:1; (○) C$_4$mim/ BF$_4$ only.*

zwitterionic imidazolium salts containing the vinyl group gave rubberlike polymers [15]. The ionic conductivity of pure zwitterionic-type imidazolium salt polymers was less than 10^{-9} S cm^{-2}, but mixed with 100 mol% LiTFSI these polymers showed good ionic conductivities of around 10^{-5} S cm^{-2} at 50°C despite their rubberlike properties.

However, radical polymerization of common vinyl monomers in situ in ionic liquids [16] produced new polymer-electrolytes called "ion gels." These ion gels have completely compatible combinations of ionic liquids and the resulting network polymers. The ion gels can softly bend, and they showed fast ionic conductivity of 10^{-2} S cm^{-2} at room temperature.

REFERENCES

1. T. Welton, *Chem. Rev.*, **99**, 2071 (1999).

2. P. Wasserscheid, T. Welton, *Ionic Liquids in Synthesis*, Wiley-VCH, 2002.

3. R. D. Rogers, K. R. Seddon, *Ionic Liquids—Industrial Applications for Green Chemistry*, ASC Symposium Series, 818 ACS, 2002.

4. P. Terech, R. G. Weiss, *Chem. Rev.*, **97**, 3133 (1997).

5. J. H. Esch, B. L. Feringa, *Angew. Chem. Int. Ed.*, **39**, 2263 (2000).

6. P. Bonhôte, A. Dias, N. Papagerorgiou, K. Kalyanasundaram, M. Grätzel, *Inorg. Chem.*, **35**, 1168 (1996).

7. K. Hanabusa, M. Matsumoto, M. Kimura, A. Kakehi, H. Shirai, *J. Colloid Interface Sci.*, **224**, 231 (2002).

8. K. Hanabusa, K. Hiratsuka, M. Kimura, H. Shirai, *Chem. Mater.*, **11**, 649 (1999).

9. K. Hanabusa, M. Yamada, M. Kimura, H. Shirai, *Angew. Chem. Int. Ed. Engl.*, **35**, 1949 (1996).

10. K. Hanabusa, J. Tange, Y. Taguchi, T. Koyama, H. Shirai, *Chem. Commun.*, 390 (1993).

11. K. Hanabusa, H. Nakayama, M. Kimura, H. Shirai, *Chem. Lett.*, 2000, 1070 (2000).

12. K. Hanabusa, K. Shimura, K. Hirose, M. Kimura, H. Shirai, *Chem. Lett.*, 1996, 885 (1996).

13. P. G. Bruce, *Solid State Elecrochemistry*, Cambridge University Press, 1995.

14. B. V. Ratnakumar, S. R. Narayanan, *Handbook Solid State Batteries Capacitors*, **1** (1995).

15. M. Yoshizawa, M. Hirao, K. Ito-Akita, H. Ohno, *J. Mater. Chem.*, **11**, 1057 (2001).

16. A. Noda, M. Watanabe, *Electochim. Acta*, **45**, 1265 (2002).

Chapter *27*

Zwitterionic Liquid/ Polymer Gels

Masahiro Yoshizawa and Hiroyuki Ohno

Typical polymer gel electrolytes are obtained by mixing a polar solvent, salt, and polymer to form a matrix. Since the polar solvent transports ions, these polymer gel electrolytes show relatively high ionic conductivity despite the film property. In addition the desirable ion conductive behavior and solid property can be attained easily by appropriately choosing the kind of polymer used as a matrix and the ratio of solvent to polymer. These polymer gel electrolytes have been used as flexible electrolyte films for dry ionics devices. However, the temperature range of this kind of polymer gel electrolyte is not wide enough because the thermal stability of polymer gel electrolytes is dependent on the solvent's properties. Although there are excellent organic solvents such as propylene carbonate and ethylene carbonate, the upper limit of the available temperature for all these remains at about 200°C. It is difficult to satisfy each parameter simultaneously, such as low viscosity, high polarity, low melting point, and high boiling point. This has prevented the application of polymer gel electrolytes in many fields. In particular, the volatility of the polar organic solvent is a serious problem (Table 27.1, type A).

Ionic liquids (ILs) are being investigated as a potential solution to this problem. ILs have attractive properties such as nonvolatility, a liquid state in a wide temperature range, and capacity to dissolve many compounds [1–3]. Furthermore ILs with relatively low viscosity show high ionic conductivity of about $10^{-2}\,\mathrm{S\,cm^{-1}}$ at room temperature [4, 5]. Although the temperature property of most of ILs is

Electrochemical Aspects of Ionic Liquids   Edited by Hiroyuki Ohno
ISBN 0-471-64851-5   Copyright © 2005 John Wiley & Sons, Inc.

TABLE 27.1 Comparison of Gel Type Polymer Electrolytes

Matrix Component	Type A Polymer, Solvent + Salt	Type B Polymer, Ionic Liquid
Target ion transport	○	×
Thermal stability	×	○

not satisfactory, there are studies made to overcome the upper limit by choosing a suitable combination of cation and anion. As reported in Chapter 26, many researchers have already succeeded in making polymer gel electrolytes using ILs [6–20]. In taking the excellent properties of ILs into account, it is easy to reach a conclusion that all these problems can be solved by using ILs as the electrolyte solutions. However, the high ionic conductivity of ILs is based on the IL itself. The low transference number of target ions such as the lithium cation and proton remains a drawback, as shown in Table 27.1, type B. To overcome this drawback, zwitterionic liquids (ZILs), which have both a cation and an anion in the same molecule, have started to be investigated (refer to Chapter 20 for details) [21]. Since ZILs do not migrate even under potential gradient, they can provide the ion conductive pathway to transport only target ions. In other words, ZILs should be suitable as the nonvolatile solvent for polymer gel electrolytes.

It is not so difficult to design a suitable structure of the IL for a particular purpose and to prepare the IL. The ease with which this can be done is a big advantage for organic ILs. Therefore the favored, IL is often called the "designer solvent" [22] or even "my solvent" [23]. This is because the matrix prepared by the polymerized IL is designed to overcome the above-mentioned problem [24, 25]. Such matrix is also designed to maximize merits of organic ILs.

The structure of a host polymer with ZILs is shown in Figure 27.1 [26]. The copolymers of vinylidene fluoride and hexafluoropropylene (P(VdF-co-HFP)) are used as the host polymer. This copolymer is used to prepare polymer gel electrolytes containing simple ILs because of their good miscibility with this component [17]. A fluorine-containing polymer generally has excellent compatibility with hydrophobic ILs. The two ZILs were used as a solvent: 1-(N-ethylimidazolio)-propane-3-sulfonate (EIm3S) and 1-(N-ethylimidazolio)-butane-4-sulfonate (EIm4S). The thermal stability of P(VdF-co-HFP), ZILs, and their mixture is

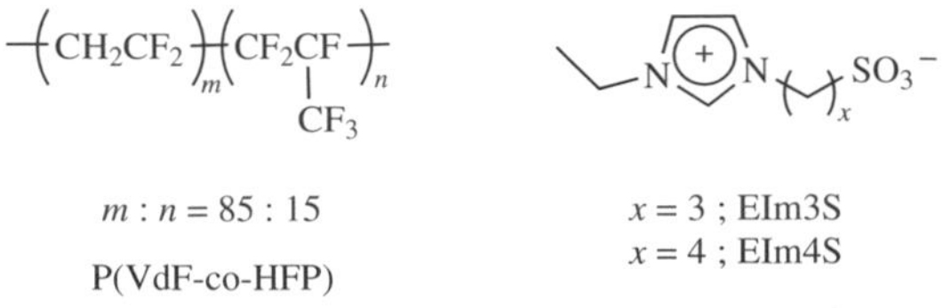

$m : n = 85 : 15$

P(VdF-co-HFP)

$x = 3$; EIm3S
$x = 4$; EIm4S

Figure 27.1 Structures of P(VdF-co-HFP) and ZILs.

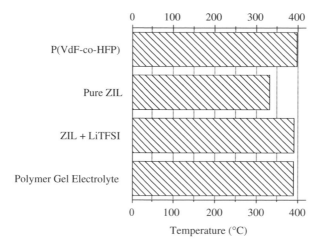

Figure 27.2 *Thermal stabilities of P(VdF-co-HFP), EIm3S, and their mixture.*

summarized in Figure 27.2. The decomposition temperature of P(VdF-co-HFP) is about 400°C, whereas that of EIm3S and EIm4S is 330°C. No weight loss was confirmed by means of TG/DTA analysis up to each decomposition temperature. Interestingly the mixture of ZILs and an equimolar amount of LiTFSI remained stable up to 390°C, which is about 60°C higher than that of pure ZILs. In the case of simple ILs, their thermal stability deeply depends on the combination of imidazolium cation and counter anion. For instance, when the imidazolium cation is combined with a TFSI anion, the decomposition temperature is generally improved at above 400°C [27, 28]. The thermal stability of ZILs is also a function of the added imidazolium cation of ZILs and the TFSI anion. The thermal stability of the polymer gel electrolyte containing the ZIL/LiTFSI mixture is about 390°C, which is consistent with that of the ZIL/LiTFSI mixture. The thermal stability of common polymer gel electrolytes is much less than this. Systems containing the ZIL/LiTFSI mixture are expected to be extremely stable gel electrolytes.

Figure 27.3 shows the temperature dependence of the ionic conductivity of these novel polymer gel electrolytes. P(VdF-co-HFP) was mixed with EIm3S/LiTFSI at a mixing ratio of 1:2 by weight. The ionic conductivity of this system is about 10^{-4} S cm^{-1} at 200°C. When the EIm4S/LiTFSI mixture was used, a similar tendency was observed for the Arrhenius plot of the polymer gel electrolyte. Although the ionic conductivity could not be measured in a higher temperature range because of an equipment limitation, the ionic conductivity of these polymer gel electrolytes is guaranteed by the results of the TG/DTA analysis. The thermal stability of this film still remains to be improved in the present system. As seen in Figure 27.3, there is an inflection point in the ionic conductivity at 150°C. This point agrees with the melting point of P(VdF-co-HFP). It means that in this case the host polymer could not keep solid state above 150°C. This drawback was overcome by using other network polymers [29].

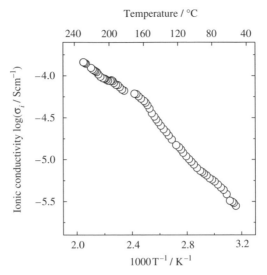

Figure 27.3 *Temperature dependence of ionic conductivity for the P(VdF-co-HFP)/equimolar mixture of EIm3S and LiTFSI. The ratio of polymer to ZIL is fixed at 1:2 by weight.*

Several polymer gel electrolytes were prepared by changing the mixing ratio of ZILs to polymer. The polymer gel electrolytes were obtained as translucent films up to 80 wt% of the ZIL/LiTFSI content. When the content of the ZIL/LiTFSI mixture was less than 40 wt%, they were obtained as flexible but white films. Figure 27.4 shows the effect of ZIL/LiTFSI mixture content on the ionic conductivity for polymer gel electrolytes. The ionic conductivity of these polymer gel

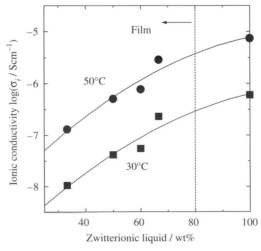

Figure 27.4 *Effect of ZIL/LiTFSI content on the ionic conductivity for the polymer gel electrolyte.*

electrolytes increased with the increasing ZIL/LiTFSI mixture content. When ZIL/LiTFSI mixture content was 66 wt%, the ionic conductivity exceeded 10^{-6} S cm^{-1} at 50°C, which is almost equivalent to that of ZIL/LiTFSI simple mixture.

We prepared a new polymer gel electrolyte of superior thermal stability by using ZILs. Several properties of the polymer gel electrolytes will be further improved through the synthesis of novel liquid ZILs at room temperature. Unfortunately, most ZILs are solid at room temperature without LiTFSI. So we found it necessary to design new ZILs in order to lower the melting point of the ZILs. We are expanding the possibility of using liquid ZILs at room temperature. Our research with liquid ZILs will be published elsewhere soon.

REFERENCES

1. P. Wasserscheid, T. Welton, eds., *Ionic Liquids in Synthesis*, Wiley-VCH, Weinheim, 2003.

2. H. Ohno, ed., *Ionic Liquids-The Front and Future of Material Development-*, CMC, Tokyo, 2003. (in Japanese).

3. T. Welton, *Chem. Rev.*, **99**, 2071 (1999).

4. P. Bonhôte, A.-P. Dias, M. Armand, N. Papageorgiou, K. Kalyanasundaram, M. Grätzel, *Inorg. Chem.*, **35**, 1168 (1996).

5. R. Hagiwara, Y. Ito, *J. Fluorine Chem.*, **105**, 221 (2000).

6. A. Noda, M. Watanabe, *Electrochim. Acta*, **45**, 1265 (2000). (b) T. Tsuda, T. Nohira, Y. Nakamori, K. Matsumoto, R. Hagiwara, Y. Ito, *Solid State Ionics*, **149**, 295 (2002).

7. H. Nakagawa, S. Izuchi, K. Kuwana, Y. Aihara, *J. Electrochem. Soc.*, **150**, A695 (2003).

8. (a) J. Sun, D. R. MacFarlane, M. Forsyth, *Solid State Ionics*, **147**, 333 (2002). (b) M. Forsyth, J. Sun, F. Zhou, D. R. MacFarlane, *Electrochim. Acta*, **48**, 2129 (2003).

9. M. Doyle, S. K. Choi, G. Proulx, *J. Electrochem. Soc.*, **147**, 34 (2000).

10. (a) J. Sun, L. R. Jordadn, M. Forsyth, D. R. MacFarlane, *Electrochim. Acta*, **46**, 1703 (2001). (b) J. Sun, D. R. MacFarlane, M. Forsyth, *Electrochim. Acta*, **48**, 1971 (2003).

11. W. Kubo, Y. Makimoto, T. Kitamura, Y. Wada, S. Yanagida, *Chem. Lett.*, 948 (2002).

12. P. Wang, S. M. Zakeeruddin, I. Exnar, M. Grätzel, *Chem. Commun.*, 2972 (2002).

13. A. Lewandowski, A. Swiderska, *Solid State Ionics*, **161**, 243 (2003).

14. M. Ue, M. Takeda, A. Toriumi, A. Kominato, R. Hagiwara, Y. Ito, *J. Electrochem. Soc.*, **150**, A499 (2003).

15. J. D. Stenger-Smith, C. K. Webber, N. Anderson, A. P. Chafin, K. Zong, R. Reynolds, *J. Electrochem. Soc.*, **149**, A973 (2002).

16. (a) W. Lu, A. G. Fadeev, B. Qi, E. Smela, B. R. Mattes, J. Ding, G. M. Spinks, J. Mazurkiewicz, D. Zhou, G. G. Wallace, D. R. MacFarlane, S. A. Forsyth, M. Forsyth, *Science*, **297**, 983 (2002). (b) D. Zhou, G. M. Spinks, G. G. Wallace, C. Tiyaboonchaiya, D. R. MacFarlane, M. Forsyth, J. Sun, *Electrochim. Acta*, **48**, 2355 (2003).

17. (a) J. Fuller, A. C. Breda, R. T. Carlin, *J. Electrochem. Soc.*, **144**, L67 (1997). (b) J. Fuller, A. C. Breda, R. T. Carlin, *J. Electroanal. Chem.*, **459**, 29 (1998).

18. C. Tiyapiboonchaiya, D. R. MacFarlane, J. Sun, M. Forsyth, *Macromol. Chem. Phys.*, **203**, 1906 (2002).

19. T. Fukushima, A. Kosaka, Y. Ishimura, T. Yamamoto, T. Takigawa, N. Ishii, T. Aida, *Science*, **300**, 2072 (2003).

20. N. Nishimura H. Ohno, *J. Mater. Chem.*, **12**, 2299 (2002).

21. M. Yoshizawa, M. Hirao, K. Ito-Akita, H. Ohno, *J. Mater. Chem.*, **11**, 1057 (2001).

22. M. Freemantle, *Chem. Eng. News*, March 30, 32 (1998).

23. H. Ohno, *Petrotech*, **23**, 556 (2000). (in Japanese)

24. (a) H. Ohno, K. Ito, *Chem. Lett.*, 751 (1998). (b) M. Hirao, K. Ito, H. Ohno, *Electrochim. Acta*, **45**, 1291 (2000). (c) H. Ohno, *Electrochim. Acta*, **46**, 1407 (2001).

25. M. Yoshizawa, W. Ogihara, H. Ohno, *Polym. Adv. Technol.*, **13**, 589 (2002).

26. H. Ohno, M. Yoshizawa, W. Ogihara, *Electrochim. Acta*, **48**, 2079 (2003).

27. H. L. Ngo, K. LeCompte, L. Hargens, A. B. McEwen, *Thermochim. Acta*, **357–358**, 97 (2000).

28. A. B. McEwen, H. L. Ngo, K. LeCompte, J. L. Goldman, *J. Electrochem. Soc.*, **146**, 1687 (1999).

29. (a) H. Ohno, M. Yoshizawa, *Polymer Preprints, Japan*, **51**, 3069 (2002). (b) M. Yoshizawa, M. Tamada, H. Ohno, *Polymer Preprints, Japan*, **52**, 2615 (2003).

Chapter 28

Ionic Liquidized DNA

Naomi Nishimura and Hiroyuki Ohno

The typical cation structure of ionic liquids is comprises heteroaromatic rings such as imidazole or pyridine, and the useful properties of these materials stem from these groups. Molecular orbital calculations suggest that imidazole is a good starting material for an onium cation based ionic liquid. Useful ionic liquids can also be prepared from other amines, and suitable amines as starting materials for ionic liquids are still sought today [1]. Many kinds of heteroaromatic rings have been examined for this purpose. Biosystems also contain many kinds of amine. In particular, nucleic acid bases composed of hetero aromatic rings promise excellent ion conductive materials. Nucleic acid bases are examined as starting materials for ionic liquids by means of neutralization. The resulting ionic liquid can be regarded as a model for ionic liquidized DNA. DNA is then converted to ionic liquid. A methodology is set out for preparing flexible DNA films having high ionic conductivity, in which DNA exhibits excellent ionic conductivity as a flexible film.

28.1 DNA

DNA is generally recognized as a typical biopolymer having genetic information. It is not so well known that the DNA is obtained easily with an inexhaustible supply in nature, if proteins are separated from the gene. Since DNA has superior characteristics compared with the synthetic polymers, DNA is a very attractive material, as shown in Figure 28.1. DNA is also readily available: salmon roe is frequently

Electrochemical Aspects of Ionic Liquids Edited by Hiroyuki Ohno
ISBN 0-471-64851-5 Copyright © 2005 John Wiley & Sons, Inc.

Figure 28.1 *Characteristics of DNA. DNA is a rigid rod-like polymer, having hetero aromatic rings, π stacked bases, inexhaustible supply in nature, and biodegradability.*

used as a Japanese dish, but considerable amounts of salmon milt, which is rich in DNA, are discarded. DNA is a kind of biomass. DNA has four kinds of nucleic acid bases, as shown in Figure 28.2, and the information is kept by hydrogen bonding between complementary base pairs. DNA has nucleic acid bases with high density. There has therefore been much interest shown in utilizing these characteristics of DNA in creating functional materials for electronics, optics, and the like [2, 3]. In particular, there have been some attempts to prepare electroconductive materials using the π–π stacking of base pairs. However, except for this research, the application of ion conductive materials using DNA has not been undertaken. We have used a composite of DNA and PEO, and we found this composite to have excellent ionic conductivity [4].

28.2 LOW MOLECULAR WEIGHT MODEL COMPOUNDS [5]

There are two methods for preparing ionic liquids: (1) quaternization of the tertiary amines with alkyl halide followed by the anion exchange, and (2) neutralization of bases with acids in pure water. Since DNA is soluble in an aqueous medium, the DNA can be ionic liquidized by a neutralization method in pure water. Before we prepared the ionic liquid from DNA, we examined four kinds of bases as possible models. Table 28.1 gives the melting point (T_m) and the glass transition

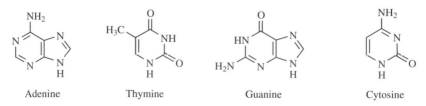

Adenine   Thymine   Guanine   Cytosine

Figure 28.2 *Structure of four kinds of nucleic acid bases.*

TABLE 28.1 T_m of A, G, C, and T Before and After Reaction with Acids

		+HBF$_4$		+HTFSI	
	T_d (°C)^a	T_m (°C)	Appearance	T_m (°C)	Appearance
Adenine	360–365	243.5	White powder	—	Glass
Cytosine	360	185.9	White powder	32.3	Liquid
Guanine	320–325	251.3	White powder	50.6^b	White powder
Thymine	335–337	317.9	White powder	220.6	White powder

$^a T_d$: decompositon temperature.
^{b}Decomposed to guanidine.

temperature (T_g) of the salts obtained from the neutralization of the four bases (adenine, **A**; guanine, **G**; cytosine, **C**; and thymine, **T**) with HBF$_4$ and *bis*(trifluoromethanesulfonyl)imide (HTFSI). The TFSI$^-$ is an excellent anion for polymer electrolyte preparation, and it can also yield many of the ionic liquids. All these bases before neutralization are white powders, and they decompose above 320°C without melting. Salts neutralized with HBF$_4$ are also obtained as white powder, but their T_m varies from 317.9° to 185.9°C depending on the base species, as shown in Table 28.1. Adenine neutralized with HTFSI had no melting point but showed a T_g at −13.2°C. Furthermore the **C**·TFSI salt obtained by the neutralization of cytosine with HTFSI became an ionic liquid with a $T_m = 32.3$°C. **A**·TFSI and **C**·TFSI were only slightly soluble in water due to the hydrophobic property of the TFSI anion. They should be classed as ionic liquids with their corresponding bases of DNA.

The pK$_b$ of **A**, **C**, **T**, and **G** were 9.8, 9.4, 4.1, and 3.2, so both **A** and **C** could be neutralized with acid. No reaction occurred when thymine was mixed with acid. Guanine was degraded by the acid, yielding a product having a lower T_m at 50°C. The IR and ^{1}H–NMR measurements indicated a cleavage of the purine rings. These results show that both adenine and cytosine can form ionic liquids.

Figure 28.3 shows the temperature dependence of the ionic conductivity of the corresponding bases after neutralization with HTFSI. **A**·TFSI and **C**·TFSI showed relatively high ionic conductivity, especially **C**·TFSI with 6.85×10^{-5} S cm^{-1} at 50°C (Figure 28.3; ■). This high ionic conductivity is probably related to the low T_g, since **A**·TFSI and **C**·TFSI had T_g's of −13.2 and −30.8°C respectively. These remarkable characteristics of the corresponding bases are the result of ionic liquid formation through neutralization with HTFSI.

To select a suitable acid for the formation of ionic liquid, cytosine was neutralized with 11 different acids. The obtained salts were studied with ^{1}H–NMR spectroscopy. Protons at the 5 and 6 positions shifted to the lower magnetic field side depending on the salt formation. Although pure cytosine decomposed above 360°C, the T_m of the thus prepared cytosine salts depended on the acid species being used. **C**·TFSI showed the lowest T_m, at 32.3°C. No other acid proved more effective in making an ionic liquid. Figure 28.4 gives the relationships between the T_m's

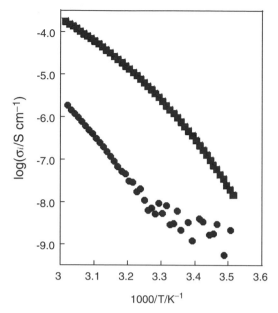

Figure 28.3 *Temperature dependence on the ionic conductivity of neutralized nucleic acid bases.* ●; A·TFSI and ■; C·TFSI.

of the resulting salt and the pK_a's of the acid used for neutralization. Since the pK_a of HTFSI is not yet determined, the T_m of this salt is indicated by the arrow in Figure 28.4. Strong acids such as $HClO_4$ and CF_3SO_3H lowered the T_m of these salts. This relationship seen in Figure 28.4 is similar to that of the imidazolium-

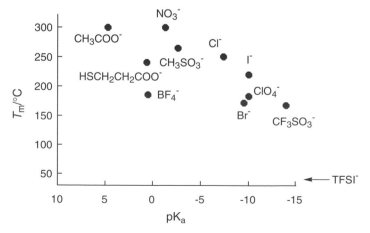

Figure 28.4 *Relationship between T_m of cytosine after neutralization and pK_a of the applied acids.*

type ionic liquid. That is, the stronger acid produced a salt with a lower T_m. However, the ionic liquids from the corresponding bases showed higher T_m's than those of the imidazolium salts.

28.3 IONIC LIQUIDIZED DNA

Since adenine and cytosine formed ionic liquids after neutralization, these subunits in DNA should also form ionic liquids after acid treatment. DNA has many bases aligned on the chain. If all these bases are converted into ionic liquids, successive ionic liquid domains should be created along the chain. According to the results in Figure 28.4, HBF_4, HTFSI, and CF_3SO_3H are strongly indicated to be effective acids for this purpose. DNA was treated with these acids. The four kinds of bases were assumed to be contained equally in the DNA. Acids of 50 mol% to the total bases were mixed with DNA to neutralize all adenines and cytosines. After neutralization with the stated acids individually in pure water, all products were obtained as precipitates. It is known that purine rings are dissociated from DNA by strong acid treatment. There was some fear of the dissociation of the purine rings occurring in the present treatments. Nevertheless, we proceeded to prepare the ionic liquid domain of adenine salts in the DNA. The resulting samples were washed with water to remove unreacted acid, dissociated purine (salts), and so forth. This process gave us an acid-treated DNA without the dissociated purines. The hydrogen bonding between complementary base pairs was broken upon neutralization, and the hydrophobic bases turned out from the helix. CD and IR spectra gave strong indications that these DNAs lose their double-strand helical structure.

The ionic conductivity of the neutralized DNA was about 1×10^{-9} S cm^{-1} at room temperature. This can be explained by the insufficient ionic liquid fraction. The weight fraction of all the bases in the solid DNA was about 40 wt%. However, if all the adenines and cytosines had yielded ionic liquids, the corresponding domain fraction would have been only about 20 wt%. A closely packed ion conduction path cannot be formed with this low fraction. C·TFSI, which had the lowest T_m of the neutralized bases, was therefore added to the neutralized DNA to assist the construction of a continuous ion conduction path. The neutralized DNA and C·TFSI were mixed in a solvent, and then they were cast to prepare films. The problem that we encountered in the processing was that DNA·TFSI is excellent matrix from the viewpoint of ionic conductivity, but it cannot be dissolved in any solvent. Since DNA·BF$_4$ can be dissolved in excess of water, we used the DNA·BF$_4$ in further experiments. When the C·TFSI content in DNA·BF$_4$ was less than 50 wt%, the ionic conductivity of the mixture was very low, at around 1×10^{-8} S cm^{-1} at 50°C (Figure 28.5; ▲). It abruptly increased when more than 70 wt% C·TFSI was added, reaching 4.76×10^{-5} S cm^{-1} at 50°C when 80 wt% of C·TFSI was added to the DNA·BF$_4$. The mixture was, however, obtained as a flexible film even with 80 wt% C·TFSI. Since the ionic conductivity of C·TFSI in the bulk was only 6.85×10^{-5} S cm^{-1}, the DNA·BF$_4$ and C·TFSI mixed film showed reasonable conductivities, indicating the formation of a successive ionic

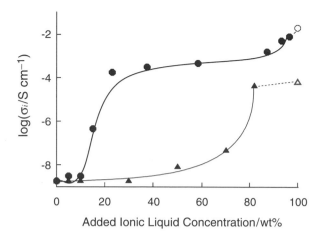

Figure 28.5 *Effects of added ionic liquid concentration on the ionic conductivities of DNA·BF₄/ C·TFSI (▲) and DNA·BF₄/EImBF₄ (●) at 50°C. Those of pure C·TFSI (△) and pure EImBF₄ (○) are also depicted as references.*

liquid phase. To improve the ionic conductivity of the DNA·BF₄ film, we had to mix the DNA·BF₄ with a more conductive ionic liquid. For this purpose ethylimidazolium tetrafluoroborate (EImBF₄) was added to the DNA·BF₄ (Figure 28.5; ●). EImBF₄ shows very high ionic conductivity of around 10^{-2} S cm^{-1} at 50°C. Since both DNA·BF₄ and EImBF₄ have the same BF₄$^-$ anion and are soluble in pure water, the DNA·BF₄/EImBF₄ mixture was conveniently prepared as a homogeneous film by casting. When the amount of the added EImBF₄ was up to 10 wt%, the mixture's ionic conductivity was the same as that of pure DNA·BF₄. However, the ionic conductivity of the film containing 15 wt% EImBF₄ was about 4.62×10^{-7} S cm^{-1}, and that of film containing 23.7 wt% EImBF₄ was 1.74×10^{-4} S cm^{-1} at 50°C. This excellent ionic conductivity was maintained up to 85 wt%. The highest ionic conductivity, 5.05×10^{-3} S cm^{-1} at 50°C, was observed when the film was prepared with 93 wt% EImBF₄. Further addition of EImBF₄ maintained high ionic conductivity, but no film was obtained. We found, quite surprisingly, that DNA with only 7 wt% ionic liquid forms a film, and a solidity with 93 wt% ionic liquid. This is because of the high molecular weight of DNA and the strong affinity of EImBF₄ for DNA·BF₄. We have already reported on the preparation of excellent ion conductive films from a mixture of native DNA and EImBF₄. There is no difference in the ionic conductivity between the DNA·BF₄/EImBF₄ film and the DNA/EImBF₄ film at high ionic liquid content (>70 wt%), as shown in Figure 28.6 [6]. However, the DNA·BF₄/EImBF₄ film displayed much higher ionic conductivity at a low EImBF₄ content. This can be explained by the formation of an effective ionic liquid pathway in the DNA matrix due to neutralization of the bases. Further the DNA·BF₄/EImBF₄ film showed high ionic conductivity and excellent flexibility over a wide ionic liquid content. Significant improvement in both the ionic conductivity and flexibility of the DNA films is

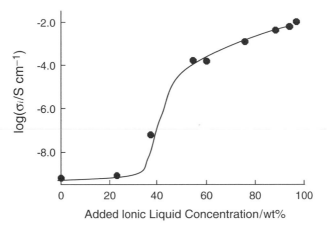

Figure 28.6 *The relationship of ionic conductivity at 50° C and added EImBF$_4$ concentration in native DNA.*

observed compared with the DNA/PEO$_{1000}$/NaClO$_4$ film [4] or the simple mixture of DNA and ionic liquid [6]. A photograph of the flexible and transparent DNA film is shown in Figure 28.7. DNA·BF$_4$ was mixed with 40 wt% EImBF$_4$, and the ionic conductivity of the film was 1.32×10^{-3} S cm^{-1} at room temperature. In other words, flexible film having high ionic conductivity can be obtained when DNA·BF$_4$ is mixed with only a small amount of EImBF$_4$. The experiments were carried out in a dry nitrogen atmosphere at room temperature. DNA·BF$_4$/EImBF$_4$ film showed excellent stability, and no leakage of ionic liquid was detected. This stability was observed continually at room temperature over several months.

Figure 28.7 *Photograph of flexible and transparent DNA film. DNA·BF$_4$/EImBF$_4$ (40 wt%).*

28.4 FUTURE OF DNA AS MATERIALS

In this chapter we reviewed our work on preparing ion conductive films by using DNA. The highest ionic conductivity of 5.05×10^{-3} S cm^{-1} was found to be at 50°C when 93 wt% EImBF$_4$ is mixed with DNA neutralized with HBF$_4$. The DNA film was obtained by casting a DNA aqueous solution. The attractive characteristics of DNA is that it is a biodegradable material that has an inexhaustible supply in nature. We showed in this chapter that several properties, such as ionic conductivity of DNA films, can be controlled by certain factors. It is our hope that our results well open the way to new applications of DNA.

REFERENCES

1. M. Hirao, H. Sugimoto, H. Ohno, *J. Electrochem. Soc.*, **147**, 4168 (2000).
2. C. J. Murphy, M. R. Arkin, Y. Jenkins, N. D. Ghatlia, S. H. Bossmann, N. J. Turro, J. K. Barton, *Science*, **262**, 1025 (1993).
3. D. Porath, A. Bezryadin, S. de Vries, C. Dekker, *Nature*, **403**, 635 (2000).
4. (a) H. Ohno, N. Takizawa, *Chem. Lett.*, 642 (2000). (b) N. Nishimura, S. Kokubo, H. Ohno, *Polym. Adv. Technol.*, **15**, 335 (2004).
5. N. Nishimura, H. Ohno, *J. Mater. Chem.*, **12**, 2299 (2002).
6. H. Ohno, N. Nishimura, *J. Electrochem. Soc.*, **148**, E168 (2001).

Part VII

Polymerized Ionic Liquids

Chapter *29*

Ion Conductive Polymers

Hiroyuki Ohno and Masahiro Yoshizawa

Because ionic liquids (ILs) consist only of ions, they offer two brilliant features: very high concentration of ions [1] and high mobility of component ions at room temperature. Because many ILs show the ionic conductivity of over 10^{-2} S cm^{-1} at room temperature [2, 3], there are plenty of possible applications as electrolyte materials, among these, for rechargeable lithium-ion batteries [4–8], fuel cells [9–12], solar cells [13–17], and capacitors [18–23].

However, so far the ILs studied as electrolyte materials have shown a fatal drawback in that their component ions migrate along with the potential gradient. As a result there is a concern about leakage of liquid ILs in battery technology as well as in other typical organic electrolyte applications. Certain kinds of polymer gel electrolytes containing ILs have been investigated because of their high ionic conductivity and good mechanical properties. For example, findings for poly(vinylidene fluoride-co-hexafluoropropylene) (P(VdF-co-HFP)) [24], Nafion [9], DNA [25], and the polymer electrolyte [26] containing ILs have been reported. Polymer-in-IL electrolytes were further prepared by in situ polymerization of vinyl monomers in ILs [27].

In this chapter, we report on the ion conductive behavior of new ion conductive polymers that polymerize the IL itself.

Electrochemical Aspects of Ionic Liquids   Edited by Hiroyuki Ohno
ISBN 0-471-64851-5   Copyright © 2005 John Wiley & Sons, Inc.

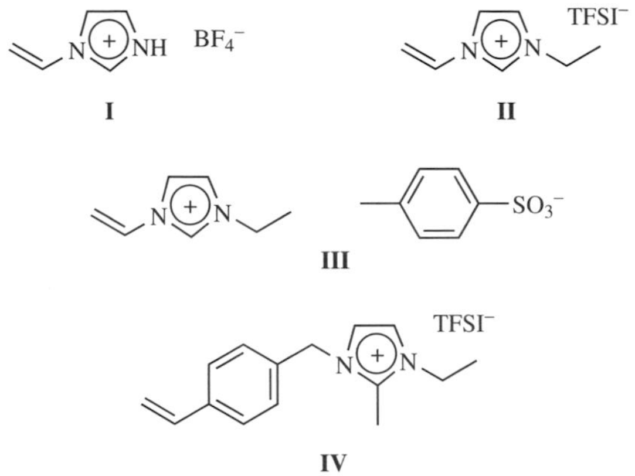

Figure 29.1 *Structure of IL monomers with different polymerizable cations.*

29.1 POLYCATIONS

The IL monomers were prepared from vinylimidazole and polymerized in order to investigate the relationship between the IL polymer structure and their ionic conductivity. Figure 29.1 shows the structures of a number of IL monomers having different polymerizable cations. Monomer I was prepared by neutralization of N-vinylimidazole and HBF_4. It has already been reported that ILs can be obtained by neutralization of the tertiary amines and organic acids (see Chapter 19) [28–30]. This procedure is very simple, and only an equimolar mixing of tertiary amines and acids provides salts without generation of by-product. Therefore this is one of the most favored procedures to prepare pure ILs and their models. Monomer I showed melting point (T_m) at 55°C, whereas it showed glass transition temperature (T_g) at −5°C after polymerization. On the other hand, although monomer I showed the ionic conductivity of about 10^{-3} S cm^{-1} at 50°C, the ionic conductivity of corresponding polymer was about 10^{-8} S cm^{-1}, which is 10,000 times lower than that of monomer, as shown in Figure 29.2. This decrease should be based on the decrease of carrier ion number and their mobility. There are two rough strategies to overcome this drop of ionic conductivity of polymer I. One is the design of polymer structure to lower the T_g. This will be mentioned in Chapter 31. The other is to add the carrier ions. In the present chapter we consider the case where Li salt was added to the polymer matrix. The ionic conductivity of polymer I containing 25 mol% of $LiBF_4$ to imidazolium cation was about 10^{-4} S cm^{-1} at 50°C, which is almost equivalent to that of monomer I. The carrier ions increased with the added $LiBF_4$, but the number of ions was not the only factor to govern the ionic conductivity. That is, when LiCl was added to the polymer, no improvement of ionic conductivity was found. The dissociation degree of the added salt as well as changes in viscosity (in terms of segmental motion) would turn out to be important factors in improving the ionic conductivity of the IL polymer.

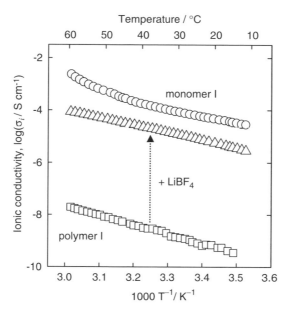

Figure 29.2 *Arrhenius plots of the ionic conductivity for monomer I, its polymer, and the polymer containing LiBF₄ (polymer: LiBF₄ = 4:1, mole fraction).*

Monomer II is also a polymerizable IL composed of quaternized imidazolium salt, as shown in Figure 29.1. This monomer is liquid at room temperature and shows a T_g only at $-70°C$. Its high ionic conductivity of about 10^{-2} S cm^{-1} at room temperature reflects a low T_g. Although the ionic conductivity of this monomer decreased after polymerization as in the case of monomer I, it was considerably improved by the addition of a small amount of LiTFSI. Figure 29.3 shows the effect of LiTFSI concentration on the ionic conductivity and lithium transference number (t_{Li}^+) for polymer II. The bulk ionic conductivity of polymer II was 10^{-5} S cm^{-1} at 50°C. When LiTFSI was added to polymer II, the ionic conductivity increased up to 10^{-3} S cm^{-1}. After that, the ionic conductivity of polymer II decreased gradually with the increasing LiTFSI concentration. On the other hand, when the LiTFSI concentration was 100 mol%, the t_{Li}^+ of this system exceeded 0.5. Because of the fixed imidazolium cations on the polymer chain, mobile anion species exist more than cation species in the polymer matrix at this concentration. Since the TFSI anions form the IL domain with the imidazolium cation, the anion can supply a successive ion conduction path for the lithium caiton. Such behavior is not observed in monomeric IL systems, and is understood to be due to the concentrated charge domains created by the polymerization.

Polymers I and II inherently showed low ionic conductivity and had poor mechanical properties. IL monomers having benzene rings were prepared to improve the properties of these polymers. This is because it is known that most imidazolium salts having benzene rings form ILs at room temperature and show relatively high ionic conductivity [30]. In addition the polymers having benzene rings are generally

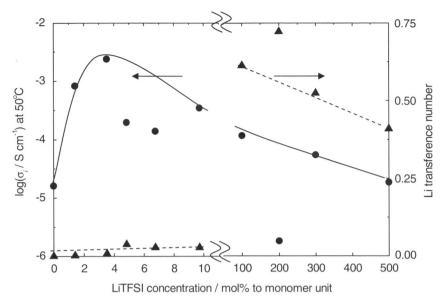

Figure 29.3 *Effects of LiTFSI concentration on the ionic conductivity and lithium transference number for polymer II.*

superior in terms of thermal stability and film preparation. Two kinds of IL mono-mers having a benzene ring on either the cation or the anion were synthesized, as shown in Figure 29.1. Monomer III was obtained by the reaction of vinylimidazole and by an equimolar amount of p-toluenesulfonic acid ethyl ester (p-TSE). This is also a simple reaction similar to the neutralized amines. Since p-TSE is cheap, var-ious ILs were prepared by this reaction [31]. The monomer IV was prepared by quaternization of N-ethylimidazole with 4-chlorostyrene.

Monomer III showed a T_m at 83°C, whereas monomer IV showed a T_g at −59°C. The ionic conductivity of monomer IV (3.07×10^{-3} S cm^{-1} at 50°C) performed four orders higher than that of monomer III (1.64×10^{-7} S cm^{-1} at 50°C), reflect-ing its improved thermal property. Monomer IV is a kind of IL, and it shows high ionic conductivity. Figure 29.4 gives the ionic conductivity of III and IV after polymerization. The effects of the LiTFSI addition on the ionic conductivities of polymer III and IV can be seen in this figure. Polymer III acted mostly as an insu-lator, so its conductivity was too low to appear on this scale. Polymer IV's ionic conductivity was about 10^{-6} S cm^{-1} at room temperature. When an equimolar amount of LiTFSI was added to the imidazolium unit, the ionic conductivity of these polymers increased considerably. Polymer III especially showed an ionic conductivity of over 10^{-4} S cm^{-1} at room temperature, which is higher than that of the corresponding monomer.

Figure 29.5 shows the effects of the LiTFSI concentration on the ionic conduc-tivities at 30°C for polymers III and IV. Although the ionic conductivity of polymer III was very poor in bulk, it improved considerably with the increasing LiTFSI concentration. When the LiTFSI concentration was 100 mol% to the imidazolium

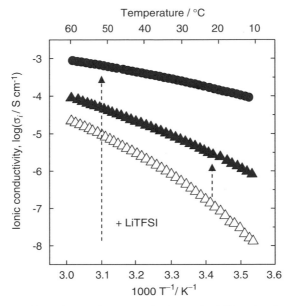

Figure 29.4 *Effects of LiTFSI addition on the ionic conductivities of polymer III and IV.* △: *polymer IV;* ▲: *polymer IV + LiTFSI;* ●: *polymer III + LiTFSI.*

unit, the ionic conductivity of polymer III reached its maximum value. After further additions, there was no obvious increase of ionic conductivity. The T_g of polymer III lowered with increasing LiTFSI concentration. The T_g fell to $-80°C$ when the LiTFSI concentration was 100 mol%. The fall of T_g was brought on by the

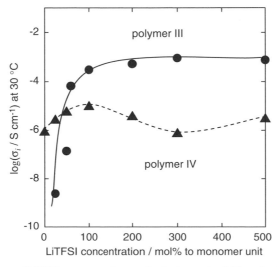

Figure 29.5 *Effects of LiTFSI concentration on the ionic conductivities of polymers III and IV at 30° C.*

increase in the number of imidazolium cation and TFSI anion pairs. However, while the addition of TFSI salt can generally be expected to produce a plasticizing effect, this effect cannot be used to fully explain the lowest T_g at the 100 mol% addition. On the other hand, the ionic conductivity of polymer IV also increased with the addition of LiTFSI. Although the increase was small, it had to do with the relatively high ionic conductivity of polymer IV in bulk. Two T_g values of polymer IV were observed at the high salt concentration, indicating that polymer IV had a micro phase separation with LiTFSI. For example, two T_g values were observed at $-4°C$ and $-49°C$ when salt concentration was 100 mol%. This phase separation would be caused by the low solubility of salts in the polymer matrix having a benzene ring.

Polycation-type ILs can provide an ion conductive path for the added salts. However, the bulk ionic conductivity of their polymers is very poor at present. It is necessary to improve their ionic conductivity for wide application.

29.2   POLYANIONS

The IL polymers were prepared from polymerizable anions and free imidazolium cations. To investigate the effect of the anion structure on the properties, four kinds of IL monomers were synthesized by neutralizing N-ethylimidazole with acrylic acid, p-styrenesulfonic acid, vinylsulfonic acid, and vinylphosphonic acid, respectively, as shown in Figure 29.6 [32]. Table 29.1 summarizes the properties of IL monomers composed of the polymerizable anion and an ethylimidazolium cation. They are liquid at room temperature and show quite low T_g's in the range of $-100°$ to $-60°C$. EImVS showed the lowest T_g of $-95°C$ among these four IL monomers. In addition it showed the highest ionic conductivity of about 10^{-2} S cm^{-1} at $30°C$, which is comparable to that of the best ILs such as the 1-ethyl-3-methylimidazolium salts.

The IL monomers composed of polymerizable acid having lower pKa show higher ionic conductivity. Table 29.1 also summarizes the properties of the polymerized IL monomers. The T_g was found to be elevated about $30°C$ after

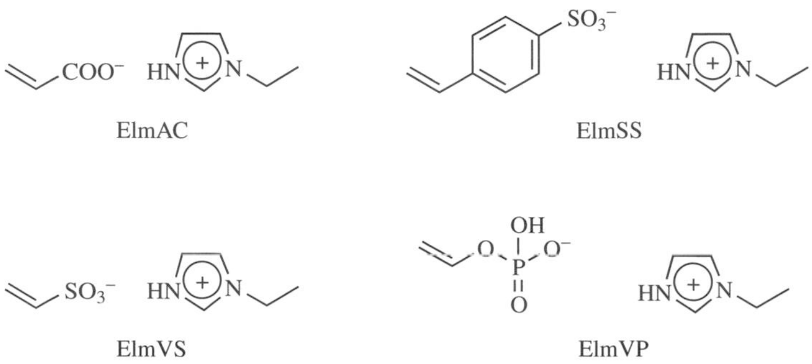

Figure 29.6   Structures of IL monomers having different polymerizable anions.

TABLE 29.1 Ionic Conductivity and T_g for Ionic Liquid Monomers and their Polymers

	T_g / °C		σ_i/S cm^{-1} (30°C)	
	Monomer	Polymer	Monomer	Polymer
EImAC	−61	−32	1.4×10^{-4}	1.2×10^{-6}
EImSS	−77	—	8.7×10^{-5}	1.1×10^{-8}
EImVS	−95	−63	9.0×10^{-3}	1.1×10^{-4}
EImVP	−78	−56	1.5×10^{-4}	2.9×10^{-5}

Note:–Not detected.

polymerization. P(EImVS) (P implies polymer) showed the lowest T_g of −63°C and the highest ionic conductivity of 1.1×10^{-4} S cm^{-1} at 30°C among these four polymers. The IL having the VS anion kept its matrix flexibility high even after polymerization. On the other hand, the ionic conductivity of P(EImSS) was extremely low. Although the T_g of P(EImSS) was not detected by the DSC measurement, the T_g of P(EImSS) might have increased considerably after polymerization. As mentioned previously, there is a certain relationship between T_g and the ionic conductivity [33], and accordingly this can be used to estimate the T_g from its ionic conductivity. When the ion concentration is comparable, the ionic conductivity can roughly be estimated from this relation. As a result the T_g of P(EImSS) can be estimated to be around 5°C based on the very low ionic conductivity it showed of about 10^{-8} S cm^{-1} at 30°C.

REFERENCES

1. P. Wasserscheid, T. Welton, eds., *Ionic Liquids in Synthesis*, Wiley-VCH, 2003.

2. R. Hagiwara, Y. Ito, *J. Fluorine Chem.*, **105**, 221 (2000).

3. P. Bonhôte, A.-P. Dias, M. Armand, N. Papageorgiou, K. Kalyanasundaram, M. Grätzel, *Inorg. Chem.*, **35**, 1168 (1996).

4. V. R. Koch, C. Nanjundiah, G. B. Appetecchi, B. Scrosati, *J. Electrochem. Soc.*, **142**, L116 (1995).

5. D. R. MacFarlane, J. Huang, M. Forsyth, *Nature*, **402**, 792 (1999).

6. H. Nakagawa, S. Izuchi, K. Kuwana, T. Nukuda, Y. Aihara, *J. Electrochem. Soc.*, **150**, A695 (2003).

7. A. Hayashi, M. Yoshizawa, C. A. Angell, F. Mizuno, T. Minami, M. Tatsumisago, *Electrochem. Solid-State Lett.*, **6**, E19 (2003).

8. H. Sakaebe, H. Matsumoto, *Electrochem. Commun.*, **5**, 594 (2003).

9. M. Doyle, S. K. Choi, G. Proulx, *J. Electrochem. Soc.*, **147**, 34 (2000).

10. (a) J. Sun, D. R. MacFarlane, M. Forsyth, *Electrochim. Acta*, **46**, 1673 (2001). (b) J. Sun, L. R. Jordan, M. Forsyth, D. R. MacFarlane, *Electrochim. Acta*, **46**, 1703 (2001).

11. (a) A. Noda, M. A. B. H. Susan, K. Kudo, S. Mitsushima, K. Hayamizu, M. Watanabe, *J. Phys. Chem. B*, **107**, 4024 (2003). (b) M. A. B. H. Susan, A. Noda, S. Mitsushima, M. Watanabe, *Chem. Commun.*, 938 (2003).

12. M. Yoshizawa, W. Xu, C. A. Angell, *J. Am. Chem. Soc.*, **125**, 15411 (2003).

13. N. Papageogiou, Y. Athanassov, M. Armand, P. Bonhôte, H. Pettersson, A. Azam, M. Grätzel, *J. Electrochem. Soc.*, **143**, 3099 (1996).

14. R. Kawano, M. Watanabe, *Chem. Commun.*, 330 (2002).

15. W. Kubo, T. Kitamura, K. Hanabusa, Y. Wada, S. Yanagida, *Chem. Commun.*, 374 (2002).

16. P. Wang, S. M. Zakeeruddin, I. Exnar, M. Grätzel, *Chem. Commun.*, 2972 (2002).

17. H. Matsumoto, T. Matsuda, T. Tsuda, R. Hagiwara, Y. Ito, Y. Miyazaki, *Chem. Lett.*, 26 (2001).

18. A. Lewandowski, A. Swiderska, *Solid State Ionics*, **161**, 243 (2003).

19. M. Ue, M. Takeda, A. Toriumi, A. Kominato, R. Hagiwara, Y. Ito, *J. Electrochem. Soc.*, **150**, A499 (2003).

20. J. D. Stenger-Smith, C. K. Webber, N. Anderson, A. P. Chafin, K. Zong, R. Reynolds, *J. Electrochem. Soc.*, **149**, A973 (2002).

21. A. B. McEwen, H. L. Ngo, K. LeCompte, J. L. Goldman, *J. Electrochem. Soc.*, **146**, 1687 (1999).

22. A. B. McEwen, S. F. McDevitt, V. R. Koch, *J. Electrochem. Soc.*, **144**, L84 (1997).

23. C. Nanjundiah, S. F. McDevitt, V. R. Koch, *J. Electrochem. Soc.*, **144**, 3392 (1997).

24. (a) J. Fuller, A. C. Breda, R. T. Carlin, *J. Electrochem. Soc.*, **144**, L67 (1997). (b) J. Fuller, A. C. Breda, R. T. Carlin, *J. Electroanal. Chem.*, **459**, 29 (1998). (c) H. Ohno, M. Yoshizawa, W. Ogihara, *Electrochim. Acta*, **48**, 2079 (2003).

25. N. Nishimura, H. Ohno, *J. Mater. Chem.*, **12**, 2299 (2002).

26. M. Forsyth, J. Sun, F. Zhou, D. R. MacFarlane, *Electrochim. Acta*, **48**, 2129 (2003).

27. (a) A. Noda, M. Watanabe, *Electrochim. Acta*, **45**, 1265 (2000). (b) C. Tiyapiboonchaiya, D. R. MacFarlane, J. Sun, M. Forsyth, *Macromol. Chem. Phys.*, **203**, 1906 (2002). (c) T. Tsuda, T. Nohira, Y. Nakamori, K. Matsumoto, R. Hagiwara, Y. Ito, *Solid State Ionics*, **149**, 295 (2002).

28. M. Hirao, H. Sugimoto, H. Ohno, *J. Electrochem. Soc.*, **147**, 4168 (2000).

29. M. Yoshizawa, W. Ogihara, H. Ohno, *Electrochem. Solid-State Lett.*, **4**, E25 (2001).

30. H. Ohno, M. Yoshizawa, *Solid State Ionics*, **154–155**, 303 (2002).

31. J. Golding, S. Forsyth, D. R. MacFarlane, M. Forsyth, G. B. Deacon, *Green Chem.*, **4**, 223 (2002).

32. H. Ohno, M. Yoshizawa, W. Ogihara, *Electrochim. Acta*, **50**, 255 (2004).

33. (a) Y. Nakai, K. Ito, H. Ohno, *Solid State Ionics*, **113–115**, 199 (1998). (b) Y. Tominaga, N. Takizawa, H. Ohno, *Electrochim. Acta*, **45**, 1285 (2000). (c) M. Yoshizawa, K. Ito-Akita, H. Ohno, *Electrochim. Acta*, **45**, 1617 (2000).

Chapter 30

Amphoteric Polymers

Hiroyuki Ohno, Masahiro Yoshizawa, and Wataru Ogihara

The polymerized ionic liquid (IL) shows great promise for diverse applications. Some polymerization methods have already been oriented toward specific applications. Polymerized ILs are useful in polar environments or where there are ion species for transport in the matrix. Amphoteric polymers that contain no carrier ions are being considered for several porposes in polymer electrolytes. Zwitterionic liquids (ZILs) were introduced in Chapter 20 as ILs in which component ions cannot move with the potential gradient. ZILs can provide ion conductive paths upon addition of salt to the matrix. It is therefore possible to realize selective ion transport in an IL matrix. If the resulting matrix can form solid film over a wide temperature range, many useful ionic devices can be realized. This chapter focuses on the preparation and characteristics of amphoteric IL polymers.

Amphoteric IL polymers contain no carrier ions, since the migration of component ions is completely excluded by fixing them on the polymer chain. They can therefore transport only added target ions in the matrix. Research into solid polymer electrolytes has so far mainly involved the use of polyether matrices, since these polyethers can dissolve salts and transport the dissociated ions along with intra- and intermolecular motion of these polyether chains [1, 2]. Polymer matrices having both high polarity and low glass transition temperature are rare. However, the ionic conductivity of certain polyether systems is about $10^{-4}\,\mathrm{S\,cm^{-1}}$ at room temperature [3], and it is difficult to improve cationic conductivity. This is a consequence of the ion–dipole interaction of polyethers to target ions, known as strong solvation of polyethers to cations. Alternative matrices have therefore been sought in order to overcome the problem. Some polyampholytes, in which both

Electrochemical Aspects of Ionic Liquids   Edited by Hiroyuki Ohno
ISBN 0-471-64851-5   Copyright © 2005 John Wiley & Sons, Inc.

cation and anion are fixed in the polymer main chain and have low glass transition temperature, have been studied as possible polyether-alternative ion conductive matrix. These amphoteric polymers can provide ion conductive paths for target ions without migration of the IL itself. There are two main categories of polyampholytes: copolymers and poly(zwitterion)s.

30.1 COPOLYMERS

Copolymers can be obtained by the polymerization of IL monomers synthesized by the neutralization of acid and base which both have vinyl groups [4]. In the case of copolymer systems, the ionic conductivity of IL copolymers should be quite low, since there are no carrier ions in the matrix. However, IL copolymers are expected to provide ion conductive paths for target ions upon adding appropriate salts. Figure 30.1 shows the structure of IL monomers where both cation and anion have vinyl groups. To compare the effect of alkyl spacer on the properties of IL copolymers, monomer **2** was also synthesized and polymerized, having alkyl spacer between the vinyl group and sulfonate group.

Table 30.1 summarizes the effect of alkyl spacer on the properties of IL copolymers. Both monomers had only a glass transition temperature (T_g), and were colorless transparent liquids at room temperature. The ionic conductivity of the monomers was 10^{-4}–10^{-3} S cm^{-1} at room temperature. This value reflects the quite low T_g values of $-83°$ and $-73°$C, respectively. However, **1** did not show T_g after polymerization, whereas polymerized **2** had $T_g = -31°$C, implying that **2** maintains high mobility of charged units. Both copolymers displayed low ionic conductivity,

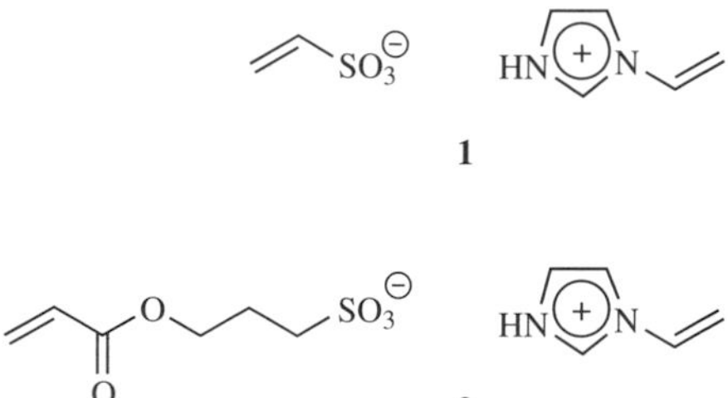

1

2

Figure 30.1 Structures of IL monomers composed of cation and anion both having vinyl group.

TABLE 30.1 Thermal Property and Ionic Conductivity of IL Monomers and their Polymers

	T_g/°C		σ_i at 30°C/S cm^{-1}	
	Monomer	Polymer	Monomer	Polymer
1	−83	—	3.5×10^{-3}	$<10^{-9}$
2	−73	−31	6.5×10^{-4}	$<10^{-9}$

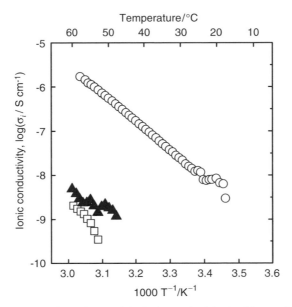

Figure 30.2 *Temperature dependence of the ionic conductivity for P1 containing an equimolar amount of LiX.* ○: X = TFSI; ▲: BF₄; □: CF₃SO₃.

however. It was confirmed that neither contains any carrier ions that can support long-distance migration.

Li salts were added to **P1** (**P1** implies polymerized **1**) so as to generate carrier ions, and their ionic conductivity was measured. Three distinct Li salts (LiTFSI, LiBF$_4$, and LiCF$_3$SO$_3$) were used as additives. It is well known that these anions can form ILs at room temperature after mixing with imidazolium cations. Figure 30.2 shows the temperature dependence of the ionic conductivity for **P1** containing an amount of Li salt equimolar to the imidazolium unit. When LiTFSI was added to **P1**, it had ionic conductivity of 7.2×10^{-7} S cm^{-1} at 50°C. In contrast, the ionic conductivity of **P1** containing the other Li salts was about 10^{-9} S cm^{-1}, which is almost insulating. The ionic conductivity of **P1** improved significantly upon adding LiTFSI even though other anion species have the ability to form ILs. This difference will be based on the plasticizing effect of the TFSI anion [5]. A similar effect has been observed in ZILs [6]. Unfortunately, their T_g has not previously been settled, although DSC measurement has been carried out. Addition of the TFSI anion is effective in improving the ionic conductivity of IL copolymers.

An amount of LiTFSI equimolar to the imidazolium cation unit was added to **P1** and **P2** in order to study the effect of alkyl spacer on the ionic conductivity. Figure 30.3 shows Arrhenius plots of the ionic conductivity for the copolymers after addition of salt. Within the measured temperature regime, **P2** with alkyl spacer had ionic conductivity one order higher than **P1** without spacer. There are two possible explanations. An increase in the length of the alkyl spacer causes an increase of free volume and maintains the high mobility of the IL domain. Although

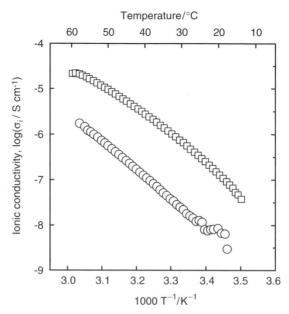

Figure 30.3 *Temperature dependences of the ionic conductivities for P1 (○) and P2 (□) containing an equimolar amount of LiTFSI.*

the spacer unit is effective in improving the ionic conductivity of copolymers, a spacer of appropriate length should be used because it is difficult to form successive domains for ion conduction with a longer spacer.

Proton conducting copolymers of vinylphosphonic acid and 4-vinylimidazole have been reported recently by Bozkurt et al. [7]. Since the imidazole ring can act as a proton hopping site in the polymer matrix [8–10], these copolymers had ionic conductivity of about 10^{-6} S cm^{-1} at 60°C without solvent or salt. To realize fast proton transport in copolymer systems, it is essential to design an ion conductive paths that uses the IL domain.

30.2 POLY(ZWITTERIONIC LIQUID)S

ZIL monomers, having both cation and anion in the same monomer unit, were prepared in order to give different types of ampholyte polymers. Figure 30.4 shows the structure of the monomers used here [6]. Four different monomers were synthesized to study the relation between structure and such properties as ionic conductivity and T_g.

Polymers having a ZIL unit should provide an excellent ion conductive matrix for added target ions alone, as stated in Chapter 20.3. Accordingly poly(ZIL)s were synthesized, and their ionic conductivity and thermal properties were studied. **P6** showed the lowest T_g, both before and after addition of salt. This matrix proved to have higher ionic conductivity than the other polymers. Figure 30.5 shows the

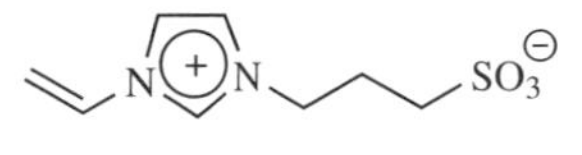

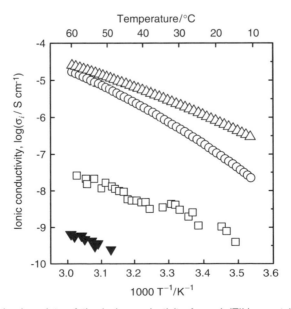

3

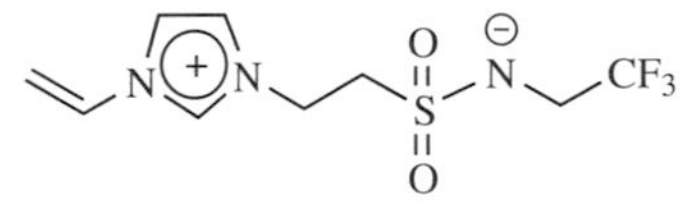

4

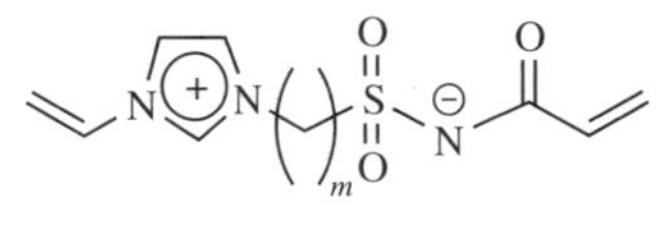

5; $m = 2$
6; $m = 3$

Figure 30.4 *Structures of zwitterionic monomers.*

temperature dependence of the ionic conductivity for poly(ZIL)s containing equimolar amounts of LiTFSI. Two groups of poly(ZIL)s were synthesized, with an imidazolium cation and with a sulfonamide group on the main chain, to study the effect of freedom of the imidazolium cation. The ionic conductivity of polymers having an imidazolium cation on the main chain, **P3** and **P4**, containing 100 mol% LiTFSI to imidazolium salt unit, was about $10^{-8} - 10^{-9}$ S cm^{-1}. Poly(ZIL)s having

Figure 30.5 *Arrhenius plots of the ionic conductivity for poly(ZIL)s containing LiTFSI.* □: *P3;* ▼: *P4;* ○: *P5;* △: *P6.*

an immobilized counter anion on the main chain (**P5** and **P6** containing LiTFSI) had relatively high ionic conductivity of about 10^{-5} S cm^{-1} at 50°C despite their rubberlike properties. **P6**, with a long alkyl spacer (m = 3), showed slightly greater ionic conductivity.

The difference in ionic conductivity can be attributed to the difference in freedom of the imidazolium cation. This effect has already been confirmed empirically in our laboratory [11, 12]. For a simple system such as 1-ethyl-3-vinylimidazolium TFSI, the ionic conductivity decreased by about four orders upon polymerization. The ionic conductivity of IL polymer brushes having PEO or hydrocarbon chains as spacers between the vinyl group and imidazolium salt are almost the same before and after polymerization. In spite of their rubberlike physical properties, IL polymer brushes had excellent ionic conductivity, around 10^{-4} S cm^{-1}, at room temperature. Details of the IL polymer brushes are given in Chapter 31. The ionic conductivity of **P3** and **P4**, in which the imidazolium cation is fixed to the main chain, was very low even after salts were added to the matrix. **P5** and **P6** having the counter anion on the main chain displayed an ionic conductivity four orders higher than **P3** or **P4** (about 10^{-5} S cm^{-1} at 50°C). The distance between the vinyl polymer and the imidazolium cation is important for the high ionic conductivity in IL polymers.

Poly(ZIL)s composed of aliphatic ammonium cation have also been reported by Galin et al. [13, 14]. These authors studied the ionic conductivity and dipole moment of poly(ZIL)s. The dipole moment increased with increasing distance between the cation and anion fixed on the polymer chain. This observation suggests that these poly(ZIL)s have different solubility from the added salts, which is reflected in their polarity. However, the ionic conductivities of **P4** and **P5** were very different, though the spacer distance between the cation and anion was nearly identical. This result suggests that the distance between the vinyl polymer and imidazolium cation is important in maintaining the high ionic conductivity of IL polymers. Although other factors muddy the comparison, the ionic conductivity of **P6** is nevertheless 100 times higher than that of the ammonium systems prepared by Galin et al. We attribute this difference not only to the fixed position of the cation but also the use of a suitable anion to form the ILs.

Ampholyte-type polymers should become attractive materials once successive and effective ion conductive paths are successfully designed in the matrix.

REFERENCES

1. J. R. MacCallum, C. A. Vincent, eds., in *Polymer Electrolyte Reviews 1 and 2*, Elsevier Applied Science, 1987 and 1989.
2. P. G. Bruce, ed., in *Solid State Electrochemistry*, Cambridge University Press, 1995.
3. R. Hooper, L. J. Lyons, M. K. Mapes, D. Schumacher, D. A. Moline, R. West, *Macromolecules*, **34**, 931 (2001).
4. M. Yoshizawa, W. Ogihara, H. Ohno, *Polym. Adv. Technol.*, **13**, 589 (2002).
5. S. Besner, A. Vallée, G. Bouchard, J. Prud'homme, *Macromolecules*, **25**, 6480 (1992).

6. M. Yoshizawa, M. Hirao, K. Ito-Akita, H. Ohno, *J. Mater. Chem.*, **11**, 1057 (2001).

7. A. Bozkurt, W. H. Meyer, J. Gutmann, G. Wegner, *Solid State Ionics*, **164**, 169 (2003).

8. (a) M. A. B. H. Susan, A. Noda, S. Mitsushima, M. Watanabe, *Chem. Commun.*, 938 (2003). (b) A. Noda, M. A. B. H. Susan, K. Kudo, S. Mitsushima, K. Hayamizu, M. Watanabe, *J. Phys. Chem. B*, **107**, 4024 (2003).

9. M. Yoshizawa, W. Xu, C. A. Angell, *J. Am. Chem. Soc.*, **125**, 15411 (2003).

10. (a) H. G. Herz, K. D. Kreuer, J. Maier, G. Scharfenberger, M. F. H. Schuster, W. H. Meyer, *Electrochim. Acta*, **48**, 2165 (2003). (b) K. D. Kreuer, A. Fuchs, M. Ise, M. Spaeth, J. Maier, *Electrochim. Acta*, **43**, 1281 (1998). (c) M. Schuster, W. H. Meyer, G. Wegner, H. G. Herz, M. Ise, M. Schuster, K. D. Kreuer, J. Maier, *Solid State Ionics*, **145**, 85 (2001).

11. H. Ohno, K. Ito, *Chem. Lett.*, 751 (1998).

12. M. Yoshizawa, H. Ohno, *Electrochim. Acta*, **46**, 1723 (2001).

13. M. Galin, A. Mathis, J.-C. Galin, *Polym. Adv. Technol.*, **12**, 574 (2001).

14. M. Galin, E. Marchal, A. Mathis, J.-C. Galin, *Polym. Adv. Technol.*, **8**, 75 (1997).

Chapter 31

Polymer Brushes

Masahiro Yoshizawa and Hiroyuki Ohno

When ionic liquids (ILs) are applied as electrolyte materials, the major mobile ions are the component ions, and therefore any specific target ion transport is impaired. Target ion transport in ILs has already been discussed in Chapters 20 and 30 [1, 2]. Polymerized derivatives of ILs are required for applications in ionics devices [3]. So far many researchers have reported work on polymer systems containing ILs for ionics devices with applications such as rechargeable lithium-ion batteries [4], fuel cells [5], solar cells [6, 7], and capacitors [8–11]. However, when an IL unit is polymerized, ion transport becomes selective [12]. In the polymer matrix the fixed cation or anion cannot move along with the potential gradient, only the counter ion can move. Therefore IL polymers are expected to be novel single ion conductive matrices. Still, as mentioned in Chapter 29, the ionic conductivity of ILs composed of vinylimidazolium cation decrease considerably after polymerization [12]. It is therefore necessary to design new IL polymers to maintain a low T_g. Certain IL-type polymer brushes having a flexible spacer between polymerizable group and IL moiety have been developed to minimize this drawback [13–17].

Figure 31.1 shows the structure of IL monomers having a flexible spacer. IL monomers having an ethylene oxide (EO) or a hydrocarbon (HC) spacer between the vinyl group and the imidazolium cation were synthesized, and the effects of the structure and length of the spacer groups on both the ionic conductivity and thermal properties were investigated.

First, the effect of the EO spacer on the physical properties for IL-type polymer brushes is discussed. Both IL monomers, a_8Cl and a_8TFSI, are liquids at room temperature and have very low T_g's at $-74°C$ and $-72°C$, respectively. These

Electrochemical Aspects of Ionic Liquids Edited by Hiroyuki Ohno
ISBN 0-471-64851-5 Copyright © 2005 John Wiley & Sons, Inc.

(a)

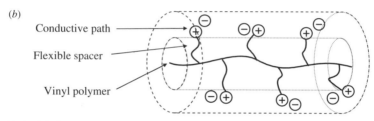

$A^- = Cl^-$ or $TFSI^-$

$a_m A$; $m = 2$ or 8

$A^- = Br^-$ or $TFSI^-$

$b_n A$; $R_1 = Me$, $R_2 = H$, $n = 6$
$c_n A$; $R_1 = Et$, $R_2 = H$, $n = 2, 3, 6, 9,$ or 12
$d_n A$; $R_1 = Bu$, $R_2 = H$, $n = 6$
$e_n A$; $R_1 = Et$, $R_2 = Me$, $n = 6$

(b)

Conductive path

Flexible spacer

Vinyl polymer

Figure 31.1 (a) Structures of IL monomers having flexible spacer between vinyl group and imidazolium cation. (b) Model structure of IL polymer brushes.

monomers showed very high ionic conductivity at room temperature, as shown in Figure 31.2. The ionic conductivity of $a_8 TFSI$ was 5.51×10^{-4} S cm^{-1} at 30°C, which was 20 times higher than that with Cl$^-$ as counter anion. This improvement can be attributed to the IL formation by the combination of the imidazolium cation and TFSI anion and to a much weaker interaction force of ether oxygens toward TFSI$^-$ than that to Cl$^-$. Since ion radius of TFSI$^-$ (325 pm) [18] is larger than that of Cl$^-$ (181 pm), the delocalized surface charge density of TFSI$^-$ was much smaller, permitting the weaker interaction.

Then ionic conductivity for these samples was measured after polymerization. The IL polymer brushes having EO spacer are rubberlike solids regardless of the anion species. The T_g's of these polymers $P(a_8 Cl)$ and $P(a_8 TFSI)$ (P implies polymer) are $-62°C$ and $-64°C$, respectively. From comparing the T_g's for both monomers and polymers, it was confirmed that the T_g was kept low even after polymerization. Despite their rubberlike properties these IL polymer brushes showed high ionic conductivity, reflecting their low T_g. That is, the flexibility of the side-chain containing the imidazolium salts was effective in maintaining the high ionic conductivity. In particular, when the counter anion specie of the polymer was TFSI$^-$, excellent ionic conductivity (1.20×10^{-4} S cm^{-1} at 30°C) was observed.

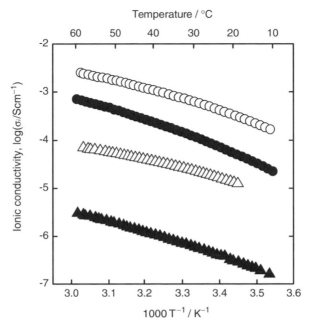

Figure 31.2 *Arrhenius plots of the ionic conductivity for IL monomers having EO spacer a_8A and their polymers. Open plots: monomers; closed plots: polymers. $\bigcirc, \bullet$: TFSI$^-$; $\triangle, \blacktriangle$: Cl$^-$.*

The EO-tethering of the IL and the polymer matrix was effective in suppressing the decrease of ionic conductivity after polymerization.

Next Figure 31.3 shows the effect of HC spacer length (n) on the ionic conductivity and T_g for $c_n TFSI$. Note that all monomers show almost the same T_g between $-80°$ and $-70°$C. There is a good relationship between ionic conductivity and the T_g's of ordinary ILs [19]. In other words, ILs having lower T_g's show higher ionic conductivity. In the ionic conductivities of these IL monomers decreased monotonically with the increasing spacer chain length. No increase in the ionic conductivity could be detected with the drop in T_g. This may be caused by the decrease of ion density and the increase of viscosity with the increasing HC length. Figure 31.4 shows the relationship between the carbon number (n) of the HC spacer and the ionic conductivity or T_g for $P(c_n TFSI)$. The ionic conductivity of P(1-ethyl-3-vinylimidazolium TFSI) (*P(EVImTFSI)*) is also depicted as a reference, with $n = 0$ for comparison [12a].

P(EVImTFSI) was obtained as a glassy solid at room temperature. The ionic conductivity decreased more than four orders, compared to the system before polymerization. However, the, *P(c_n TFSI)s* were sticky and rubbery, and their ionic conductivity was almost the same before and after polymerization. The decrease of ionic conductivity caused by polymerization was minimized in these cases within about one order. The T_g of *P(c_n TFSI)s* was almost the same value, around $-60°$C except for $n = 2$ ($-47.8°$C). Nearly equal ionic conductivity of a series of

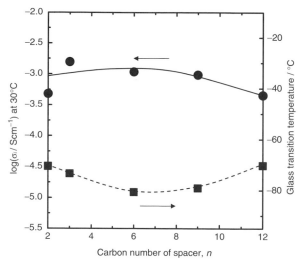

Figure 31.3 *Effects of HC spacer length (n) on the ionic conductivities and T_g's for IL monomers $c_n TFSI$.*

$P(c_n TFSI)$ can be explained by the excellent T_g values. It was also confirmed that the HC spacer is as effective in maintaining high ionic conductivity, even after polymerization, as the effect of the EO spacer between the vinyl group and the imidazolium cation [15].

The IL monomers tethering different cation structures on the spacer end were synthesized to study the effect of the cation structure on the ionic conductivity.

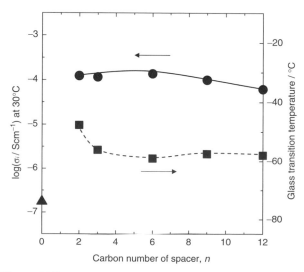

Figure 31.4 *Effects of HC spacer length (n) on both the ionic conductivities (▲: the data of P(EVImTFSI) was also shown as reference) and T_g's for IL polymer brushes $P(c_n TFSI)$.*

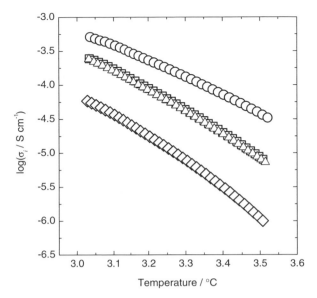

Figure 31.5 *Arrhenius plots of the ionic conductivity for various polycation-type ILs P(x₆TFSI).*
○: x = c; △: x = d; □: x = b; ◇: x = e.

Figure 31.5 shows the Arrhenius plots of the ionic conductivity for $P(x_6TFSI)$.
The $P(c_6TFSI)$ having an ethylimidazolium cation showed the highest ionic conductivity among these polymer brushes. It showed ionic conductivity of over
10^{-4} S cm^{-1} at room temperature, which is one order higher than that of $P(e_6TFSI)$
having the 1-ethyl-2-methylimidazolium cation. It appears that the ethylimidazolium cation is suitable for higher ionic conductivity for the IL polymer brushes
as well as for simple ILs [20]. The T_g's of $P(x_6TFSI)$ are $-53.2°$ ($x = b$), $-59.1°(x = c)$,
$-51.1°(x = d)$, and $-42.3°$C ($x = e$). The system with lower T_g's showed the higher
ionic conductivity.

In addition a thermogravimetric analysis of these polymers was carried out.
Their thermal stability was evaluated in the following sequence of the onset decomposition temperature: $P(c_6TFSI)$ ($T = 389°$C) > $P(d_6TFSI)$ ($T = 382°$C) > $P(e_6TFSI)$
($T = 381°$C) > $P(b_6TFSI)$ ($T = 371°$C). From these results the IL-type polymer
brushes were determined to be thermally stable up to around 400°C. This is an
excellent property for application to electrochemical devices.

Although the $P(c_6TFSI)$, which showed the highest ionic conductivity among
all systems of $P(x_6TFSI)$, had ionic conductivity of over 10^{-4} S cm^{-1} at 30°C,
the sticky characteristics of this polymer made it difficult to handle. To surmount
this problem, the $P(c_6TFSI)$ were polymerized with various cross-linkers, as shown
in Figure 31.6. First, the CLf_3 was used as a cross-linker so that the effects of
various amounts of the cross-linker on their properties could be examined. The
sticky rubbery property of IL polymers was improved by the addition of
the cross-linkers. While they were sticky rubbers without cross-linkers, they formed

CLf$_x$; $x = 2$ or 3

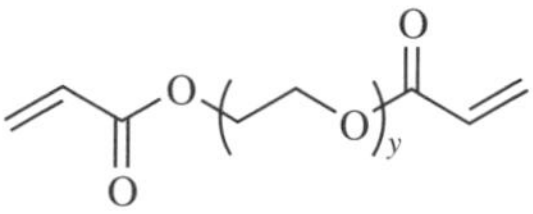

CLg$_y$; $y = 4$ or 13–14

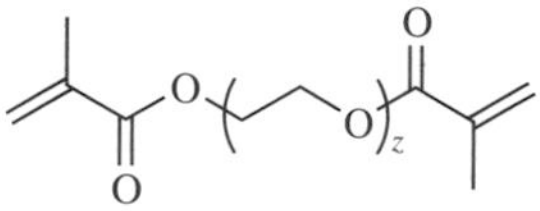

CLh$_z$; $z = 2$, 4 or 9

Figure 31.6 Structures of various cross-linkers used for obtaining film electrolytes.

flexible films after the addition of the cross-linker. In addition they could be obtained as films by adding just a small amount of the cross-linker. Figure 31.7 shows a typical photograph of a flexible film prepared by the polymerization of c_6TFSI with 3 mol% CLf_3 as the cross-linker. It was confirmed a small amount of cross-linker of about 1 mol% was enough to obtain the IL polymers as flexible films.

Figure 31.8 shows the effect of the cross-linker (CLf_3) concentration on the ionic conductivity and the T_g for the IL-type network polymer $P(c_6TFSI–CLf_3)$. After cross-linking, the T_g of the network was increased about 5°C, compared to that of the linear polymer, and the ionic conductivity decreased by about one-third.

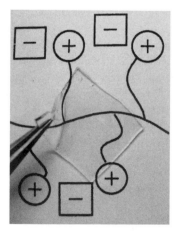

Figure 31.7 Photograph of film electrolyte composed of polymerized IL (cross-linker CLf_3 3 mol% to c_6TFSI).

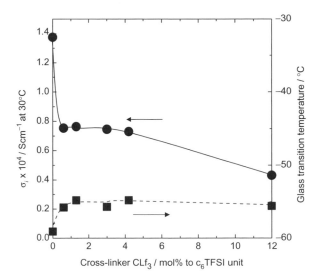

Figure 31.8 *Effects of cross-linker concentration CLf$_3$ on the ionic conductivity and T$_g$ of the IL-type network polymer P(c$_6$ TFSI-CLf$_3$).*

Within the range of the amount of cross-linker from 0.5 to 5 mol%, the ionic conductivity was almost the same. Generally, cross-linking restricts the motion of the polymer backbone. Therefore higher T_g should be found with increasing degrees of cross-linking. The ionic conductivities of network polymer electrolytes are usually much lower than those of the corresponding linear polymers [21]. There is a report that ionic conductivity remained almost the same as that for the linear polymers when monomers containing the EO spacer were polymerized with the cross-linkers (up to 5 mol%) [22]. This is due to the successive ion conduction in the domain even after cross-linking despite a remarkable increase in viscosity. Because the IL structure is tethered to the end of the flexible spacer, almost no change in T_g is found after polymerization in the presence of small amount of cross-linker. This seems to be the mechanism that maintains the high ionic conductivity.

Using various EO cross-linkers (*CLf$_x$*, *CLg$_y$*, and *CLh$_z$*), we next synthesized network polymers to investigate the effect of the spacer length and the polymerizable group of EO cross-linkers on ionic conductivity. The added amount was fixed at 1.3 mol%, because the addition of about 1 mol% *CLf$_3$* has been confirmed to be enough to obtain these polymers as films.

Although all the obtained network systems showed almost the same T_g's around $-55°$ to $-60°$C, there appeared some differences in the ionic conductivity, as shown in Figure 31.9. In the *CLf$_3$* system, the network exhibited higher ionic conductivity than the network of *CLf$_2$*, and in the *CLg$_y$* system, the network of *CLg$_4$* exhibited higher ionic conductivity than the network of *CLg$_{13-14}$*. There was a suitable length in the spacer of the cross-linker to maintain the high ionic conductivity despite the small amount of cross-linkers. Besides the *CLh$_z$* system,

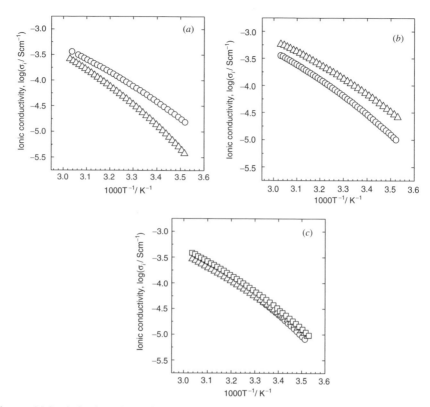

Figure 31.9 *Arrhenius plots of the ionic conductivity for IL-type network polymers. (a)* *P(c₆TFSI-CLfₓ); (b): P(c₆TFSI-CLgᵧ); (c) P(c₆TFSI-CLhᵤ).*

the temperature dependence of the ionic conductivity was the same regardless of the spacer's length. These results indicate that an appropriate spacer length must be found for every cross-linker.

When *CLf₃* was used as the cross-linker, a minimum amount of 1 mol% was necessary to obtain a film. However, the minimum amount depended on the structure of the cross-linker. Among these, *CLgᵧ*, *CLh₄*, and *CLh₉* were effective in forming films after the addition of only 0.5 mol%. In particular, *CLg₄* was the best and induced the highest ionic conductivity. The ionic conductivity of the transparent and flexible film obtained by adding 0.5 mol% of *CLg₄* was 1.1×10^{-4} S cm^{-1} at 30°C, which is almost equivalent to that of the monomer.

To investigate the effect of the flexible spacer on the physical properties of the polyanion-type ILs, we synthesized polyanion-type ILs having a hydrocarbon or an EO spacer. Figure 31.10 shows the structure of the IL monomers having a flexible spacer between polymerizable group and anion. The effect of the HC spacer on the ionic conductivity of the polyanion-type ILs was analyzed. Figure 31.11 shows the Arrhenius plots of the ionic conductivity for *EImVS*, *EImC3S*, and their

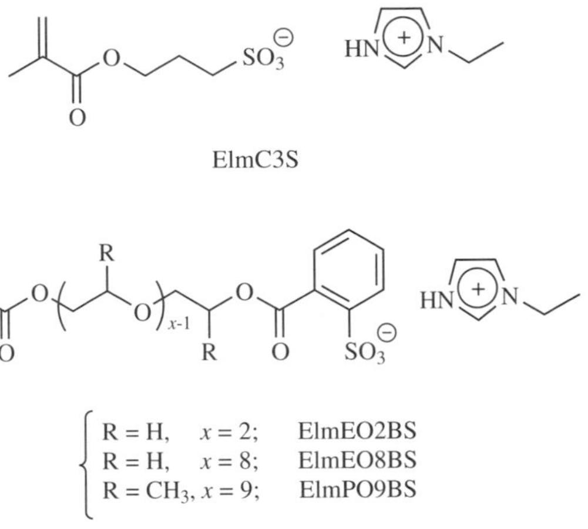

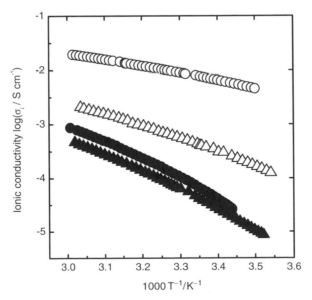

ElmC3S

$\begin{cases} R = H, & x = 2; & ElmEO2BS \\ R = H, & x = 8; & ElmEO8BS \\ R = CH_3, x = 9; & ElmPO9BS \end{cases}$

Figure 31.10 *Structures of IL monomers having flexible spacer between polymerizable group and anion structure.*

polymers. The ionic conductivity of *ElmC3S* was about 10^{-3} S cm^{-1} at room temperature, which is one order lower than that of *ElmVS*. This difference could be due to the large size anion of *ElmC3S*. After polymerization, both systems showed an ionic conductivity of about 10^{-4} S cm^{-1} at room temperature. Although the alkyl spacer group was introduced between the vinyl group and the sulfonate,

Figure 31.11 *Temperature dependences of the ionic conductivities for ElmVS, ElmC3S, and their polymers.* ○: *ElmVS;* ●: *P(ElmVS);* △: *ElmC3S;* ▲: *P(ElmC3S).*

suppression of the drop in ionic conductivity was not observed. The flexibility of the imidazolium cation is inherently high in the polyanion-type IL containing sulfonate.

Higher ionic conductivity of the polymerized ILs was expected when the motion of imidazolium cation was coupled with that of anion containing the tethered vinyl group. For this reason some IL monomers were prepared, as shown in Figure 31.10. Figure 31.12 shows the effect of the spacer length on ionic conductivity and the T_g's for *EImEO2BS*, *EImEO8BS*, *EImPO9BS*, and their polymers. *EImSS* was used as the reference with a spacer length of 0. The ionic conductivity

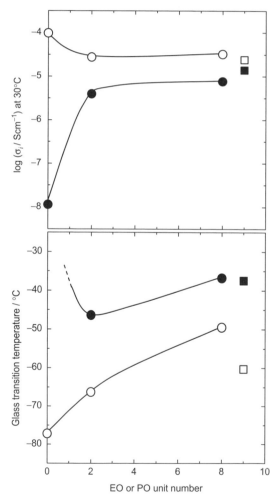

Figure 31.12 *Effects of spacer length and structure on the ionic conductivities and T_g's for IL monomers and polymers containing polyether spacers. Structures are shown in Figure 31.10.* ○, ●: *EO spacer*; □, ■: *PO spacer. Open plots: monomers; closed plots: polymers.*

of *EImSS* was about 10^{-4} S cm^{-1} at room temperature. However, its polymer showed a low ionic conductivity of about 10^{-8} S cm^{-1}. As mentioned above, the lowering of ionic conductivity after polymerization seems to reflect the increase of T_g. On the other hand, polyanion-type ILs having an EO or a PO spacer showed ionic conductivity of over 10^{-6} S cm^{-1} at room temperature. When the repeating unit number of spacer was 2, suppression of the ionic conductivity was diminished. In particular, the ionic conductivity of the PO system was about 10^{-5} S cm^{-1}, which is nearly equal to that of the corresponding monomer. This effect should be due to the weaker interaction between the imidazolium cation and the PO units than with the EO units. The ionic conductivity of polyanion-type ILs containing benzene sulfonate was improved by the introduction of the polyether spacer, whereas their T_g's increased with the increasing spacer unit number. Generally, the increase of T_g causes a considerable drop in ionic conductivity [3]. Although the T_g's of the polyanion-type ILs having the polyether spacer increased with the increasing spacer unit number, they maintained ionic conductivities of over 10^{-6} S cm^{-1} at room temperature. These T_g values are based on the polyether segment, and not an the ionic liquid part of the spacer end. It is suggested that the polyether spacer does not work as an ion conductive path, that is, the transport of imidazolium cation does not depend on the mobility of polyether segment but rather on the flexibility of the charged site. So the PO unit is a favorable spacer for keeping the ionic conductivity of the polymerized ILs high.

Recently a new composite material based on the IL-type polymer brush and the single-walled carbon nanotube (SWNT) was reported by Aida et al. [23]. The film of the IL polymer brush including SWNT (3.8 wt%) showed a nearly 400% increase in dynamic hardness compared with that of the pure IL polymer film. It is thought that the good dynamic hardness was obtained because of a strong SWNT–imidazolium cation interaction. In addition the film showed conductivity of 0.56 S cm^{-1} at room temperature, which is equal to that of typical conductive polymers whose films are being engineered as coating materials, antistatic materials, novel electronic devices, and so on.

REFERENCES

1. (a) M. Yoshizawa, M. Hirao, K. Ito-Akita, H. Ohno, *J. Mater. Chem.*, **11**, 1057 (2001).
   (b) M. Yoshizawa, A. Narita, H. Ohno, *Aust. J. Chem.*, **57**, 139 (2004).
2. M. Yoshizawa, W. Ogihara, H. Ohno, *Polym. Adv. Technol.*, **13**, 589 (2002).
3. J. R. MacCallum, C. A. Vincent, eds., *Polymer Electrolyte Reviews*, Elsevier Applied Science., vol. 1, 1987, and vol. 2, 1989.
4. H. Nakagawa, S. Izuchi, K. Kuwana, T. Nukuda, Y. Aihara, *J. Electrochem. Soc.*, **150**, A695 (2003).
5. M. Doyle, S. K. Choi, G. Proulx, *J. Electrochem. Soc.*, **147**, 34 (2000).
6. W. Kubo, T. Kitamura, K. Hanabusa, Y. Wada, S. Yanagida, *Chem. Commun.*, 374 (2002).
7. P. Wang, S. M. Zakeeruddin, I. Exnar, M. Grätzel, *Chem. Commun.*, 2972 (2002).

8. A. Lewandowski, A. Swiderska, *Solid State Ionics*, **161**, 243 (2003).

9. M. Ue, M. Takeda, A. Toriumi, A. Kominato, R. Hagiwara, Y. Ito, *J. Electrochem. Soc.*, **150**, A499 (2003).

10. J. D. Stenger-Smith, C. K. Webber, N. Anderson, A. P. Chafin, K. Zong, R. Reynolds, *J. Electrochem. Soc.*, **149**, A973 (2002).

11. C. Nanjundiah, S. F. McDevitt, V. R. Koch, *J. Electrochem. Soc.*, **144**, 3392 (1997).

12. (a) H. Ohno, K. Ito, *Chem. Lett.*, 751 (1998). (b) M. Hirao, K. Ito, H. Ohno, *Electrochim. Acta*, **45**, 1291 (2000). (c) M. Hirao, K. Ito-Akita, H. Ohno, *Polym. Adv. Technol.*, **11**, 534 (2000).

13. M. Yoshizawa, H. Ohno, *Chem. Lett.*, 889 (1999).

14. H. Ohno, *Electrochim. Acta*, **46**, 1407 (2001).

15. M. Yoshizawa, H. Ohno, *Electrochim. Acta*, **46**, 1723 (2001).

16. S. Washiro, M. Yoshizawa, H. Nakajima, H. Ohno, *Polymer*, **45**, 1577 (2004).

17. H. Ohno, M. Yoshizawa, W. Ogihara, *Electrochim. Acta*, **50**, 255 (2004).

18. M. Ue, *J. Electrochem. Soc.*, **141**, 3336 (1994).

19. M. Hirao, H. Sugimoto, H. Ohno, *J. Electrochem. Soc.*, **147**, 4168 (2000).

20. H. Ohno, M. Yoshizawa, *Solid State Ionics*, **154–155**, 303 (2002).

21. (a) W. H. Meyer, *Adv. Mater.*, **10**, 439 (1998). (b) M. Watanabe, T. Hirakimoto, S. Mutoh, A. Nishimoto, *Solid State Ionics*, **148**, 399 (2002).

22. J. M. G. Cowie, K. Sadaghianizadeh, *Polymer*, **30**, 509 (1989).

23. T. Fukushima, A. Kosaka, Y. Ishimura, T. Yamamoto, T. Takigawa, N. Ishii, T. Aida, *Science*, **300**, 2072 (2003).

Chapter *32*

Future Prospects

Hiroyuki Ohno

Ionic liquids are a repository of a wealth of creative possibility. In the present book we did our best to introduce the latest electrochemical trends using ionic liquids. We expect the future of ionic liquids to exceed our anticipated applications.

The variety of ions is endless. One gets unlimited species of ions when organic compounds are ionized because there are millions of organic molecules, and in principle, their combinations of cation and anion can provide more than trillions of salts. Furthermore there are endless combinations of mixed salts. Their combinations can be designed on demand. We see at least two paths of development in the future. Already novel ionic liquids are being vigorously developed by scientists, but the next step should be the fine-tuning of their functionalities, by mixing two or more ionic liquids just like wines, colors, and/or other cultural materials. Because blending of ILs can generate unexpected functions and abilities, an evolution of ionic liquids can be envisioned. Some functional groups could take the form, for example, of advanced deodorant liquids that absorb smells and decompose them into odorless molecules will be synthesized. Liquidized metal complexes should be effective for this kind of task. Surface modifications with some ionic liquids may provide maintenance-free surfaces. Some equipments will always be clean and smooth because surface-modified ionic liquids are harnessed to decompose hazardous molecules. Such a system may be even applicable to domestic life. Because catalytic activities are supported electrochemically, the electrochemical characteristics of ionic liquids will continue to find meaningful applications. Solvents that possess catalytic activity are intelligent solvents. They show physico-chemical properties such as polarity, basicity, acidity, solution temperature, and

Electrochemical Aspects of Ionic Liquids   Edited by Hiroyuki Ohno
ISBN 0-471-64851-5   Copyright © 2005 John Wiley & Sons, Inc.

TABLE 32.1 Some Expected Functions and Applications of Ionic Liquids

Functions	Devices	Features
Ion conduction	Battery	Liquid
Electro-conduction	Capacitor	Liquid crystal
Thermal conduction	Actuator	Liquid (micro or nano) particle
Magnetism	Luminescence	Solid
Emission	Sensory system	Gel
Luminosity	Nano-switch	Metal complex
Refractive-index control	Biochip	Supramolecular assembly
Polarity	Catalyst	Nano-sheet
Solubility	Waveguide	Polymer
Chirality	Optical switch	Film
Lubrication	Solvents (for separation, extraction, isolation, etc.)	Rubber
Degradability	Drug carrier	Membrane
Degradation switching	Biocompatible surface	Interface material
Edibility	Cultivation of cells	
	Column	
	Biodegradable	
	Liquid brake	
	Liquid switch	

even reaction step by a color change. A liquid brake in which the viscosity can be controlled by a given potential may be prepared soon. A conductive liquid can transport not only ions but also electrons to supply organic solder, information display liquid, microchips, and so on. A liquid battery can be prepared by coating three layers of electrodes and electrolyte. Unlimited applications can be imagined with functionalized ionic liquids for human consumption such as finely-blended alcoholic drinks. Some of expected features of future ionic liquids are summarized by key words in Table 32.1. Not only liquid systems but also system with polymerized features will become attractive and be developed as useful sheets. The combination of polymers and ionic liquids could become a key ssystem for the expanding technologies. Some applications could require the electrochemical regulation of their functions by other ionic liquids.

With advances in the science of ionic liquids, there are bound to be some excellent treatments for ionic liquid decomposition. Ionic liquids could be developed to have electrolytes that carry a decomposition switch. Both developments would make ionic liquids more "green" and useful electrolyte materials. The recovering process, recycling process, and other surrounding technologies should be developed along with the progress made in the functional design of ionic liquids.

These are but a handful of suggestions on the prospects for ionic liquids discussed in this book. Readers may be inspired to come up with their own ideas for the development of ionic liquids based on their unique backgrounds.

Index

Electrochemical Aspects of Ionic Liquids   Edited by Hiroyuki Ohno
ISBN 0-471-64851-5   Copyright © 2005 John Wiley & Sons, Inc.